The Noisy Brain
Stochastic Dynamics
as a Principle of
Brain Function

The Noisy Brain
Stochastic Dynamics as a Principle of Brain Function

Edmund T. Rolls
Oxford Centre for Computational Neuroscience
Oxford, England

Gustavo Deco
Institucio Catalana de Recerca i Estudis Avancats (ICREA)
Universitat Pompeu Fabra
Barcelona, Spain

OXFORD
UNIVERSITY PRESS

OXFORD
UNIVERSITY PRESS

Great Clarendon Street, Oxford OX2 6DP

Oxford University Press is a department of the University of Oxford.
It furthers the University's objective of excellence in research, scholarship,
and education by publishing worldwide in

Oxford New York

Auckland Cape Town Dar es Salaam Hong Kong Karachi
Kuala Lumpur Madrid Melbourne Mexico City Nairobi
New Delhi Shanghai Taipei Toronto

With offices in

Argentina Austria Brazil Chile Czech Republic France Greece
Guatemala Hungary Italy Japan Poland Portugal Singapore
South Korea Switzerland Thailand Turkey Ukraine Vietnam

Oxford is a registered trade mark of Oxford University Press
in the UK and in certain other countries

Published in the United States
by Oxford University Press Inc., New York

© Edmund T. Rolls and Gustavo Deco, 2010

British Library Cataloguing in Publication Data
Data available

Library of Congress Cataloging in Publication Data
Data available

Typeset by Edmund T. Rolls
Printed in Great Britain
on acid-free paper by the
MPG Books Group, Bodmin and King's Lynn

ISBN 978-0-19-958786-5

10 9 8 7 6 5 4 3 2 1

Preface

The relatively random spiking times of individual neurons produce a source of noise in the brain. The aim of this book is to consider the effects of this and other noise on brain processing. We show that in cortical networks this noise can be an advantage, for it leads to probabilistic behaviour that is advantageous in decision-making, by preventing deadlock, and is important in signal detectability. We show how computations can be performed through stochastic dynamical effects, including the role of noise in enabling probabilistic jumping across barriers in the energy landscape describing the flow of the dynamics in attractor networks. The results obtained in neurophysiological studies of decision-making and signal detectability are modelled by the stochastical neurodynamics of integrate-and-fire networks of neurons with probabilistic neuronal spiking. We describe how these stochastic neurodynamical effects can be analyzed, and their importance in many aspects of brain function, including decision-making, perception, memory recall, short-term memory, and attention. We show how instabilities in these brain dynamics may contribute to the cognitive symptoms in aging and in psychiatric states such as schizophrenia, and how overstability may contribute to the symptoms in obsessive-compulsive disorder.

This is a new approach to the dynamics of neural processing, in which we show that noise breaks deterministic computations, and has many advantages. These principles need to be analyzed in order to understand brain function and behaviour, and it is an aim of this book to elucidate the stochastic, that is probabilistic, dynamics of brain processing, and its advantages. The book describes approaches that provide a foundation for this understanding, including integrate-and-fire models of brain and cognitive function that incorporate the stochastic spiking-related dynamics, and mean-field analyses that are consistent in terms of the parameters with these, but allow formal analysis of the networks. A feature of the treatment of the mean-field approach is that we introduce new ways in which it can be extended to include some of the effects of noise on the operation of the system. The book thus describes the underpinnings in physics of this new approach to the probabilistic functioning of the brain. However, at the same time, most of the concepts of the book and the principles of the stochastic operation of the brain described in the book, can be understood by neuroscientists and others interested in brain function who do not have expertise in mathematics or theoretical physics, and the book has been written with this in mind.

We believe that the principles of the stochastic dynamics of brain function described in this book are important, for brain function can not be understood as a deterministic noiseless system.

To understand how the brain works, including how it functions in memory, attention, and decision-making, it is necessary to combine different approaches, including neural computation. Neurophysiology at the single neuron level is needed because this is the level at which information is exchanged between the computing elements of the brain, the neurons. Evidence from the effects of brain damage, including that available from neuropsychology, is needed to help understand what different parts of the system do, and indeed what each part is necessary for. Functional neuroimaging is useful to indicate where in the human brain different processes take place, and to show which functions can be dissociated from each other. Knowledge of the biophysical and synaptic properties of neurons is essential to understand how the computing elements of the brain work, and therefore what the building blocks of biologically realistic computational models should be. Knowledge of the anatomical and functional architecture of the cortex is needed to show what types of neuronal network actually perform the computation. The approach of neural computation is also needed, as this is required to link together all the empirical evidence to produce an understanding of how the system actually works. But an understanding of the role of noise in brain computation is also crucial, as we show in this book. This book utilizes evidence from all these approaches to develop an understanding of how different types of memory, perception, attention, and decision-making are implemented by processing in the brain, and are influenced by the effects of noise.

We emphasize that to understand memory, perception, attention, and decision-making in the brain, we are dealing with large-scale computational systems with interactions between the parts, and that this understanding requires analysis at the computational and global level of the operation of many neurons to perform together a useful function. Understanding at the molecular level is important for helping to understand how these large-scale computational processes are implemented in the brain, but will not by itself give any account of what computations are performed to implement these cognitive functions. Instead, understanding cognitive functions such as memory recall, attention, and decision-making requires single neuron data to be closely linked to computational models of how the interactions between large numbers of neurons and many networks of neurons allow these cognitive problems to be solved. The single neuron level is important in this approach, for the single neurons can be thought of as the computational units of the system, and is the level at which the information is exchanged by the spiking activity between the computational elements of the brain. The single neuron level is therefore, because it is the level at which information is communicated between the computing elements of the brain, the fundamental level of information processing, and the level at which the information can be read out (by recording the spiking activity) in order to understand what information is being represented and processed in each brain area. Moreover, the probabilistic spiking of individual neurons is an important source of noise in the brain, and must be taken into account to understand brain function.

A test of whether one's understanding is correct is to simulate the processing on a computer, and to show whether the simulation can perform the tasks of memory systems in the brain, and whether the simulation has similar properties to the real brain. The approach of neural computation leads to a precise definition of how the computation is performed, and to precise and quantitative tests of the theories produced. How memory systems in the brain work is a paradigm example of this approach, because memory-like operations which involve altered functionality as a result of synaptic modification are at the heart of how all computations in the brain are performed. It happens that attention and decision-making can be understood in terms of interactions between and fundamental operations of networks that implement computations that implement memory operations in the brain, and therefore it is natural to treat these areas of cognitive neuroscience as well as memory in this book. The same fundamental concepts based on the operation of neuronal circuitry can be applied to all these functions, as is shown in this book.

One of the distinctive properties of this book is that it links the neural computation approach not only firmly to neuronal neurophysiology, which provides much of the primary data about how the brain operates, but also to psychophysical studies (for example of attention); to psychiatric studies of patients; to functional magnetic resonance imaging (fMRI) (and other neuroimaging) approaches; and to approaches influenced by theoretical physics about how the operation of large scale systems can be understood as a result of statistical effects in its components, in this case the neurons. The empirical evidence that is brought to bear is largely from non-human primates and from humans, because of the considerable similarity of their memory and related systems, and the overall aims to understand how memory, attention, decision-making and related functions are implemented in the human brain, and the disorders that can arise.

The overall aims of the book are developed further, and the plan of the book is described, in Chapter 1, Section 1.1.

Part of the material described in the book reflects work performed in collaboration with many colleagues, whose tremendous contributions are warmly appreciated. The contributions of many will be evident from the references cited in the text. Especial appreciation is due to Alessandro Treves, Marco Loh, and Simon M. Stringer, who have contributed greatly in an always interesting and fruitful research collaboration on computational aspects of brain function, and to many neurophysiology and functional neuroimaging colleagues who have contributed to the empirical discoveries that provide the foundation to which the computational neuroscience must always be closely linked, and whose names are cited throughout the text. Much of the work described would not have been possible without financial support from a number of sources, particularly the Medical Research Council of the UK, the Human Frontier Science Program, the Wellcome Trust, and the James S. McDonnell Foundation. The book was typeset by the Edmund Rolls using LaTex and WinEdt, and Gustavo Deco took primary responsibility for the Appendix.

The covers show part of the picture *Ulysses and the Sirens* painted in 1909 by Herbert James Draper. The version on the back cover has noise added, and might be called *Ulysses and the Noisy Sirens*. The metaphors are of noise: sirens, and stormy, irregular, water; of waves and basins of attraction: the waves on the horizon; of decision-making: the rational conscious in Ulysses resisting the gene-based emotion-related attractors; and of Ulysses the explorer (the Greek Odysseus of Homer), always and indefatigably (like the authors) seeking new discoveries about the world (and how it works).

Updates to some of the publications cited in this book are available at **http://www.oxcns.org**.

We dedicate this work to the overlapping group: our families, friends, and many colleagues whose contributions are greatly appreciated – *in salutem praesentium, in memoriam absentium*. In addition, Gustavo Deco thanks and dedicates this book to his family, Maria Eugenia, Nikolas, Sebastian, Martin, and Matthias. We remember too a close colleague and friend, the theoretical physicist Daniel Amit, who contributed much to the analysis of attractor networks (Amit 1989, Brunel and Amit 1997).

Contents

1 Introduction: Neuronal, Cortical, and Network foundations

1.1 Introduction and overview

To understand how the brain works, including how it functions in memory, decision-making, and attention, it is necessary to combine different approaches, including neural computation. Neurophysiology at the single neuron level is needed because this is the level at which information is exchanged between the computing elements of the brain. Evidence from the effects of brain damage, including that available from neuropsychology, is needed to help understand what different parts of the system do, and indeed what each part is necessary for. Functional neuroimaging is useful to indicate where in the human brain different processes take place, and to show which functions can be dissociated from each other. Knowledge of the biophysical and synaptic properties of neurons is essential to understand how the computing elements of the brain work, and therefore what the building blocks of biologically realistic computational models should be. Knowledge of the anatomical and functional architecture of the cortex is needed to show what types of neuronal network actually perform the computation. And finally the approach of neural computation is needed, as this is required to link together all the empirical evidence to produce an understanding of how the system actually works, including how the noise generated within the brain by the stochastic firing of its neurons affects the brain's functioning. This book utilizes evidence from all these disciplines to develop an understanding of how different types of memory, attention, decision-making, and related functions are implemented by processing in the brain, and how noise contributes, often usefully, to the functions being performed.

A test of whether one's understanding is correct is to simulate the processing on a computer, and to show whether the simulation can perform the tasks of memory systems, decision-making, and attention in the brain, and whether the simulation has similar properties to the real brain. The approach of neural computation leads to a precise definition of how the computation is performed, and to precise and quantitative tests of the theories produced. How memory systems in the brain work is a paradigm example of this approach, because memory-like operations which involve altered functionality as a result of synaptic modification, and how noise affects the way in which the system settles into a state and maintains it, are at the heart of how all computations in the brain are performed. It happens that attention and decision-making can be understood in terms of interactions between, and fundamental operations of, networks that implement computations that implement memory operations in the brain, and therefore it is natural to treat these areas of cognitive neuroscience as well as memory in this book. The same fundamental concepts based on the operation of neuronal circuitry can be applied to all these functions, as is shown in this book.

One of the distinctive properties of this book is that it links the neural computation approach, including that related to theoretical physics, not only firmly to neuronal neurophysiology, which provides much of the primary data about how the brain operates, but also to psychophysical studies (for example of attention); to studies of patients with brain damage and psychiatric disorders; and to functional magnetic resonance imaging (fMRI) (and other neuroimaging) approaches. The empirical evidence that is brought to bear is largely from non-human primates and from humans, because of the considerable similarity of their memory and related systems, and the overall aims to understand how memory and related functions are implemented in the human brain, and the disorders that can arise.

The overall plan of the book is as follows.

Chapter 1 provides an introduction to information processing in neural systems in the brain.

Chapter 2 describes a computational approach to brain function in which the effects of noise on the operation of attractor networks involved in short-term memory, attention, and decision-making are described. The focus is on integrate-and-fire models of brain and cognitive function that incorporate the stochastic spiking-related dynamics, and mean-field analyses that are consistent in terms of the parameters with these, but allow formal analysis of the networks. A feature of the treatment of the mean-field approach is that we introduce new ways in which it can be extended to include some of the effects of noise on the operation of the system, in this Chapter, and in Appendix A.

Chapter 3 describes how short-term memory systems in the brain are implemented by attractor networks, and how noise produced by the probabilistic spiking of individual neurons affects the operation and stability of short-term memory.

Chapter 4 develops an understanding of attentional mechanisms that involves the subject of attention being held in a short-term memory, and the activity in this short-term memory acting as the source of a bias to influence the competition between representations in perceptual and memory areas of the brain. Interactions between the short-term memory and perceptual networks are key in understanding the dynamical, temporal processing involved in attention, and how noise produced by the probabilistic firing of neurons influences the maintenance of attention. The concepts are applied to understanding attentional disorders.

Chapter 5 shows how probabilistic decision-making can be understood in terms of the way in which an attractor network settles from spontaneous activity into a state that represents a decision in a way that depends on the probabilistic spiking of finite numbers of neurons in the network.

Chapter 6 shows how this stochastic dynamical approach to decision-making provides an understanding at the computational, neuronal, functional neuroimaging, and psychophysical levels of confidence in decision-making, and how this is implemented in the brain.

Chapter 7 shows how the probabilistic detection of signals close to threshold is influenced by the stochastic noise generated by the probabilistic spiking of neurons in the networks involved in perception.

Chapter 8 shows how the concept of probabilistic operation of networks in the brain has applications to understanding a myriad of brain functions, including the non-deterministic properties of brain function, perceptual rivalry, avoiding predators, creative thought, the cognitive changes in aging, and the substance of dreams. Chapter 8 also shows how too little stability in attractor networks, caused by the effects of noise caused by neuronal spiking affecting attractor networks with a reduced depth of the basins of attraction, can destabilize attractor networks involved in short-term memory and attention, and may contribute to the symptoms of schizophrenia. Chapter 8 also shows how too much stability in attractor networks, caused by an increased depth of the basins of attraction, so that the active networks cannot be shifted by another stimulus and noise, may contribute to the symptoms of obsessive-compulsive disorder (OCD).

Appendix A provides a summary and derivation of mean-field approaches that are consistent with integrate-and-fire spiking networks, and which therefore help to provide an analytic understanding of how noise influences brain function.

We emphasize that to understand memory, perception, attention, and decision-making in the brain, we are dealing with large-scale computational systems with interactions between the parts, and that this understanding requires analysis at the computational and global level of the operation of many neurons to perform together a useful function. Understanding at the molecular level is important for helping to understand how these large-scale computational processes are implemented in the brain, but will not by itself give any account of what computations are performed to implement these cognitive functions. Instead, understanding cognitive functions such as object recognition, memory recall, attention, and decision-making requires single neuron data to be closely linked to computational models of how the interactions between large numbers of neurons and many networks of neurons allow these cognitive problems to be solved. The single neuron level is important in this approach, for the single neurons can be thought of as the computational units of the system, and is the level at which the information is exchanged by the spiking activity between the computational elements of the brain. The single neuron level is therefore, because it is the level at which information is communicated between the computing elements of the brain, the fundamental level of information processing, and the level at which the information can be read out (by recording the spiking activity) in order to understand what information is being represented and processed in each brain area (Rolls, Grabenhorst and Franco 2009b).

Because of this importance of being able to analyze the activity of single neurons and populations of neurons in order to understand brain function, Section 1.11 describes rigorous approaches to understanding how information is represented by neurons, and summarizes evidence on how the information is actually represented. This is important for understanding the noise that is generated by a population of neurons, for the sparseness of the representation, and the distribution of the firing rates of the neurons when any stimulus or event is being represented, are important for understanding the noise generated by the spiking of the neurons.

In the rest of this chapter, we introduce some of the background for understanding brain computation, such as how single neurons operate; how some of the essential features of this can be captured by simple formalisms; some of the biological background to what it can be taken happens in the nervous system, such as synaptic modification based on information available locally at each synapse; the nature and operation of attractor networks; and how information is represented by neuronal firing, as this has implications for understanding the statistics of the noise generated by neuronal firing in the brain.

1.2 Neurons in the brain, and their representation in neuronal networks

Neurons in the vertebrate brain typically have, extending from the cell body, large dendrites which receive inputs from other neurons through connections called synapses. The synapses operate by chemical transmission. When a synaptic terminal receives an all-or-nothing action potential from the neuron of which it is a terminal, it releases a transmitter that crosses the synaptic cleft and produces either depolarization or hyperpolarization in the postsynaptic neuron, by opening particular ionic channels. (A textbook such as Kandel, Schwartz and Jessell (2000) gives further information on this process.) Summation of a number of such depolarizations or excitatory inputs within the time constant of the receiving neuron, which is typically 15–25 ms, produces sufficient depolarization that the neuron fires an action potential. There are often 5,000–20,000 inputs per neuron. Examples of cortical neurons are shown in Fig. 1.1, and further examples are shown in Shepherd (2004) and Rolls and Treves (1998). Once firing is initiated in the cell body (or axon initial segment of the cell body), the action potential is conducted in an all-or-nothing way to reach the synaptic terminals of the neuron, whence it may affect other neurons. Any inputs the neuron receives that cause it to

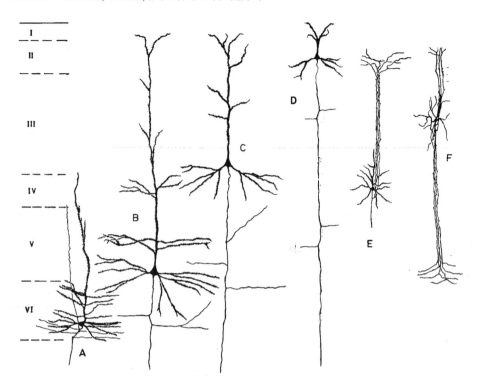

Fig. 1.1 Examples of neurons found in the brain. Cell types in the cerebral neocortex are shown. The different laminae of the cortex are designated I–VI, with I at the surface. Cells A–D are pyramidal cells in the different layers. Cell E is a spiny stellate cell, and F is a double bouquet cell. (After Jones 1981; see Jones and Peters 1984, p. 7.)

become hyperpolarized make it less likely to fire (because the membrane potential is moved away from the critical threshold at which an action potential is initiated), and are described as inhibitory. The neuron can thus be thought of in a simple way as a computational element that sums its inputs within its time constant and, whenever this sum, minus any inhibitory effects, exceeds a threshold, produces an action potential that propagates to all of its outputs. This simple idea is incorporated in many neuronal network models using a formalism of a type described in the next section.

1.3 Synaptic modification

For a neuronal network to perform useful computation, that is to produce a given output when it receives a particular input, the synaptic weights must be set up appropriately. This is often performed by synaptic modification occurring during learning.

A simple learning rule that was originally presaged by Donald Hebb (1949) proposes that synapses increase in strength when there is conjunctive presynaptic and postsynaptic activity. The Hebb rule can be expressed more formally as follows

$$\delta w_{ij} = \alpha y_i x_j \qquad (1.1),$$

where δw_{ij} is the change of the synaptic weight w_{ij} which results from the simultaneous (or conjunctive) presence of presynaptic firing x_j and postsynaptic firing y_i (or strong depolarization), and α is a learning rate constant that specifies how much the synapses alter on any one pairing. The presynaptic and postsynaptic activity must be present approximately simultaneously (to within perhaps 100–500 ms in the real brain).

The Hebb rule is expressed in this multiplicative form to reflect the idea that both presynaptic and postsynaptic activity must be present for the synapses to increase in strength. The multiplicative form also reflects the idea that strong pre- and postsynaptic firing will produce a larger change of synaptic weight than smaller firing rates. The Hebb rule thus captures what is typically found in studies of associative Long-Term Potentiation (LTP) in the brain, described in Section 1.4.

One useful property of large neurons in the brain, such as cortical pyramidal cells, is that with their short electrical length, the postsynaptic term, h_i (see Equation 1.19), is available on much of the dendrite of a cell. The implication of this is that once sufficient postsynaptic activation has been produced, any active presynaptic terminal on the neuron will show synaptic strengthening. This enables associations between coactive inputs, or correlated activity in input axons, to be learned by neurons using this simple associative learning rule.

If, in contrast, a group of coactive axons made synapses close together on a small dendrite, then the local depolarization might be intense, and only these synapses would modify onto the dendrite. (A single distant active synapse might not modify in this type of neuron, because of the long electrotonic length of the dendrite.) The computation in this case is described as Sigma-Pi ($\Sigma\Pi$), to indicate that there is a local product computed during learning; this allows a particular set of locally active synapses to modify together, and then the output of the neuron can reflect the sum of such local multiplications (see Rumelhart and McClelland (1986), Koch (1999)). Sigma-Pi neurons are not used in most of the networks described in this book. There has been some work on how such neurons, if present in the brain, might utilize this functionality in the computation of invariant representations (Mel, Ruderman and Archie 1998, Mel and Fiser 2000).

1.4 Long-term potentiation and long-term depression as models of synaptic modification

Long-term potentiation (LTP) and long-term depression (LTD) provide useful models of some of the synaptic modifications that occur in the brain. The synaptic changes found appear to be synapse-specific, and to depend on information available locally at the synapse. LTP and LTD may thus provide a good model of the biological synaptic modifications involved in real neuronal network operations in the brain. Some of the properties of LTP and LTD are described next, together with evidence that implicates them in learning in at least some brain systems. Even if they turn out not to be the basis for the synaptic modifications that occur during learning, they have many of the properties that would be needed by some of the synaptic modification systems used by the brain.

Long-term potentiation is a use-dependent and sustained increase in synaptic strength that can be induced by brief periods of synaptic stimulation. It is usually measured as a sustained increase in the amplitude of electrically evoked responses in specific neural pathways following brief trains of high-frequency stimulation (see Fig. 1.2b). For example, high frequency stimulation of the Schaffer collateral inputs to the hippocampal CA1 cells results in a larger response recorded from the CA1 cells to single test pulse stimulation of the pathway. LTP is long-lasting, in that its effect can be measured for hours in hippocampal slices, and in chronic *in vivo* experiments in some cases it may last for months. LTP becomes evident rapidly, typically in less than one minute. LTP is in some brain systems associative. This is illustrated in Fig. 1.2c, in which a weak input to a group of cells (e.g. the commissural input

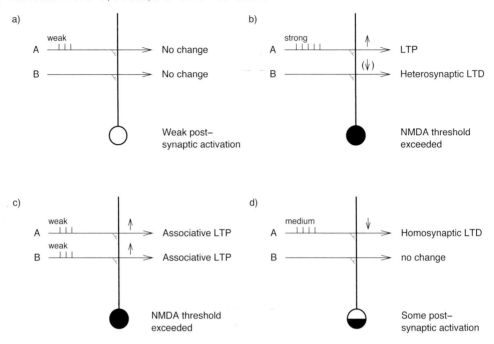

Fig. 1.2 Schematic illustration of synaptic modification rules as revealed by long-term potentiation (LTP) and long-term depression (LTD). The activation of the postsynaptic neuron is indicated by the extent to which its soma is black. There are two sets of inputs to the neuron: A and B. (a) A weak input (indicated by three spikes) on the set A of input axons produces little postsynaptic activation, and there is no change in synaptic strength. (b) A strong input (indicated by five spikes) on the set A of input axons produces strong postsynaptic activation, and the active synapses increase in strength. This is LTP. It is homosynaptic in that the synapses that increase in strength are the same as those through which the neuron is activated. LTP is synapse-specific, in that the inactive axons, B, do not show LTP. They either do not change in strength, or they may weaken. The weakening is called heterosynaptic LTD, because the synapses that weaken are other than those through which the neuron is activated (hetero- is Greek for other). (c) Two weak inputs present simultaneously on A and B summate to produce strong postsynaptic activation, and both sets of active synapses show LTP. (d) Intermediate strength firing on A produces some activation, but not strong activation, of the postsynaptic neuron. The active synapses become weaker. This is homosynaptic LTD, in that the synapses that weaken are the same as those through which the neuron is activated (homo- is Greek for same).

to CA1) does not show LTP unless it is given at the same time as (i.e. associatively with) another input (which could be weak or strong) to the cells. The associativity arises because it is only when sufficient activation of the postsynaptic neuron to exceed the threshold of NMDA receptors (see later) is produced that any learning can occur. The two weak inputs summate to produce sufficient depolarization to exceed the threshold. This associative property is shown very clearly in experiments in which LTP of an input to a single cell only occurs if the cell membrane is depolarized by passing current through it at the same time as the input arrives at the cell. The depolarization alone or the input alone is not sufficient to produce the LTP, and the LTP is thus associative. Moreover, in that the presynaptic input and the postsynaptic depolarization must occur at about the same time (within approximately 500 ms), the LTP requires temporal contiguity. LTP is also synapse-specific, in that, for example, an inactive input to a cell does not show LTP even if the cell is strongly activated by other inputs (Fig. 1.2b, input B).

These spatiotemporal properties of long term potentiation can be understood in terms

Pre-synaptic terminal   Post-synaptic membrane   Synaptic potentials

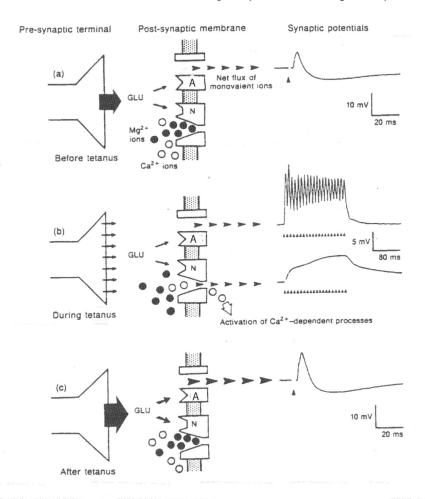

Fig. 1.3 The mechanism of induction of LTP in the CA1 region of the hippocampus. (a) Neurotransmitter (e.g. L-glutamate, GLU) is released and acts upon both AMPA (A) and NMDA (N) receptors. The NMDA receptors are blocked by magnesium and the excitatory synaptic response (EPSP) is therefore mediated primarily by ion flow through the channels associated with AMPA receptors. (b) During high-frequency activation ('tetanus'), the magnesium block of the ion channels associated with NMDA receptors is released by depolarization. Activation of the NMDA receptor by transmitter now results in ions moving through the channel. In this way, calcium enters the postsynaptic region to trigger various intracellular mechanisms which eventually result in an alteration of synaptic efficacy. (c) Subsequent low-frequency stimulation results in a greater EPSP. See text for further details. (After Collingridge and Bliss 1987.)

of actions of the inputs on the postsynaptic cell, which in the hippocampus has two classes of receptor, NMDA (N-methyl-D-aspartate) and AMPA (alpha-amino-3-hydroxy-5-methyl-isoxasole-4-propionic acid), activated by the glutamate released by the presynaptic terminals. The NMDA receptor channels are normally blocked by Mg^{2+}, but when the cell is strongly depolarized by strong tetanic stimulation of the type necessary to induce LTP, the Mg^{2+} block is removed, and Ca^{2+} entering via the NMDA receptor channels triggers events that lead to the potentiated synaptic transmission (see Fig. 1.3). Part of the evidence for this is that NMDA antagonists such as AP5 (D-2-amino-5-phosphonopentanoate) block LTP. Further, if the postsynaptic membrane is voltage clamped to prevent depolarization by a strong input, then LTP does not occur. The voltage-dependence of the NMDA receptor channels introduces a threshold and thus a non-linearity that contributes to a number of the phenomena of some types of LTP, such as cooperativity (many small inputs together produce sufficient depolarization to allow the NMDA receptors to operate); associativity (a weak input alone will not produce sufficient depolarization of the postsynaptic cell to enable the NMDA receptors to be activated, but the depolarization will be sufficient if there is also a strong input); and temporal contiguity between the different inputs that show LTP (in that if inputs occur non-conjunctively, the depolarization shows insufficient summation to reach the required level, or some of the inputs may arrive when the depolarization has decayed). Once the LTP has become established (which can be within one minute of the strong input to the cell), the LTP is expressed through the AMPA receptors, in that AP5 blocks only the establishment of LTP, and not its subsequent expression (Bliss and Collingridge 1993, Nicoll and Malenka 1995, Fazeli and Collingridge 1996, Lynch 2004, Collingridge and Bliss 1987).

There are a number of possibilities about what change is triggered by the entry of Ca^{2+} to the postsynaptic cell to mediate LTP. One possibility is that somehow a messenger reaches the presynaptic terminals from the postsynaptic membrane and, if the terminals are active, causes them to release more transmitter in future whenever they are activated by an action potential. Consistent with this possibility is the observation that, after LTP has been induced, more transmitter appears to be released from the presynaptic endings. Another possibility is that the postsynaptic membrane changes just where Ca^{2+} has entered, so that AMPA receptors become more responsive to glutamate released in future. Consistent with this possibility is the observation that after LTP, the postsynaptic cell may respond more to locally applied glutamate (using a microiontophoretic technique).

The rule that underlies associative LTP is thus that synapses connecting two neurons become stronger if there is conjunctive presynaptic and (strong) postsynaptic activity. This learning rule for synaptic modification is sometimes called the Hebb rule, after Donald Hebb of McGill University who drew attention to this possibility, and its potential importance in learning, in 1949.

In that LTP is long-lasting, develops rapidly, is synapse-specific, and is in some cases associative, it is of interest as a potential synaptic mechanism underlying some forms of memory. Evidence linking it directly to some forms of learning comes from experiments in which it has been shown that the drug AP5 infused so that it reaches the hippocampus to block NMDA receptors blocks spatial learning mediated by the hippocampus (Morris 1989, Martin, Grimwood and Morris 2000, Morris 2003, Morris 2006, Wang and Morris 2010). The task learned by the rats was to find the location relative to cues in a room of a platform submerged in an opaque liquid (milk). Interestingly, if the rats had already learned where the platform was, then the NMDA infusion did not block performance of the task. This is a close parallel to LTP, in that the learning, but not the subsequent expression of what had been learned, was blocked by the NMDA antagonist AP5. Although there is still some uncertainty about the experimental evidence that links LTP to learning (see for example Martin, Grimwood and Morris (2000) and Lynch (2004)), there is a need for a synapse-specific modifiability of synaptic strengths on neurons if neuronal networks are to learn. If LTP is not always an exact model of the synaptic modification that occurs during learning, then something with many of the properties of LTP is nevertheless needed, and is likely to be present in the brain given the functions known to be implemented in many brain regions (see Rolls and Treves (1998)).

In another model of the role of LTP in memory, Davis (2000) has studied the role of the amygdala in learning associations to fear-inducing stimuli. He has shown that blockade of NMDA synapses in the amygdala interferes with this type of learning, consistent with the idea that LTP also provides a useful model of this type of learning (see further Rolls (2008d), Chapter 3).

Long-Term Depression (LTD) can also occur. It can in principle be associative or non-associative. In associative LTD, the alteration of synaptic strength depends on the pre- and post-synaptic activities. There are two types. Heterosynaptic LTD occurs when the postsynaptic neuron is strongly activated, and there is low presynaptic activity (see Fig. 1.2b input B). Heterosynaptic LTD is so called because the synapse that weakens is other than (hetero) the one through which the postsynaptic neuron is activated. Heterosynaptic LTD is important in associative neuronal networks, and in competitive neuronal networks (see Rolls (2008d)). In competitive neural networks it would be helpful if the degree of heterosynaptic LTD depended on the existing strength of the synapse, and there is some evidence that this may be the case (Rolls 2008d). Homosynaptic LTD occurs when the presynaptic neuron is strongly active, and the postsynaptic neuron has some, but low, activity (see Fig. 1.2d). Homosynaptic LTD is so-called because the synapse that weakens is the same as (homo) the one that is active. Heterosynaptic and homosynaptic LTD are found in the neocortex (Artola and Singer 1993, Singer 1995, Fregnac 1996) and hippocampus (Christie 1996), and in many cases are dependent on activation of NMDA receptors (see also Fazeli and Collingridge (1996)). LTD in the cerebellum is evident as weakening of active parallel fibre to Purkinje cell synapses when the climbing fibre connecting to a Purkinje cell is active (Ito 1984, Ito 1989, Ito 1993b, Ito 1993a).

An interesting time-dependence of LTP and LTD has been observed, with LTP occurring especially when the presynaptic spikes precede by a few milliseconds (ms) the postsynaptic activation, and LTD occurring when the presynaptic spikes follow the postsynaptic activation by a few milliseconds (Markram, Lubke, Frotscher and Sakmann 1997, Bi and Poo 1998, Bi and Poo 2001, Senn, Markram and Tsodyks 2001, Dan and Poo 2004, Dan and Poo 2006). This is referred to as spike timing-dependent plasticity, STDP. This type of temporally asymmetric Hebbian learning rule, demonstrated in the hippocampus and neocortex, can induce associations over time, and not just between simultaneous events. Networks of neurons with such synapses can learn sequences (Minai and Levy 1993), enabling them to predict the future state of the postsynaptic neuron based on past experience (Abbott and Blum 1996) (see further Koch (1999), Markram, Pikus, Gupta and Tsodyks (1998) and Abbott and Nelson (2000)). This mechanism, because of its apparent time-specificity for periods in the range of ms or tens of ms, could also encourage neurons to learn to respond to temporally synchronous presynaptic firing (Gerstner, Kreiter, Markram and Herz 1997, Gutig and Sompolinsky 2006), and indeed to decrease the synaptic strengths from neurons that fire at random times with respect to the synchronized group. This mechanism might also play a role in the normalization of the strength of synaptic connection strengths onto a neuron. Further, there is accumulating evidence (Sjostrom, Turrigiano and Nelson 2001) that a more realistic description of the protocols for inducing LTP and LTD probably requires a combination of dependence on spike timing – to take into account the effects of the backpropagating action potential – and dependence on the sub-threshold depolarization of the postsynaptic neuron. However, these spike timing dependent synaptic modifications may be evident primarily at low firing rates rather than those that often occur in the brain (Sjostrom, Turrigiano and Nelson 2001), and may not be especially reproducible in the cerebral neocortex. Under the somewhat steady-state conditions of the firing of neurons in the higher parts of the ventral visual system on the 10-ms timescale that are observed not only when single stimuli are presented for 500 ms (see Fig. 8.5), but also when macaques have found a search target and are looking at it (Rolls, Aggelopoulos and Zheng 2003a, Aggelopoulos and Rolls 2005, Rolls 2008d), the average of the presynaptic and postsynaptic rates are likely to be the important determinants of synaptic modification. Part of the reason for this is that correlations between the firing of simultaneously recorded inferior temporal cortex neurons are not common, and if present are not very strong or typically restricted to a short time window in the order of 10 ms (Rolls 2008d). This point is also made in the context that each neuron has thousands of inputs, several tens of which are normally likely to be active when a cell is firing above its spontaneous firing rate and is strongly depolarized. This may make it unlikely statistically that there will be a strong correlation between a particular presynaptic spike and postsynaptic firing, and thus that this is likely to be a main determinant of synaptic strength under these natural conditions.

1.5 Neuronal biophysics

An important part of the processing by a single neuron is given by the difference of potential across the neuronal cell membrane. The difference of potential across the membrane results from currents produced by ion flows. The main ions involved in this process are the positively charged cations: sodium (Na^+), potassium (K^+), calcium (Ca^+), and the negatively charged anion: chloride (Cl^-). The cell membrane consists of two layers of lipid macromolecules that separate the interior from the exterior of cell, with both sides containing solutions in which the ions can move freely. In the membrane there are pores, which are called ion channels because the different ions can flow through these channels. This ion flow can change the net charge on each side of the membrane, and thus cause a difference of potential across the cell membrane. There are also negatively charged macromolecules (M^-) that are confined to the interior of the cell because they cannot flow through these ion channels.

There are two different types of ion channel, passive and active. Passive ion channels let the ions flow freely through the membrane, with their flow influenced by Fick's law (which states that the ions flow so that the difference of potential is compensated), and by the diffusion law (which states that the ions flow so that the difference of concentration is compensated). The equilibrium between these two sources of movement is called the Nernst equilibrium and is described by the Nernst equation:

$$V_T = \frac{RT}{zF} \ln \frac{[T]_{\text{out}}}{[T]_{\text{in}}} \tag{1.2}$$

where V_T is the potential difference, $[T]_{\text{out}}$ and $[T]_{\text{in}}$ are the concentrations of the ions T inside and outside the cell, respectively, R is the universal gas constant (8,315 mJ/(K°.mol)), T is the absolute temperature in degrees Kelvin, z is the valence (number of charges) of the ion, and F is Faraday's constant (96,480 coulombs/mol). At equilibrium the difference in concentration causes a Nernst potential. If the voltage difference ΔV is smaller than the value of the Nernst potential V_T, ions flow into the cell. If the voltage is larger than the Nernst potential V_T, ions flow out of the cell. Thus the direction of the current is reversed when the voltage ΔV passes the Nernst potential, and therefore the Nernst potential V_T of a T ion-channel is called its reversal potential. When not in equilibrium, the net ionic current is governed by Ohm's law which states that the current I_T is proportional to the difference of potential through the membrane V, i.e.:

$$I_T = g_T(V - V_T) \tag{1.3}$$

where g_T is the membrane conductance for the ion T.

The value at which the net inward and outward currents balance is called the resting potential. If there are only ions of type T, than the resting potential of the neuron would be given by the reversal potential V_T. However, neurons have more than one ion type, so the resting potential results from the global equilibrium due to all ion currents. Typically, if the main currents are due to Na^+, K^+, Ca^+, and Cl^-, the resulting expression for the resting potential is given by:

$$V_{rest} = \frac{g_{Na}V_{Na} + g_{Ca}V_{Ca} + g_K V_K + g_{Cl}V_{Cl}}{g_{Na} + g_{Ca} + g_K + g_{Cl}}. \qquad (1.4)$$

On the other hand, active ion channels influence the flow of the ions through the membrane. There are three main types of active ion channel: voltage-dependent ion channels, chemically-dependent ion channels, and ion pumps.

The ion pumps are responsible for maintaining a difference of concentration of some types of ion across the cell membrane. The most relevant ion pump is the so-called Na^+–K^+ pump which transports out three Na^+ ions for every two K^+ ions pumped in. Due to this gradient, the reversal potentials of sodium and potassium are different from zero, and therefore the resting potential of the cell is also different from zero.

The chemically-gated ion channels are channels whose conductance is dependent on the presence of a specific chemical near the channel, for example, a neurotransmitter. In Chapter 2 we will describe the most typical synaptically activated ion channels, which are activated by NMDA, AMPA, and GABA receptors.

The voltage-gated ion channels are responsible for the generation of an action potential or spike. In this case, the conductance of the channel depends on the membrane potential. There are two different types of gate, namely, gates that open the channel, and gates that close the channel. Hodgkin and Huxley (1952) described mathematically how these channels work. They assumed that the conductance of these channels can be described by two gating variables p and q, in the following way:

$$g_T = p^a q^b \qquad (1.5)$$

where a, b are the number of activation or deactivation gates required for that channel. Further, they assumed that the gating variables follow a simple kinetic equation of the type:

$$\frac{dp}{dt} = \frac{p_\infty(V(t)) - p(t)}{\tau(V(t))} \qquad (1.6)$$

where the asymptotic activation value p_∞ and the time constant τ can be measured experimentally. A similar equation is used for q. We will see in Section 1.6 how these voltage-dependent ion-channels explain the dynamics of a single neuron.

1.6 Action potential dynamics

Developing what has just been described about voltage-dependent ion channels, a point-like neuron can generate spikes by altering the membrane potential V according to continuous equations of the Hodgkin–Huxley type:

$$C\frac{dV}{dt} = -I_L - I_K - I_{Na} - I_{syn}, \qquad (1.7)$$

where I_{syn} is the current produced through the synaptically activated ion channels, as described in Chapter 2.

The ionic currents (leak current I_L, sodium current I_{Na}, and potassium current I_K) are given by:

$$I_L = g_L(V - V_L), \qquad (1.8)$$

$$I_{\text{Na}} = g_{\text{Na}} m^3 h (V - V_{\text{Na}}),\tag{1.9}$$

$$I_{\text{K}} = g_{\text{K}} n^4 (V - V_{\text{K}}).\tag{1.10}$$

The gating variables h, m and n satisfy the usual type of differential equation:

$$\frac{dw}{dt} = \alpha_w(V)(1 - w) - \beta_w(V)w = \frac{w_\infty(V) - w}{\tau_w(V)},\tag{1.11}$$

where w represents h, m or n. In general, $w_\infty(V) = (\alpha_w(V))/(\alpha_w(V) + \beta_w(V))$, and $\tau_w(V) = [\alpha_w(V) + \beta_w(V)]^{-1}$. The specific $\alpha_w(V)$ and $\beta_w(V)$ functions are defined as

$$\alpha_h(V) = 0.128 \exp(-(50 + V)/18),\tag{1.12}$$

$$\beta_h(V) = 4/(1 + \exp(-(V + 27)/5)),\tag{1.13}$$

$$\alpha_n(V) = 0.032(V + 52)/(1 - \exp(-(V + 52)/5)),\tag{1.14}$$

$$\beta_n(V_j) = 0.5 \exp(-(57 + V)/40),\tag{1.15}$$

$$\alpha_m(V) = 0.32(V + 50)/(1 - \exp(-(V + 50)/4)),\tag{1.16}$$

$$\beta_m(V) = 0.28(V + 27)/(\exp((V + 27)/5) - 1)\tag{1.17}$$

The changes in the membrane potential, driven by the input current I, interact with the opening and closing of intrinsic voltage-gated conductances (here a sodium conductance, whose channels are gated by the 'particles' m and h, and a potassium conductance, whose channels are gated by n (Hodgkin and Huxley 1952)). These equations provide an effective description, phenomenological but broadly based on physical principles, of the conductance changes underlying action potentials.

1.7 Systems-level analysis of brain function

To understand the neuronal network operations of any one brain region, it is useful to have an idea of the systems-level organization of the brain, in order to understand how the networks in each region provide a particular computational function as part of an overall computational scheme. In the context of vision, it is very useful to appreciate the different processing streams, and some of the outputs that each has. Some of the processing streams are shown in Fig. 1.4. Some of these regions are shown in the drawings of the primate brain in the next few figures. Each of these routes is described in turn. The description is based primarily on studies in non-human primates, for they have well-developed cortical areas that in many cases correspond to those found in humans, and it has been possible to analyze their connectivity and their functions by recording the activity of neurons in them.

Information in the 'ventral or what' visual cortical processing stream projects after the primary visual cortex, area V1, to the secondary visual cortex (V2), and then via area V4 to the posterior and then to the anterior inferior temporal visual cortex (see Figs. 1.4, 1.5, and 1.6).

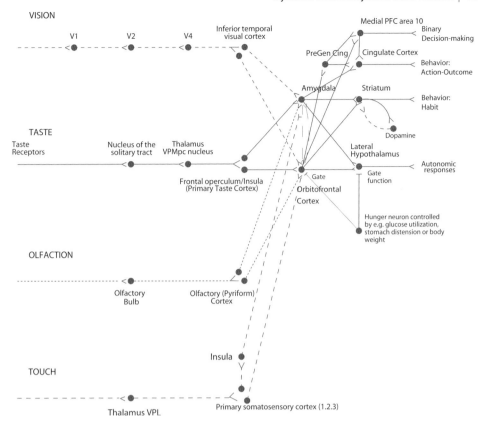

Fig. 1.4 The pathways involved in some different systems described in the text. The top pathway, also shown in Fig. 1.5, shows the connections in the 'ventral or what visual pathway' from V1 to V2, V4, the inferior temporal visual cortex, etc., with some connections reaching the amygdala and orbitofrontal cortex. There are onward projections from the orbitofrontal cortex to the pregenual cingulate cortex (PreGen Cing), to the anterior cingulate cortex, and to medial prefrontal cortex (PFC) area 10. The taste pathways project after the primary taste cortex to the orbitofrontal cortex and amygdala. The olfactory pathways project from the primary olfactory (pyriform) cortex to the orbitofrontal cortex and amygdala. The bottom pathway shows the connections from the primary somatosensory cortex, areas 1, 2 and 3, to the mid-insula, orbitofrontal cortex, and amygdala. Somatosensory areas 1, 2 and 3 also project via area 5 in the parietal cortex, to area 7b.

Information processing along this stream is primarily unimodal, as shown by the fact that inputs from other modalities (such as taste or smell) do not anatomically have significant inputs to these regions, and by the fact that neurons in these areas respond primarily to visual stimuli, and not to taste or olfactory stimuli, etc. (Rolls 2000a, Baylis, Rolls and Leonard 1987, Ungerleider 1995, Rolls and Deco 2002). The representation built along this pathway is mainly about what object is being viewed, independently of exactly where it is on the retina, of its size, and even of the angle with which it is viewed (Rolls and Deco 2002, Rolls 2008d, Rolls 2007d, Rolls and Stringer 2006), and for this reason it is frequently referred to as the 'what' visual pathway. The representation is also independent of whether the object is associated with reward or punishment, that is the representation is about objects per se (Rolls, Judge and Sanghera 1977). The computation that must be performed along this stream is thus primarily to build a representation of objects that shows invariance (Rolls 2008d). After this processing, the visual representation is interfaced to other sensory systems in areas such as

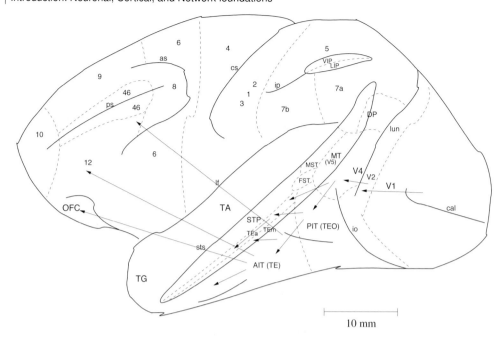

Fig. 1.5 Lateral view of the macaque brain showing the connections in the 'ventral or what visual pathway' from V1 to V2, V4, the inferior temporal visual cortex, etc., with some connections reaching the amygdala and orbitofrontal cortex. as, arcuate sulcus; cal, calcarine sulcus; cs, central sulcus; lf, lateral (or Sylvian) fissure; lun, lunate sulcus; ps, principal sulcus; io, inferior occipital sulcus; ip, intraparietal sulcus (which has been opened to reveal some of the areas it contains); sts, superior temporal sulcus (which has been opened to reveal some of the areas it contains). AIT, anterior inferior temporal cortex; FST, visual motion processing area; LIP, lateral intraparietal area; MST, visual motion processing area; MT, visual motion processing area (also called V5); OFC, orbitofrontal cortex; PIT, posterior inferior temporal cortex; STP, superior temporal plane; TA, architectonic area including auditory association cortex; TE, architectonic area including high order visual association cortex, and some of its subareas TEa and TEm; TG, architectonic area in the temporal pole; V1–V4, visual areas 1–4; VIP, ventral intraparietal area; TEO, architectonic area including posterior visual association cortex. The numbers refer to architectonic areas, and have the following approximate functional equivalence: 1, 2, 3, somatosensory cortex (posterior to the central sulcus); 4, motor cortex; 5, superior parietal lobule; 7a, inferior parietal lobule, visual part; 7b, inferior parietal lobule, somatosensory part; 6, lateral premotor cortex; 8, frontal eye field; 12, inferior convexity prefrontal cortex; 46, dorsolateral prefrontal cortex.

the orbitofrontal cortex, amygdala, and hippocampus in which simple associations must be learned between stimuli in different modalities (Rolls 2008d, Rolls 2005). The representation must thus be in a form in which the simple generalization properties of associative networks can be useful. Given that the association is about what object is present (and not where it is on the retina), the representation computed in sensory systems must be in a form that allows the simple correlations computed by associative networks to reflect similarities between objects, and not between their positions on the retina (Rolls 2008d). The way in which such invariant sensory representations could be built in the brain is described elsewhere (Rolls and Deco 2002, Rolls 2008d, Rolls and Stringer 2006, Rolls, Tromans and Stringer 2008d).

The ventral visual stream converges with other mainly unimodal information processing streams for taste, olfaction, touch, and hearing in a number of areas, particularly the amygdala and orbitofrontal cortex (see Figs. 1.4, 1.5, and 1.6). These areas appear to be necessary for learning to associate sensory stimuli with other reinforcing (rewarding or punishing) stimuli. For example, the amygdala is involved in learning associations between the sight of food and

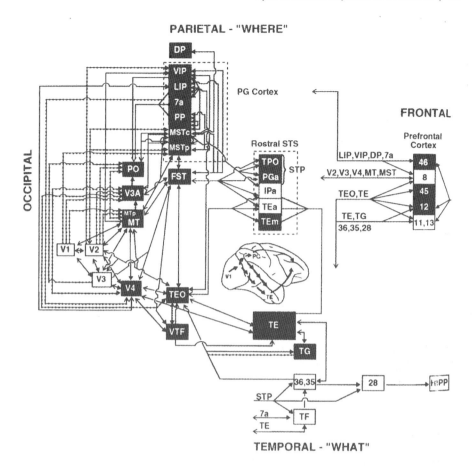

Fig. 1.6 Visual processing pathways in monkeys. Solid lines indicate connections arising from both central and peripheral visual field representations; dotted lines indicate connections restricted to peripheral visual field representations. Shaded boxes in the 'ventral (lower) or what' stream indicate visual areas related primarily to object vision; shaded boxes in the 'dorsal or where' stream indicate areas related primarily to spatial vision; and white boxes indicate areas not clearly allied with only one stream. Abbreviations: DP, dorsal prelunate area; FST, fundus of the superior temporal area; HIPP, hippocampus; LIP, lateral intraparietal area; MSTc, medial superior temporal area, central visual field representation; MSTp, medial superior temporal area, peripheral visual field representation; MT, middle temporal area; MTp, middle temporal area, peripheral visual field representation; PO, parieto-occipital area; PP, posterior parietal sulcal zone; STP, superior temporal polysensory area; V1, primary visual cortex; V2, visual area 2; V3, visual area 3; V3A, visual area 3, part A; V4, visual area 4; and VIP, ventral intraparietal area. Inferior parietal area 7a; prefrontal areas 8, 11 to 13, 45 and 46 are from Brodmann (1925). Inferior temporal areas TE and TEO, parahippocampal area TF, temporal pole area TG, and inferior parietal area PG are from Von Bonin and Bailey (1947). Rostral superior temporal sulcal (STS) areas are from Seltzer and Pandya (1978) and VTF is the visually responsive portion of area TF (Boussaoud, Desimone and Ungerleider 1991). Areas 11–13 are in the orbitofrontal cortex, and area 46 is in the dorsolateral prefrontal cortex. Areas 35 and 36 are in the perirhinal cortex, and area 28 is the entorhinal cortex. (Reprinted with permission from Ungerleider 1995.)

its taste. (The taste is a primary or innate reinforcer.) The orbitofrontal cortex is especially involved in rapidly relearning these associations, when environmental contingencies change (see Rolls (2005) and Rolls (2000c)). They thus are brain regions in which the computation

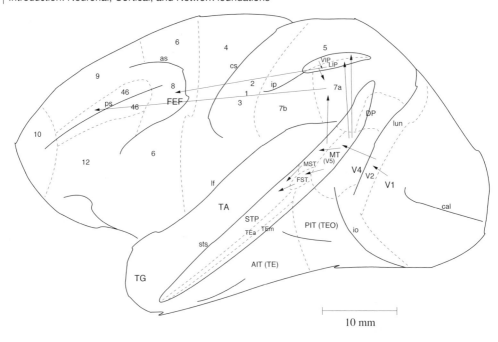

Fig. 1.7 Lateral view of the macaque brain showing the connections in the 'dorsal or where visual pathway' from V1 to V2, MST, LIP, VIP, and parietal cortex area 7a, with some connections then reaching the dorsolateral prefrontal cortex including area 46. Abbreviations as in Fig. 1.5. FEF–frontal eye field.

at least includes simple pattern association (e.g. between the sight of an object and its taste). In the orbitofrontal cortex, this association learning is also used to produce a representation of flavour, in that neurons are found in the orbitofrontal cortex that are activated by both olfactory and taste stimuli (Rolls and Baylis 1994), and in that the neuronal responses in this region reflect in some cases olfactory to taste association learning (Rolls, Critchley, Mason and Wakeman 1996b, Critchley and Rolls 1996). In these regions too, the representation is concerned not only with what sensory stimulus is present, but for some neurons, with its hedonic or reward-related properties, which are often computed by association with stimuli in other modalities. For example, many of the visual neurons in the orbitofrontal cortex respond to the sight of food only when hunger is present. This probably occurs because the visual inputs here have been associated with a taste input, which itself in this region only occurs to a food if hunger is present, that is when the taste is rewarding (Rolls 2005, Rolls 2000c, Rolls 2008d). The outputs from these associative memory systems, the amygdala and orbitofrontal cortex, project onwards to structures such as the hypothalamus, through which they control autonomic and endocrine responses such as salivation and insulin release to the sight of food; and to the striatum, including the ventral striatum, through which behaviour to learned reinforcing stimuli is produced.

The *'dorsal or where' visual processing stream* shown in Figs. 1.7, and 1.6 is that from V1 to MT, MST and thus to the parietal cortex (see Ungerleider (1995); Ungerleider and Haxby (1994); and Rolls and Deco (2002)). This 'where' pathway for primate vision is involved in representing where stimuli are relative to the animal (i.e. in egocentric space), and the motion of these stimuli. Neurons here respond, for example, to stimuli in visual space around the animal, including the distance from the observer, and also respond to optic flow or to moving stimuli. Outputs of this system control eye movements to visual stimuli (both slow pursuit and saccadic eye movements). These outputs proceed partly via the frontal eye fields, which then project to the striatum, and then via the substantia nigra reach the superior colliculus (Goldberg 2000). Other outputs of these regions are to the dorsolateral prefrontal cortex, area 46, which is important as a short-term memory for where fixation should occur next, as shown by the effects of lesions to the prefrontal cortex on saccades to remembered targets, and by neuronal activity in this region (Goldman-Rakic 1996). The dorsolateral prefrontal cortex short-term memory systems in area 46 with spatial information received from the parietal cortex play an important role in attention, by holding on-line the target being attended to, as described in Chapter 4.

The hippocampus receives inputs from both the 'what' and the 'where' visual systems (Rolls and Kesner 2006, Rolls 2008d) (see Fig. 1.6). By rapidly learning associations between conjunctive inputs in these systems, it is able to form memories of particular events occurring in particular places at particular times. To do this, it needs to store whatever is being represented in each of many cortical areas at a given time, and later to recall the whole memory from a part of it. The types of network it contains that are involved in this simple memory function are described elsewhere (Rolls and Kesner 2006, Rolls 2008d).

1.8 Introduction to the fine structure of the cerebral neocortex

An important part of the approach to understanding how the cerebral cortex could implement the computational processes that underlie memory, decision-making, and perception is to take into account as much as possible its fine structure and connectivity, as these provide important indicators of and constraints on how it computes.

1.8.1 The fine structure and connectivity of the neocortex

The neocortex consists of many areas that can be distinguished by the appearance of the cells (cytoarchitecture) and fibres or axons (myeloarchitecture), but nevertheless, the basic organization of the different neocortical areas has many similarities, and it is this basic organization that is considered here. Useful sources for more detailed descriptions of neocortical structure and function are the book *Cerebral Cortex* edited by Jones and Peters (Jones and Peters (1984) and Peters and Jones (1984)); and Douglas, Markram and Martin (2004). Approaches to quantitative aspects of the connectivity are provided by Braitenberg and Schutz (1991) and by Abeles (1991). Some of the connections described in Sections 1.8.2 and 1.8.3 are shown schematically in Fig. 1.9.

1.8.2 Excitatory cells and connections

Some of the cell types found in the neocortex are shown in Fig. 1.1. Cells A–D are pyramidal cells. The dendrites (shown thick in Fig. 1.1) are covered in spines, which receive the excitatory synaptic inputs to the cell. Pyramidal cells with cell bodies in different laminae of the cortex (shown in Fig. 1.1 as I–VI) not only have different distributions of their dendrites, but also different distributions of their axons (shown thin in Fig. 1.1), which connect both within that cortical area and to other brain regions outside that cortical area (see labelling at the bottom of Fig. 1.9).

The main information-bearing afferents to a cortical area have many terminals in layer 4. (By these afferents, we mean primarily those from the thalamus or from the preceding cortical area. We do not mean the cortico-cortical backprojections, nor the subcortical cholinergic, noradrenergic, dopaminergic, and serotonergic inputs, which are numerically minor, although they are important in setting cortical cell thresholds, excitability, and adaptation, see for example Douglas, Markram and Martin (2004).) In primary sensory cortical areas only there are spiny stellate cells in a rather expanded layer 4, and the thalamic terminals synapse onto these cells (Lund 1984, Martin 1984, Douglas and Martin 1990, Douglas, Markram and Martin 2004, Levitt, Lund and Yoshioka 1996). (Primary sensory cortical areas receive their inputs from the primary sensory thalamic nucleus for a sensory modality. An example is the primate striate cortex which receives inputs from the lateral geniculate nucleus, which in turn receives inputs from the retinal ganglion cells. Spiny stellate cells are so-called because they have radially arranged, star-like, dendrites. Their axons usually terminate within the cortical area in which they are located.) Each thalamic axon makes 1,000–10,000 synapses, not more than several (or at most 10) of which are onto any one spiny stellate cell. In addition to these afferent terminals, there are some terminals of the thalamic afferents onto pyramidal cells with cell bodies in layers 6 and 3 (Martin 1984) (and terminals onto inhibitory interneurons such as basket cells, which thus provide for a feedforward inhibition) (see Fig. 1.8). Even in layer 4, the thalamic axons provide less than 20% of the synapses. The spiny stellate neurons in layer 4 have axons which terminate in layers 3 and 2, at least partly on dendrites of pyramidal cells with cell bodies in layers 3 and 2. (These synapses are of Type I, that is are asymmetrical and are on spines, so that they are probably excitatory. Their transmitter is probably glutamate.) These layer 3 and 2 pyramidal cells provide the onward cortico-cortical projection with axons which project into layer 4 of the next cortical area. For example, layer 3 and 2 pyramidal cells in the primary visual (striate) cortex of the macaque monkey project into the second visual area (V2), layer 4.

In non-primary sensory areas, important information-bearing afferents from a preceding cortical area terminate in layer 4, but there are no or few spiny stellate cells in this layer (Lund 1984, Levitt, Lund and Yoshioka 1996). Layer 4 still looks 'granular' (due to the presence of many small cells), but these cells are typically small pyramidal cells (Lund 1984). (It may be noted here that spiny stellate cells and small pyramidal cells are similar in many ways, with a few main differences including the absence of a major apical dendrite in a spiny stellate which accounts for its non-pyramidal, star-shaped, appearance; and for many spiny stellate cells, the absence of an axon that projects outside its cortical area.) The terminals presumably make synapses with these small pyramidal cells, and also presumably with the dendrites of cells from other layers, including the basal dendrites of deep layer 3 pyramidal cells (see Fig. 1.9).

The axons of the *superficial (layer 2 and 3) pyramidal cells* have collaterals and terminals in layer 5 (see Fig. 1.9), and synapses are made with the dendrites of the layer 5 pyramidal cells (Martin 1984). The axons also typically project out of that cortical area, and on to the next cortical area in sequence, where they terminate in layer 4, forming the forward cortico-cortical projection. It is also from these pyramidal cells that projections to the amygdala arise in some sensory areas that are high in the hierarchy (Amaral, Price, Pitkanen and Carmichael 1992).

The axons of the *layer 5 pyramidal cells* have many collaterals in layer 6 (see Fig. 1.1), where synapses could be made with the layer 6 pyramidal cells (based on indirect evidence, see Fig. 13 of Martin (1984)), and axons of these cells typically leave the cortex to project to subcortical sites (such as the striatum), or back to the preceding cortical area to terminate in layer 1. It is remarkable that there are as many of these backprojections as there are forward connections between two sequential cortical areas. The possible computational significance of this connectivity is considered below in Section 1.9 and elsewhere (Rolls and Kesner 2006, Rolls 2008d).

The *layer 6 pyramidal cells* have prolific dendritic arborizations in layer 4 (see Fig. 1.1), and receive synapses from thalamic afferents (Martin 1984), and also presumably from pyramidal cells in other cortical layers. The axons of these cells form backprojections to the thalamic nucleus which projects into that cortical area, and also axons of cells in layer 6 contribute to the backprojections to layer 1 of the preceding cortical area (see Jones and Peters (1984) and Peters and Jones (1984); see Figs. 1.1 and 1.9).

Although the pyramidal and spiny stellate cells form the great majority of neocortical neurons with excitatory outputs, there are in addition several further cell types (see Peters and Jones (1984), chapter 4). Bipolar cells are found in layers 3 and 5, and are characterized by having two dendritic systems, one ascending and the other descending, which, together with the axon distribution, are confined to a narrow vertical column often less than 50 μm in diameter (Peters 1984a). Bipolar cells form asymmetrical (presumed excitatory) synapses with pyramidal cells, and may serve to emphasize activity within a narrow vertical column.

1.8.3 Inhibitory cells and connections

There are a number of types of neocortical inhibitory neurons. All are described as smooth in that they have no spines, and use GABA (gamma-amino-butyric acid) as a transmitter. (In older terminology they were called Type II.) A number of types of inhibitory neuron can be distinguished, best by their axonal distributions (Szentagothai 1978, Peters and Regidor 1981, Douglas, Markram and Martin 2004). One type is the *basket cell*, present in layers 3–6, which has few spines on its dendrites so that it is described as smooth, and has an axon that participates in the formation of weaves of preterminal axons which surround the cell bodies of pyramidal cells and form synapses directly onto the cell body, but also onto the dendritic spines (Somogyi, Kisvarday, Martin and Whitteridge 1983) (Fig. 1.8). Basket cells comprise 5–7% of the total cortical cell population, compared with approximately 72% for pyramidal cells (Sloper and Powell 1979b, Sloper and Powell 1979a). Basket cells receive synapses from the main extrinsic afferents to the neocortex, including thalamic afferents (Fig. 1.8), so that they must contribute to a feedforward type of inhibition of pyramidal cells. The inhibition is feedforward in that the input signal activates the basket cells and the pyramidal cells by independent routes, so that the basket cells can produce inhibition of pyramidal cells that does not depend on whether the pyramidal cells have already fired. Feedforward inhibition of this type not only enhances stability of the system by damping the responsiveness of the pyramidal cell simultaneously with a large new input, but can also be conceived of as a mechanism which normalizes the magnitude of the input vector received by each small region of neocortex (Rolls 2008d). In fact, the feedforward mechanism allows the pyramidal cells to be set at the appropriate sensitivity for the input they are about to receive. Basket cells can also be polysynaptically activated by an afferent volley in the thalamo-cortical projection (Martin 1984), so that they may receive inputs from pyramidal cells, and thus participate in feedback inhibition of pyramidal cells.

The transmitter used by the basket cells is gamma-amino-butyric acid (GABA), which opens chloride channels in the postsynaptic membrane. Because the reversal potential for Cl^- is approximately -10 mV relative to rest, opening the Cl^- channels does produce an inhibitory postsynaptic potential (IPSP), which results in some hyperpolarization, especially in the dendrites. This is a subtractive effect, hence it is a linear type of inhibition (Douglas and Martin 1990, Douglas, Markram and Martin 2004). However, a major effect of the opening of the Cl^- channels in the cell body is that this decreases the membrane resistance, thus producing a shunting effect. The importance of shunting is that it decreases the magnitude of excitatory postsynaptic potentials (EPSPs) (cf. Andersen, Dingledine, Gjerstad, Langmoen and Laursen (1980) for hippocampal pyramidal cells), so that the effect of shunting is to produce division (i.e. a multiplicative reduction) of the excitatory inputs received by the cell, and not just to act by subtraction (see further Bloomfield (1974), Martin (1984), Douglas and

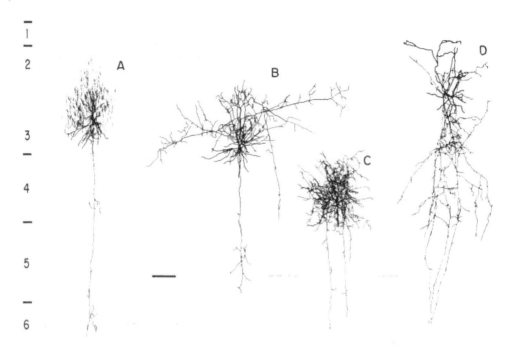

Fig. 1.8 Smooth cells from cat visual cortex. (A) Chandelier or axoaxonic cell. (B) Large basket cell of layer 3. Basket cells, present in layers 3–6, have few spines on their dendrites so that they are described as smooth, and have an axon which participates in the formation of weaves of preterminal axons which surround the cell bodies of pyramidal cells and form synapses directly onto the cell body. (C) Small basket or clutch cell of layer 3. The major portion of the axonal arbor is confined to layer 4. (D) Double bouquet cell. The axon collaterals run vertically. The cortical layers are as indicated. Bar = 100 μm. (Reproduced with permission from Douglas and Martin 1990, Fig. 12.4.)

Martin (1990)). Thus, when modelling the normalization of the activity of cortical pyramidal cells, it is common to include division in the normalization function (Rolls 2008d). It is notable that the dendrites of basket cells can extend laterally 0.5 mm or more (primarily within the layer in which the cell body is located), and that the axons can also extend laterally from the cell body 0.5–1.5 mm. Thus the basket cells produce a form of lateral inhibition which is quite spatially extensive. There is some evidence that each basket cell may make 4–5 synapses with a given pyramidal cell, that each pyramidal cell may receive from 10–30 basket cells, and that each basket cell may inhibit approximately 300 pyramidal cells (Martin 1984, Douglas and Martin 1990, Douglas, Markram and Martin 2004). The basket cells are sometimes called clutch cells.

A second type of GABA-containing inhibitory interneuron is the *axoaxonic (or 'chandelier') cell*, named because it synapses onto the initial segment of the axon of pyramidal cells. The pyramidal cells receiving this type of inhibition are almost all in layers 2 and 3, and much less in the deep cortical layers. One effect that axoaxonic cells probably produce is thus prevention of outputs from layer 2 and 3 pyramidal cells reaching the pyramidal cells in the deep layers, or from reaching the next cortical area. Up to five axoaxonic cells converge onto a pyramidal cell, and each axoaxonic cell may project to several hundred pyramidal cells scattered in a region that may be several hundred microns in length (Martin 1984, Peters 1984b). This implies that axoaxonic cells provide a rather simple device for preventing runaway overactivity of pyramidal cells, but little is known yet about the afferents to axoaxonic cells, so that the functions of these neurons are very incompletely understood.

A third type of (usually smooth and inhibitory) cell is the *double bouquet cell*, which has primarily vertically organized axons. These cells have their cell bodies in layer 2 or 3, and have an axon traversing layers 2–5, usually in a tight bundle consisting of varicose, radially oriented collaterals often confined to a narrow vertical column 50 μm in diameter (Somogyi and Cowey 1984). Double bouquet cells receive symmetrical, type II (presumed inhibitory) synapses, and also make type II synapses, perhaps onto the apical dendrites of pyramidal cells, so that these neurons may serve, by this double inhibitory effect, to emphasize activity within a narrow vertical column.

Another type of GABA-containing inhibitory interneuron is the smooth and sparsely spinous non-pyramidal (multipolar) neuron with local axonal plexuses (Peters and Saint Marie 1984). In addition to extrinsic afferents, these neurons receive many type I (presumed excitatory) terminals from pyramidal cells, and have inhibitory terminals on pyramidal cells, so that they may provide for the very important function of feedback or recurrent lateral inhibition (Rolls 2008d).

1.8.4 Quantitative aspects of cortical architecture

Some quantitative aspects of cortical architecture are described, because, although only preliminary data are available, they are crucial for developing an understanding of how the neocortex could work. Further evidence is provided by Braitenberg and Schutz (1991), and by Abeles (1991). Typical values, many of them after Abeles (1991), are shown in Table 1.1. The figures given are for a rather generalized case, and indicate the order of magnitude. The number of synapses per neuron (20,000) is an estimate for monkeys; those for humans may be closer to 40,000, and for the mouse, closer to 8,000. The number of 18,000 excitatory synapses made by a pyramidal cell is set to match the number of excitatory synapses received by pyramidal cells, for the great majority of cortical excitatory synapses are made from axons of cortical, principally pyramidal, cells.

Microanatomical studies show that pyramidal cells rarely make more than one connection with any other pyramidal cell, even when they are adjacent in the same area of the cerebral cortex. An interesting calculation takes the number of local connections made by a pyramidal cell within the approximately 1 mm of its local axonal arborization (say 9,000), and the number of pyramidal cells with dendrites in the same region, and suggests that the probability that a pyramidal cell makes a synapse with its neighbour is low, approximately 0.1 (Braitenberg and Schutz 1991, Abeles 1991). This fits with the estimate from simultaneous recording of nearby pyramidal cells using spike-triggered averaging to monitor time-locked EPSPs (Abeles 1991, Thomson and Deuchars 1994).

Now the implication of the pyramidal cell to pyramidal cell connectivity just described is that within a cortical area of perhaps 1 mm^2, the region within which typical pyramidal cells have dendritic trees and their local axonal arborization, there is a probability of excitatory-to-excitatory cell connection of 0.1. Moreover, this population of mutually interconnected neurons is served by 'its own' population of inhibitory interneurons (which have a spatial receiving and sending zone in the order of 1 mm^2), enabling local threshold setting and optimization of the set of neurons with 'high' (0.1) connection probability in that region. Such an architecture is effectively recurrent or re-entrant. It may be expected to show some of the properties of recurrent networks, including the fast dynamics described by Rolls and Treves (1998) and Rolls (2008d). Such fast dynamics may be facilitated by the fact that cortical neurons in the awake behaving monkey generally have a low spontaneous rate of firing (personal observations; see for example Rolls and Tovee (1995), Rolls, Treves, Tovee and Panzeri (1997b), and Franco, Rolls, Aggelopoulos and Jerez (2007)), which means that even any small additional input may produce some spikes sooner than would otherwise have

Table 1.1 Typical quantitative estimates for neocortex (partly after Abeles (1991) and reflecting estimates in macaques)

Neuronal density	20,000–40,000/mm^3
Neuronal composition:	
Pyramidal	75%
Spiny stellate	10%
Inhibitory neurons, for example smooth stellate, chandelier	15%
Synaptic density	8 x 10^8/mm^3
Numbers of synapses on pyramidal cells:	
Excitatory synapses from remote sources onto each neuron	9,000
Excitatory synapses from local sources onto each neuron	9,000
Inhibitory synapses onto each neuron	2,000
Pyramidal cell dendritic length	10 mm
Number of synapses made by axons of pyramidal cells	18,000
Number of synapses on inhibitory neurons	2,000
Number of synapses made by inhibitory neurons	300
Dendritic length density	400 m/mm^3
Axonal length density	3,200 m/mm^3
Typical cortical thickness	2 mm
Cortical area:	
human (assuming 3 mm for cortical thickness)	300,000 mm^2
macaque (assuming 2 mm for cortical thickness)	30,000 mm^2
rat (assuming 2 mm for cortical thickness)	300 mm^2

occurred, because some of the neurons may be very close to a threshold for firing. It might also show some of the autoassociative retrieval of information typical of autoassociation networks, if the synapses between the nearby pyramidal cells have the appropriate (Hebbian) modifiability. In this context, the value of 0.1 for the probability of a connection between nearby neocortical pyramidal cells is of interest, for the connection probability between hippocampal CA3 pyramidal is approximately 0.02–0.04, and this is thought to be sufficient to sustain associative retrieval (Rolls and Treves 1998, Rolls 2008d).

In the neocortex, each 1-mm^2 region within which there is a relatively high density of recurrent collateral connections between pyramidal cells probably overlaps somewhat continuously with the next. This raises the issue of modules in the cortex, described by many authors as regions of the order of 1 mm^2 (with different authors giving different sizes), in which there are vertically oriented columns of neurons that may share some property (for example, responding to the same orientation of a visual stimulus), and that may be anatomically marked (for example Powell (1981), Mountcastle (1984), see Douglas, Mahowald and Martin (1996)). The anatomy just described, with the local connections between nearby (1 mm) pyramidal cells, and the local inhibitory neurons, may provide a network basis for starting to understand the columnar architecture of the neocortex, for it implies that local recurrent connectivity on this scale implementing local re-entrancy is a feature of cortical computation. We can note that the neocortex could not be a single, global, autoassociation network, because the number of memories that could be stored in an autoassociation network, rather than increasing with the number of neurons in the network, is limited by the number of recurrent connections per neuron, which is in the order of 10,000 (see Table 1.1), or less, depending on the species, as pointed out by O'Kane and Treves (1992). This would be an impossibly small capacity for the whole cortex. It is suggested that instead a principle of cortical design is that it does have in part local connectivity, so that each part can have its own processing and storage, which may be triggered by other modules, but is a distinct operation from that which occurs simultaneously in other modules.

An interesting parallel between the hippocampus and any small patch of neocortex is the allocation of a set of many small excitatory (usually non-pyramidal, spiny stellate or granular) cells at the input side. In the neocortex this is layer 4, in the hippocampus the dentate gyrus. In both cases, these cells receive the feedforward inputs and relay them to a population of pyramidal cells (in layers 2–3 of the neocortex and in the CA3 field of the hippocampus) which have extensive recurrent collateral connections. In both cases, the pyramidal cells receive inputs both as relayed by the preprocessing array and directly. Such analogies might indicate that the functional roles of neocortical layer 4 cells and of dentate granule cells could be partially the same (see Rolls and Treves (1998)).

The short-range high density of connectivity may also contribute to the formation of cortical topographic maps, as described by Rolls (2008d). This may help to ensure that different parameters of the input space are represented in a nearly continuous fashion across the cortex, to the extent that the reduction in dimensionality allows it; or when preserving strict continuity is not possible, to produce the clustering of cells with similar response properties, as illustrated for example by colour 'blobs' in striate cortex, or by the local clustering of face cells in the temporal cortical visual areas (Rolls 2007d, Rolls 2008b, Rolls 2009c).

1.8.5   Functional pathways through the cortical layers

Because of the complexity of the circuitry of the cerebral cortex, some of which is summarized in Fig. 1.9, there are only preliminary indications available now of how information is processed by the cortex. In primary sensory cortical areas, the main extrinsic 'forward' input is from the thalamus, and ends in layer 4, where synapses are formed onto spiny stellate cells. These in turn project heavily onto pyramidal cells in layers 3 and 2, which in turn send projections forward to the next cortical area. The situation is made more complex than this by the fact that the thalamic afferents also synapse onto the basal dendrites in or close to the layer 2 pyramidal cells, as well as onto layer 6 pyramidal cells and inhibitory interneurons. Given that the functional implications of this particular architecture are not fully clear, it would be of interest to examine the strength of the functional links between thalamic afferents and different classes of cortical cell using cross-correlation techniques, to determine which neurons are strongly activated by thalamic afferents with monosynaptic or polysynaptic delays. Given that this is technically difficult, an alternative approach has been to use electrical stimulation of the thalamic afferents to classify cortical neurons as mono- or polysynaptically driven, then to examine the response properties of the neuron to physiological (visual) inputs, and finally to fill the cell with horseradish peroxidase so that its full structure can be studied (see for example Martin (1984)). Using these techniques, it has been shown in the cat visual cortex that spiny stellate cells can indeed be driven monosynaptically by thalamic afferents to the cortex. Further, many of these neurons have S-type receptive fields, that is they have distinct on and off regions of the receptive field, and respond with orientation tuning to elongated visual stimuli (Martin 1984) (see Rolls and Deco (2002)). Further, consistent with the anatomy just described, pyramidal cells in the deep part of layer 3, and in layer 6, could also be monosynaptically activated by thalamic afferents, and had S-type receptive fields (Martin 1984). Also consistent with the anatomy just described, pyramidal cells in layer 2 were di- (or poly-) synaptically activated by stimulation of the afferents from the thalamus, but also had S-type receptive fields.

Inputs could reach the layer 5 pyramidal cells from the pyramidal cells in layers 2 and 3,

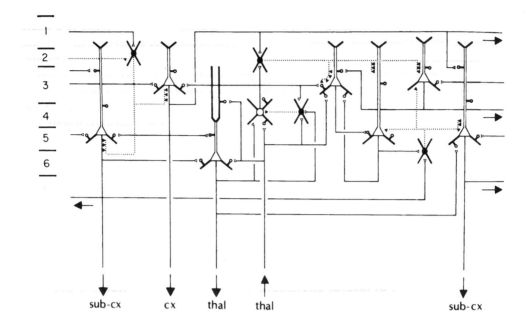

Fig. 1.9 Basic circuit for visual cortex. Excitatory neurons, which are spiny and use glutamate as a transmitter, and include the pyramidal and spiny stellate cells, are indicated by open somata; their axons are indicated by solid lines, and their synaptic boutons by open symbols. Inhibitory (smooth, GABAergic) neurons are indicated by black (filled) somata; their axons are indicated by dotted lines, and their synaptic boutons by solid symbols. thal, thalamus; cx, cortex; sub-cx, subcortex. Cortical layers 1–6 are as indicated. (Reproduced with permission from Douglas and Martin 1990, Fig. 12.7.)

the axons of which ramify extensively in layer 5, in which the layer 5 pyramidal cells have widespread basal dendrites (see Fig. 1.1), and also perhaps from thalamic afferents. Many layer 5 pyramidal cells are di- or trisynaptically activated by stimulation of the thalamic afferents, consistent with them receiving inputs from monosynaptically activated deep layer 3 pyramidal cells, or from disynaptically activated pyramidal cells in layer 2 and upper layer 3 (Martin 1984). Interestingly, many of the layer 5 pyramidal cells had C-type receptive fields, that is they did not have distinct on and off regions, but did respond with orientation tuning to elongated visual stimuli (Martin 1984) (see Rolls and Deco (2002)).

Studies on the function of inhibitory pathways in the cortex are also beginning. The fact that basket cells often receive strong thalamic inputs, and that they terminate on pyramidal cell bodies where part of their action is to shunt the membrane, suggests that they act in part as a feedforward inhibitory system that normalizes the thalamic influence on pyramidal cells by dividing their response in proportion to the average of the thalamic input received (Rolls 2008d). The smaller and numerous smooth (or sparsely spiny) non-pyramidal cells that are inhibitory may receive inputs from pyramidal cells as well as inhibit them, so that these neurons could perform the very important function of recurrent or feedback inhibition (Rolls 2008d). It is only feedback inhibition that can take into account not only the inputs received by an area of cortex, but also the effects that these inputs have, once multiplied by the synaptic weight vector on each neuron, so that recurrent inhibition is necessary for competition and contrast enhancement (Rolls 2008d).

Another way in which the role of inhibition in the cortex can be analyzed is by applying a drug such as bicuculline using iontophoresis (which blocks GABA receptors to a single neuron), while examining the response properties of the neuron (see Sillito (1984)). With this technique, it has been shown that in the visual cortex of the cat, layer 4 simple cells lose their orientation and directional selectivity. Similar effects are observed in some complex cells, but the selectivity of other complex cells may be less affected by blocking the effect of endogenously released GABA in this way (Sillito 1984). One possible reason for this is that the inputs to complex cells must often synapse onto the dendrites far from the cell body, and distant synapses will probably be unaffected by the GABA receptor blocker released near the cell body. The experiments reveal that inhibition is very important for the normal selectivity of many visual cortex neurons for orientation and the direction of movement. Many of the cells displayed almost no orientation selectivity without inhibition. This implies that not only is the inhibition important for maintaining the neuron on an appropriate part of its activation function, but also that lateral inhibition between neurons is important because it allows the responses of a single neuron (which need not be markedly biased by its excitatory input) to have its responsiveness set by the activity of neighbouring neurons (Rolls 2008d).

1.8.6   The scale of lateral excitatory and inhibitory effects, and the concept of modules

The forward cortico-cortical afferents to a cortical area sometimes have a columnar pattern to their distribution, with the column width 200–300 μm in diameter (see Eccles (1984)). Similarly, individual thalamo-cortical axons often end in patches in layer 4 which are 200–300 μm in diameter (Martin 1984). The dendrites of spiny stellate cells are in the region of 500 μm in diameter, and their axons can distribute in patches 200–300 μm across, separated by distances of up to 1 mm (Martin 1984). The dendrites of layer 2 and 3 pyramidal cells can be approximately 300 μm in diameter, but after this the relatively narrow column appears to become less important, for the axons of the superficial pyramidal cells can distribute over 1 mm or more, both in layers 2 and 3, and in layer 5 (Martin 1984). Other neurons that may contribute to the maintenance of processing in relatively narrow columns are the double bouquet cells, which because they receive inhibitory inputs, and themselves produce inhibition, all within a column perhaps 50 μm across (see above), would tend to enhance local excitation. The bipolar cells, which form excitatory synapses with pyramidal cells, may also serve to emphasize activity within a narrow vertical column approximately 50 μm across. These two mechanisms for enhancing local excitation operate against a much broader-ranging set of lateral inhibitory processes, and could it is suggested have the effect of increasing contrast between the firing rates of pyramidal cells 50 μm apart, and thus be very important in competitive interactions between pyramidal cells. Indeed, the lateral inhibitory effects are broader than the excitatory effects described so far, in that for example the axons of basket cells spread laterally 500 μm or more (see earlier) (although those of the small, smooth non-pyramidal cells are closer to 300 μm—see Peters and Saint Marie (1984)).

Such short-range local excitatory interactions with longer-range inhibition not only provide for contrast enhancement and for competitive interactions, but also can result in the formation of maps in which neurons with similar responses are grouped together and neurons with dissimilar response are more widely separated (Rolls 2008d). Thus these local interactions are consistent with the possibilities that cortical pyramidal cells form a competitive network, and that cortical maps are formed at least partly as a result of local interactions of this kind in a competitive network (Rolls 2008d).

Some of the advantages of the locality of neocortical connectivity are described in Section 8.18 on page 253.

The type of genetic specification that could provide the fundamental connectivity rules between cortical areas, which would then self-organize the details of the exact synaptic connectivity, have been considered by Rolls and Stringer (2000). They compared the connectivity of different cortical areas, thereby suggested a set of rules that the genes might be specifying, and then simulated using genetic algorithms the selection of the appropriate rules for solving particular types of computational problem, including pattern association memory, autoassociation memory, and competitive learning.

In contrast to the relatively localized terminal distributions of forward cortico-cortical and thalamo-cortical afferents, the cortico-cortical backward projections that end in layer 1 have a much wider horizontal distribution, of up to several millimetres (mm). It is suggested that this enables the backward-projecting neurons to search over a larger number of pyramidal cells in the preceding cortical area for activity that is conjunctive with their own (see later).

1.9 Theoretical significance of backprojections in the neocortex

Cortical processing takes place through a hierarchy of cortical stages (Rolls 2008d). Convergence and competition are key aspects of the processing. This processing could act in a feedforward manner, and indeed, in experiments on backward masking, it is shown that there is insufficient time for top-down processing to occur when objects can just be recognized (Rolls 2008d, Rolls 2003, Rolls 2006a). (Neurons in each cortical stage respond for 20–30 ms when an object can just be seen, and given that the time for processing to travel from V1 to the inferior temporal visual cortex (IT) is approximately 50 ms, there is insufficient time for a return projection from IT to reach V1, influence processing there, and in turn for V1 to project up to IT to alter processing there.) Nevertheless, backprojections are a major feature of cortical connectivity, and we next consider hypotheses about their possible function. Some of their functions, as described below, are in memory recall, in forming attractor networks in which there are associatively modifiable synaptic connections in both the feedforward and feedback connections between adjacent levels in the hierarchy (Section 1.9.4), and in top down attentional modulation implemented by biased competition (Chapter 4). In all these situations, noise generated by the neuronal firing will be an important aspect of how these functions operate.

1.9.1 Architecture

The forward and backward projections in the neocortex that will be considered are shown in Fig. 1.10 (for further anatomical information see Jones and Peters (1984) and Peters and Jones (1984)). As described earlier, in primary sensory cortical areas, the main extrinsic 'forward' input is from the thalamus and ends in layer 4, where synapses are formed onto spiny stellate cells. These in turn project heavily onto pyramidal cells in layers 3 and 2, which in turn send projections forwards to terminate strongly in layer 4 of the next cortical layer (on small pyramidal cells in layer 4 or on the basal dendrites of the layer 2 and 3 (superficial) pyramidal cells, and also onto layer-6 pyramidal cells and inhibitory interneurons). Inputs reach the layer 5 (deep) pyramidal cells from the pyramidal cells in layers 2 and 3 (Martin 1984), and it is the deep pyramidal cells that send backprojections to end in layer 1 of the preceding cortical area (see Fig. 1.10), where there are apical dendrites of pyramidal cells. It is important to note that in addition to the axons and their terminals in layer 1 from the succeeding cortical stage, there are also axons and terminals in layer 1 in many stages of the cortical hierarchy from the amygdala and (via the subiculum, entorhinal cortex, and parahippocampal cortex)

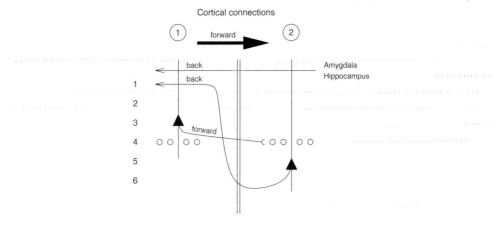

Fig. 1.10 Schematic diagram of forward and backward connections between adjacent neocortical areas. (Area 1 projects forwards to area 2 in the diagram. Area 1 would in sensory pathways be closest to the sense organs.) The superficial pyramidal cells (shown schematically as filled triangles) in layers 2 and 3 project forwards (in the direction of the arrow) to the next cortical area. The deep pyramidal cells in layer 5 project backwards to end in layer 1 of the preceding cortical area (on the apical dendrites of pyramidal cells). The hippocampus and amygdala are also sources of backprojections that end mainly in layer 1 of the higher association cortical areas. Spiny stellate cells are represented by small circles in layer 4. See text for further details.

from the hippocampal formation (see Fig. 1.10) (Van Hoesen 1981, Turner 1981, Amaral and Price 1984, Amaral 1986, Amaral 1987, Amaral, Price, Pitkanen and Carmichael 1992, Rolls 2008d).

A feature of the cortico-cortical forward and backprojection connectivity shown schematically in Fig. 1.10 is that it is 'keyed', in that the origin and termination of the connections between cortical areas provide evidence about which one is forwards or higher in the hierarchy. The reasons for the asymmetry (including the need for backprojections not to dominate activity in preceding cortical areas, see Rolls and Stringer (2001)) are described later, but the nature of the asymmetry between two cortical areas provides additional evidence about the hierarchical nature of processing in visual cortical areas especially in the ventral stream, for which there is already much other evidence (Rolls 2008d, Panzeri, Rolls, Battaglia and Lavis 2001).

One point made in Fig. 1.4 is that the orbitofrontal cortex, amygdala and hippocampus are stages of information processing at which the different sensory modalities (such as vision, hearing, touch, taste, and smell for the orbitofrontal cortex and amygdala) are brought together, so that correlations between inputs in different modalities can be detected in these regions, but not at prior cortical processing stages in each modality, as these cortical processing stages are mainly unimodal. As a result of bringing together any two sensory modalities, significant correspondences between the two modalities can be detected. One example might be that a particular visual stimulus is associated with the taste of food. Another example might be that another visual stimulus is associated with painful touch. Thus at these limbic (and orbitofrontal cortex, see Rolls (2005) and Rolls (2008d)) stages of processing, but not before, the significance of, for example, visual and auditory stimuli can be detected and signalled. Sending this information back to the neocortex could thus provide a signal that indicates to the cortex that information should be stored. Even more than this, the backprojection pathways could provide patterns of firing that could help the neocortex to store the information efficiently, one of the possible functions of backprojections within and to the neocortex considered next.

1.9.2 Recall

Evidence that during recall neural activity does occur in cortical areas involved in the original processing comes, for example, from investigations that show that when humans are asked to recall visual scenes in the dark, blood flow is increased in visual cortical areas, stretching back from association cortical areas as far as early (possibly even primary) visual cortical areas (Roland and Friberg 1985, Kosslyn 1994). Recall is a function that could be produced by cortical backprojections.

If in Fig. 1.10 only the backprojection input is presented after the type of learning just described, then the neurons originally activated by the forward-projecting input stimuli are activated. This occurs because the synapses from the backprojecting axons onto the pyramidal cells have been associatively modified only where there was conjunctive forward and backprojected activity during learning. This thus provides a mechanism for recall. The crucial requirement for recall to operate in this way is that, in the backprojection pathways, the backprojection synapses would need to be associatively modifiable, so that the backprojection input when presented alone could operate effectively as a pattern associator to produce recall. Some aspects of neocortical architecture consistent with this hypothesis (Rolls 1989a, Rolls 1989c, Rolls and Treves 1998) are as follows. First, there are many NMDA receptors on the apical dendrites of cortical pyramidal cells, where the backprojection synapses terminate. These receptors are implicated in associative modifiability of synapses, and indeed plasticity is very evident in the superficial layers of the cerebral cortex (Diamond, Huang and Ebner 1994). Second, the backprojection synapses in ending on the apical dendrite, quite far from the cell body, might be expected to be sufficient to dominate the cell firing when there is no forward input close to the cell body. In contrast, when there is forward input to the neuron, activating synapses closer to the cell body than the backprojecting inputs, this would tend to electrically shunt the effects received on the apical dendrites. This could be beneficial during the original learning, in that during the original learning the forward input would have the stronger effect on the activation of the cell, with mild guidance then being provided by the backprojections.

An example of how this recall could operate is provided next. Consider the situation when in the visual system the sight of food is forward projected onto pyramidal cells in higher order visual cortex, and conjunctively there is a backprojected representation of the taste of the food from, for example, the amygdala or orbitofrontal cortex. Neurons which have conjunctive inputs from these two stimuli set up representations of both, so that later if only the taste representation is backprojected, then the visual neurons originally activated by the sight of that food will be activated. In this way many of the low-level details of the original visual stimulus might be recalled. Evidence that during recall relatively early cortical processing stages are activated comes from cortical blood flow studies in humans, in which it has been found, for example, that quite posterior visual areas are activated during recall of visual (but not auditory) scenes (Kosslyn 1994). The backprojections are probably in this situation acting as pattern associators.

The quantitative analysis of the recall that could be implemented through the hippocampal backprojection synapses to the neocortex, and then via multiple stages of cortico-cortical backprojections, makes it clear that the most important quantitative factor influencing the number of memories that can be recalled is the number of backprojecting synapses onto each cortical neuron in the backprojecting pathways (Treves and Rolls 1994, Treves and Rolls 1991, Rolls 2008d). This provides an interpretation of why there are in general as many backprojecting synapses between two adjacent cortical areas as forward connections. The number of synapses on each neuron devoted to the backprojections needs to be large to recall as many memories as possible, but need not be larger than the number of forward inputs

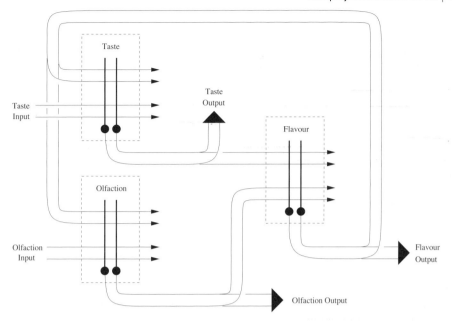

Fig. 1.11 A two-layer set of competitive nets in which feedback from layer 2 can influence the categories formed in layer 1. Layer 2 could be a higher cortical visual area with convergence from earlier cortical visual areas. In the example, taste and olfactory inputs are received by separate competitive nets in layer 1, and converge into a single competitive net in layer 2. The categories formed in layer 2 (which may be described as representing 'flavour') may be dominated by the relatively orthogonal set of a few tastes that are received by the net. When these layer 2 categories are fed back to layer 1, they may produce in layer 1 categories in, for example, the olfactory network that reflect to some extent the flavour categories of layer 2, and are different from the categories that would otherwise be formed to a large set of rather correlated olfactory inputs. A similar principle may operate in any multilayer hierarchical cortical processing system, such as the ventral visual system, in that the categories that can be formed only at later stages of processing may help earlier stages to form categories relevant to what can be identified at later stages.

to each neuron, which influences the number of possible classifications that the neuron can perform with its forward inputs (Rolls 2008d).

An implication of these ideas is that if the backprojections are used for recall, as seems likely as just discussed, then this would place severe constraints on their use for functions such as error backpropagation (Rolls 2008d). It would be difficult to use the backprojections in cortical architecture to convey an appropriate error signal from the output layer back to the earlier, hidden, layers if the backprojection synapses are also to be set up associatively to implement recall.

1.9.3 Attention

In the same way, attention could operate from higher to lower levels, to selectively facilitate only certain pyramidal cells by using the backprojections. Indeed, the backprojections described could produce many of the 'top-down' influences that are common in perception (Rolls 2008f), and implement top-down attention (Chapter 4).

1.9.4 Backprojections, attractor networks, and constraint satisfaction

If the forward connections from one cortical area to the next, and the return backprojections, are both associatively modifiable, then the coupled networks could be regarded as, effectively, an autoassociative network. (Autoassociation networks are described in Section 1.10.) A pattern of activity in one cortical area would be associated with a pattern in the next that occurred regularly with it. This could enable higher cortical areas to influence the state of earlier cortical areas, and could be especially influential in the type of situation shown in Fig. 1.11 in which some convergence occurs at the higher area. For example, if one of the earlier stages (for example the olfactory stage in Fig. 1.11) had a noisy input on a particular occasion, its representation could be cleaned up if a taste input normally associated with it was present. The higher cortical area would be forced into the correct pattern of firing by the taste input, and this would feed back as a constraint to affect the state into which the olfactory area settled. This could be a useful general effect in the cerebral cortex, in that constraints arising only after information has converged from different sources at a higher level could feed back to influence the representations that earlier parts of the network settle into. This is a way in which top-down processing could be implemented (Rolls and Deco 2002, Rolls 2008d).

The autoassociative effect between two forward and backward connected cortical areas could also be used in short-term memory (see Chapter 3), in attention (see Chapter 4), and in decision making (see Chapter 5).

1.10 Autoassociation or attractor memory

Formal models of neural networks are needed in order to provide a basis for understanding the memory, attention, and decision-making functions performed by real neuronal networks in the brain. The aims of this section are to describe formal models of attractor networks in the brain and their dynamics as implemented by neurons, to introduce the concept of stability in these networks, and to introduce the concept of noise produced by statistical fluctuations of the spiking of neurons, and how this noise influences the stability of these networks. For pedagogical reasons, the description we provide in this Section (1.10) on attractor networks uses simplified neurons. In later Chapters we use biologically realistic attractor networks with spiking neurons so that we can address the dynamical properties of these networks and their stochastic properties, a main theme of this book.

More detailed descriptions of some of the quantitative aspects of storage in autoassociation networks are provided in the Appendices of Rolls and Treves (1998) *Neural Networks and Brain Function*. Another book that provides a clear and quantitative introduction to some of these networks is Hertz, Krogh and Palmer (1991) *Introduction to the Theory of Neural Computation*, and other useful sources include Dayan and Abbott (2001), Amit (1989) (for attractor networks), Koch (1999) (for a biophysical approach), Wilson (1999) (on spiking networks), Gerstner and Kistler (2002) (on spiking networks), Rolls and Deco (2002), and Rolls (2008d) *Memory, Attention, and Decision-Making*.

An introduction to linear algebra that is useful in understanding attractor neuronal networks, and a description of other important biologically plausible neuronal networks including pattern association networks and competitive networks, are provided in Rolls (2008d) *Memory, Attention, and Decision-Making*, Rolls and Deco (2002) *Computational Neuroscience of Vision*, and Rolls and Treves (1998) *Neural Networks and Brain Function*.

Some of the background to the operation of the type of neuronal network described here is provided earlier in this Chapter, including a brief review of the evidence on neuronal structure and function, on the presence of a highly developed system of excitatory recurrent collateral connections between nearby pyramidal cells which is a hallmark of neocortical design, and on synaptic plasticity and the rules by which synaptic strength is modified as shown for example by studies of long-term potentiation.

Autoassociative memories, or attractor neural networks, store memories, each one of which is represented by a pattern of neural activity. The memories are stored in the recurrent synaptic connections between the neurons of the network, for example in the recurrent collateral connections between cortical pyramidal cells. Autoassociative networks can then recall the appropriate memory from the network when provided with a fragment of one of the memories. This is called completion. Many different memories can be stored in the network and retrieved correctly. A feature of this type of memory is that it is content addressable; that is, the information in the memory can be accessed if just the contents of the memory (or a part of the contents of the memory) are used. This is in contrast to a conventional computer, in which the address of what is to be accessed must be supplied, and used to access the contents of the memory. Content addressability is an important simplifying feature of this type of memory, which makes it suitable for use in biological systems. The issue of content addressability will be amplified later.

An autoassociation memory can be used as a short-term memory, in which iterative processing round the recurrent collateral connection loop keeps a representation active by continuing neuronal firing. The short-term memory reflected in continuing neuronal firing for several hundred milliseconds after a visual stimulus is removed which is present in visual cortical areas such as the inferior temporal visual cortex (Rolls 2008d) is probably implemented in this way. This short-term memory is one possible mechanism that contributes to the implementation of the trace memory learning rule that can help to implement invariant object recognition (Rolls 2008d, Rolls and Stringer 2006).

Autoassociation memories also appear to be used in a short-term memory role in the prefrontal cortex. In particular, the temporal visual cortical areas have connections to the ventrolateral prefrontal cortex which help to implement the short-term memory for visual stimuli (in for example delayed match to sample tasks, and visual search tasks, as described in Section 3.2). In an analogous way the parietal cortex has connections to the dorsolateral prefrontal cortex for the short-term memory of spatial responses (see Section 3.2). These short-term memories provide a mechanism that enables attention to be maintained through backprojections from prefrontal cortex areas to the temporal and parietal areas that send connections to the prefrontal cortex, as described in Chapter 4.

Autoassociation networks implemented by the recurrent collateral synapses between cortical pyramidal cells also provide a mechanism for constraint satisfaction and also noise reduction whereby the firing of neighbouring neurons can be taken into account in enabling the network to settle into a state that reflects all the details of the inputs activating the population of connected neurons, as well as the effects of what has been set up during developmental plasticity as well as later experience.

Attractor networks are also effectively implemented by virtue of the forward and backward connections between cortical areas (see Sections 1.9 and 3.2).

An autoassociation network with rapid synaptic plasticity can learn each memory in one trial. Because of its 'one-shot' rapid learning, and ability to complete, this type of network is well suited for episodic memory storage, in which each past episode must be stored and recalled later from a fragment, and kept separate from other episodic memories (Rolls 2008d, Treves and Rolls 1994, Rolls and Kesner 2006, Rolls 2008e).

p. 39

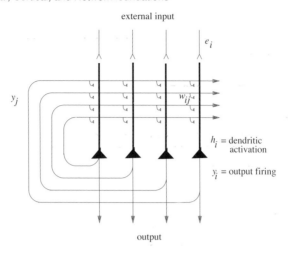

Fig. 1.12 The architecture of an autoassociative neural network. e_i is the external input to each output neuron i, and y_j is the presynaptic firing rate to synapse w_{ij}. The filled triangles represent the pyramidal cell bodies, and the thick lines above each cell the dendrites of the cell. The thin lines represent the axons.

1.10.1 Architecture and operation

The prototypical architecture of an autoassociation memory is shown in Fig. 1.12. The external input e_i is applied to each neuron i by unmodifiable synapses. This produces firing y_i of each neuron (or a vector **y** of the firing of the output neurons where each element in the vector is the firing rate of a single neuron). Each output neuron i is connected by a recurrent collateral connection to the other neurons in the network, via modifiable connection weights w_{ij}. This architecture effectively enables the output firing vector **y** to be associated during learning with itself. Later on, during recall, presentation of part of the external input will force some of the output neurons to fire, but through the recurrent collateral axons and the modified synapses, other neurons in **y** can be brought into activity. This process can be repeated a number of times, and recall of a complete pattern may be perfect. Effectively, a pattern can be recalled or recognized because of associations formed between its parts. This of course requires distributed representations.

Next we introduce a more precise and detailed description of this, and describe the properties of these networks. Ways to analyze formally the operation of these networks are introduced in Appendix A4 of Rolls and Treves (1998) and by Amit (1989).

1.10.1.1 Learning

The firing of every output neuron i is forced to a value y_i determined by the external input e_i. Then a Hebb-like associative local learning rule is applied to the recurrent synapses in the network:

$$\delta w_{ij} = \alpha y_i y_j \tag{1.18}$$

where δw_{ij} is the change in the synaptic strength or weight, y_i is the postsynaptic firing rate, y_j is the presynaptic firing rate, and α is a learning rate constant. It is notable that in a fully connected network, this will result in a symmetric matrix of synaptic weights, that is the strength of the connection from neuron 1 to neuron 2 will be the same as the strength of the connection from neuron 2 to neuron 1 (both implemented via recurrent collateral synapses).

It is a factor that is sometimes overlooked that there must be a mechanism for ensuring that during learning y_i does approximate e_i, and must not be influenced much by activity in the recurrent collateral connections, otherwise the new external pattern e will not be stored in the network, but instead something will be stored that is influenced by the previously stored memories. It is thought that in some parts of the brain, such as the hippocampus, there are processes that help the external connections to dominate the firing during learning (Treves and Rolls 1992, Rolls and Treves 1998, Rolls 2008d).

1.10.1.2 Recall

During recall, the external input e_i is applied, and produces output firing, operating through the non-linear activation function described later. The firing is fed back by the recurrent collateral axons shown in Fig. 1.12 to produce activation of each output neuron through the modified synapses on each output neuron. The activation h_i produced by the recurrent collateral effect on the ith neuron is, in the standard way, the sum of the activations produced in proportion to the firing rate of each axon y_j operating through each modified synapse w_{ij}, that is,

$$h_i = \sum_j y_j w_{ij} \qquad (1.19)$$

where $\sum_j$ indicates that the sum is over the C input axons to each neuron, indexed by j. (The activation of the neuron is equivalent to depolarization in neurophysiology.)

The output firing y_i is a function of the activation produced by the recurrent collateral effect (internal recall, h_i) and by the external input (e_i):

$$y_i = f(h_i + e_i) \qquad (1.20)$$

The activation function should be non-linear, and may be for example binary threshold, linear threshold, sigmoid, etc. (Rolls 2008d). The threshold at which the activation function operates is set in part by the effect of the inhibitory neurons in the network (not shown in Fig. 1.12). The connectivity is that the pyramidal cells have collateral axons that excite the inhibitory interneurons, which in turn connect back to the population of pyramidal cells to inhibit them by a mixture of shunting (divisive) and subtractive inhibition using GABA (gamma-amino-butyric acid) terminals, as described in Section 2.2. There are many fewer inhibitory neurons than excitatory neurons (in the order of 5–10%, see Table 1.1) and of connections to and from inhibitory neurons (see Table 1.1), and partly for this reason the inhibitory neurons are considered to perform generic functions such as threshold setting, rather than to store patterns by modifying their synapses. The non-linear activation function can minimize interference between the pattern being recalled and other patterns stored in the network, and can also be used to ensure that what is a positive feedback system remains stable. The network can be allowed to repeat this recurrent collateral loop a number of times. Each time the loop operates, the output firing becomes more like the originally stored pattern, and this progressive recall is usually complete within 5–15 iterations.

1.10.2 Introduction to the analysis of the operation of autoassociation networks

With complete connectivity in the synaptic matrix, and the use of a Hebb rule, the matrix of synaptic weights formed during learning is symmetric. The learning algorithm is fast, 'one-shot', in that a single presentation of an input pattern is all that is needed to store that pattern.

During recall, a part of one of the originally learned stimuli can be presented as an external input. The resulting firing is allowed to iterate repeatedly round the recurrent collateral system, gradually on each iteration recalling more and more of the originally learned pattern. Completion thus occurs. If a pattern is presented during recall that is similar but not identical to any of the previously learned patterns, then the network settles into a stable recall state in which the firing corresponds to that of the previously learned pattern. The network can thus generalize in its recall to the most similar previously learned pattern. The activation function of the neurons should be non-linear, since a purely linear system would not produce any categorization of the input patterns it receives, and therefore would not be able to effect anything more than a trivial (i.e. linear) form of completion and generalization.

Recall can be thought of in the following way. The external input $\mathbf{e}$ is applied, produces firing $\mathbf{y}$, which is applied as a recall cue on the recurrent collaterals as $\mathbf{y}^T$. (The notation $\mathbf{y}^T$ signifies the transpose of $\mathbf{y}$, which is implemented by the application of the firing of the neurons $\mathbf{y}$ back via the recurrent collateral axons as the next set of inputs to the neurons.) The activity on the recurrent collaterals is then multiplied by the synaptic weight vector stored during learning on each neuron to produce the new activation h_i which reflects the similarity between $\mathbf{y}^T$ and one of the stored patterns. Partial recall has thus occurred as a result of the recurrent collateral effect. The activations h_i result after thresholding (which helps to remove interference from other memories stored in the network, or noise in the recall cue) in firing y_i, or a vector of all neurons $\mathbf{y}$, which is already more like one of the stored patterns than, at the first iteration, the firing resulting from the recall cue alone, $\mathbf{y} = f(\mathbf{e})$. This process is repeated a number of times to produce progressive recall of one of the stored patterns.

Autoassociation networks operate by effectively storing associations between the elements of a pattern. Each element of the pattern vector to be stored is simply the firing of a neuron. What is stored in an autoassociation memory is a set of pattern vectors. The network operates to recall one of the patterns from a fragment of it. Thus, although this network implements recall or recognition of a pattern, it does so by an association learning mechanism, in which associations between the different parts of each pattern are learned. These memories have sometimes been called autocorrelation memories (Kohonen 1977), because they learn correlations between the activity of neurons in the network, in the sense that each pattern learned is defined by a set of simultaneously active neurons. Effectively each pattern is associated by learning with itself. This learning is implemented by an associative (Hebb-like) learning rule.

The system formally resembles spin glass systems of magnets analyzed quantitatively in statistical mechanics. This has led to the analysis of (recurrent) autoassociative networks as dynamical systems made up of many interacting elements, in which the interactions are such as to produce a large variety of basins of attraction of the dynamics. Each basin of attraction corresponds to one of the originally learned patterns, and once the network is within a basin it keeps iterating until a recall state is reached that is the learned pattern itself or a pattern closely similar to it. (Interference effects may prevent an exact identity between the recall state and a learned pattern.) This type of system is contrasted with other, simpler, systems of magnets (e.g. ferromagnets), in which the interactions are such as to produce only a limited number of related basins, since the magnets tend to be, for example, all aligned with each other. The states reached within each basin of attraction are called attractor states, and the analogy between autoassociator neural networks and physical systems with multiple attractors was drawn by Hopfield (1982) in a very influential paper. He was able to show that the recall state can be thought of as the local minimum in an energy landscape, where the energy would be defined as

$$E = -\frac{1}{2} \sum_{i,j} w_{ij} (y_i - <y>)(y_j - <y>).$$ (1.21)

This equation can be understood in the following way. If two neurons are both firing above their mean rate (denoted by $<y>$), and are connected by a weight with a positive value, then the firing of these two neurons is consistent with each other, and they mutually support each other, so that they contribute to the system's tendency to remain stable. If across the whole network such mutual support is generally provided, then no further change will take place, and the system will indeed remain stable. If, on the other hand, either of our pair of neurons was not firing, or if the connecting weight had a negative value, the neurons would not support each other, and indeed the tendency would be for the neurons to try to alter ('flip' in the case of binary units) the state of the other. This would be repeated across the whole network until a situation in which most mutual support, and least 'frustration', was reached. What makes it possible to define an energy function and for these points to hold is that the matrix is symmetric (see Hopfield (1982), Hertz, Krogh and Palmer (1991), Amit (1989)).

Physicists have generally analyzed a system in which the input pattern is presented and then immediately removed, so that the network then 'falls' without further assistance (in what is referred to as the unclamped condition) towards the minimum of its basin of attraction. A more biologically realistic system is one in which the external input is left on contributing to the recall during the fall into the recall state. In this clamped condition, recall is usually faster, and more reliable, so that more memories may be usefully recalled from the network. The approach using methods developed in theoretical physics has led to rapid advances in the understanding of autoassociative networks, and its basic elements are described in Appendix A4 of Rolls and Treves (1998), and by Hertz, Krogh and Palmer (1991) and Amit (1989).

1.10.3 Properties

The internal recall in autoassociation networks involves multiplication of the firing vector of neuronal activity by the vector of synaptic weights on each neuron. This inner product vector multiplication allows the similarity of the firing vector to previously stored firing vectors to be provided by the output (as effectively a correlation), if the patterns learned are distributed. As a result of this type of 'correlation computation' performed if the patterns are distributed, many important properties of these networks arise, including pattern completion (because part of a pattern is correlated with the whole pattern), and graceful degradation (because a damaged synaptic weight vector is still correlated with the original synaptic weight vector). Some of these properties are described next.

1.10.3.1 Completion

Perhaps the most important and useful property of these memories is that they complete an incomplete input vector, allowing recall of a whole memory from a small fraction of it. The memory recalled in response to a fragment is that stored in the memory that is closest in pattern similarity (as measured by the dot product, or correlation). Because the recall is iterative and progressive, the recall can be perfect.

This property and the associative property of pattern associator neural networks are very similar to the properties of human memory. This property may be used when we recall a part of a recent memory of a past episode from a part of that episode. The way in which this could be implemented in the hippocampus is described elsewhere (Rolls 2008d, Treves and Rolls 1994, Rolls and Kesner 2006).

1.10.3.2 Generalization

The network generalizes in that an input vector similar to one of the stored vectors will lead to recall of the originally stored vector, provided that distributed encoding is used. The principle by which this occurs is similar to that described for a pattern associator.

1.10.3.3 Graceful degradation or fault tolerance

If the synaptic weight vector $\mathbf{w}_i$ on each neuron (or the weight matrix) has synapses missing (e.g. during development), or loses synapses (e.g. with brain damage or ageing), then the activation h_i (or vector of activations $\mathbf{h}$) is still reasonable, because h_i is the dot product (correlation) of $\mathbf{y}^T$ with $\mathbf{w}_i$. The same argument applies if whole input axons are lost. If an output neuron is lost, then the network cannot itself compensate for this, but the next network in the brain is likely to be able to generalize or complete if its input vector has some elements missing, as would be the case if some output neurons of the autoassociation network were damaged.

1.10.3.4 Speed

The recall operation is fast on each neuron on a single iteration, because the pattern $\mathbf{y}^T$ on the axons can be applied simultaneously to the synapses $\mathbf{w}_i$, and the activation h_i can be accumulated in one or two time constants of the dendrite (e.g. 10–20 ms). If a simple implementation of an autoassociation net such as that described by Hopfield (1982) is simulated on a computer, then 5–15 iterations are typically necessary for completion of an incomplete input cue e. This might be taken to correspond to 50–200 ms in the brain, rather too slow for any one local network in the brain to function. However, it has been shown that if the neurons are treated not as McCulloch–Pitts neurons which are simply 'updated' at each iteration, or cycle of timesteps (and assume the active state if the threshold is exceeded), but instead are analyzed and modelled as 'integrate-and-fire' neurons in real continuous time, then the network can effectively 'relax' into its recall state very rapidly, in one or two time constants of the synapses (see Section 2.2 and Treves (1993), Battaglia and Treves (1998), and Appendix A5 of Rolls and Treves (1998)). This corresponds to perhaps 20 ms in the brain.

One factor in this rapid dynamics of autoassociative networks with brain-like 'integrate-and-fire' membrane and synaptic properties is that with some spontaneous activity, some of the neurons in the network are close to threshold already before the recall cue is applied, and hence some of the neurons are very quickly pushed by the recall cue into firing, so that information starts to be exchanged very rapidly (within 1–2 ms of brain time) through the modified synapses by the neurons in the network. The progressive exchange of information starting early on within what would otherwise be thought of as an iteration period (of perhaps 20 ms, corresponding to a neuronal firing rate of 50 spikes/s) is the mechanism accounting for rapid recall in an autoassociative neuronal network made biologically realistic in this way. Further analysis of the fast dynamics of these networks if they are implemented in a biologically plausible way with 'integrate-and-fire' neurons is provided in Section 2.2, in Appendix A5 of Rolls and Treves (1998), and by Treves (1993). *The general approach applies to other networks with recurrent connections, not just autoassociators, and the fact that such networks can operate much faster than it would seem from simple models that follow discrete time dynamics is probably a major factor in enabling these networks to provide some of the building blocks of brain function.*

Learning is fast, 'one-shot', in that a single presentation of an input pattern e (producing $\mathbf{y}$) enables the association between the activation of the dendrites (the postsynaptic term h_i) and the firing of the recurrent collateral axons $\mathbf{y}^T$, to be learned. Repeated presentation with small variations of a pattern vector is used to obtain the properties of prototype extraction, extraction of central tendency, and noise reduction, because these arise from the averaging process produced by storing very similar patterns in the network.

1.10.3.5   Local learning rule

The simplest learning used in autoassociation neural networks, a version of the Hebb rule, is (as in Equation 1.18)

$$\delta w_{ij} = \alpha y_i y_j.$$

The rule is a local learning rule in that the information required to specify the change in synaptic weight is available locally at the synapse, as it is dependent only on the presynaptic firing rate y_j available at the synaptic terminal, and the postsynaptic activation or firing y_i available on the dendrite of the neuron receiving the synapse. This makes the learning rule biologically plausible, in that the information about how to change the synaptic weight does not have to be carried to every synapse from a distant source where it is computed. As with pattern associators, since firing rates are positive quantities, a potentially interfering correlation is induced between different pattern vectors. This can be removed by subtracting the mean of the presynaptic activity from each presynaptic term, using a type of long-term depression. This can be specified as

$$\delta w_{ij} = \alpha y_i (y_j - z) \tag{1.22}$$

where α is a learning rate constant. This learning rule includes (in proportion to y_i) increasing the synaptic weight if $(y_j - z) > 0$ (long-term potentiation), and decreasing the synaptic weight if $(y_j - z) < 0$ (heterosynaptic long-term depression). This procedure works optimally if z is the average activity $< y_j >$ of an axon across patterns.

Evidence that a learning rule with the general form of Equation 1.18 is implemented in at least some parts of the brain comes from studies of long-term potentiation, described in Section 1.4. One of the important potential functions of heterosynaptic long-term depression is its ability to allow in effect the average of the presynaptic activity to be subtracted from the presynaptic firing rate (see Appendix A3 of Rolls and Treves (1998), and Rolls and Treves (1990)).

Autoassociation networks can be trained with the error-correction or delta learning rule (Rolls 2008d). Although a delta rule is less biologically plausible than a Hebb-like rule, a delta rule can help to store separately patterns that are very similar (see McClelland and Rumelhart (1988), Hertz, Krogh and Palmer (1991)).

1.10.3.6   Capacity

One measure of storage capacity is to consider how many orthogonal patterns could be stored, as with pattern associators. If the patterns are orthogonal, there will be no interference between them, and the maximum number p of patterns that can be stored will be the same as the number N of output neurons in a fully connected network. Although in practice the patterns that have to be stored will hardly be orthogonal, this is not a purely academic speculation, since it was shown how one can construct a synaptic matrix that effectively orthogonalizes any set of (linearly independent) patterns (Kohonen 1977, Kohonen 1989, Personnaz, Guyon and Dreyfus 1985, Kanter and Sompolinsky 1987). However, this matrix cannot be learned with a local, one-shot learning rule, and therefore its interest for autoassociators in the brain is limited. The more general case of random non-orthogonal patterns, and of Hebbian learning rules, is considered next.

With non-linear neurons used in the network, the capacity can be measured in terms of the number of input patterns **y** (produced by the external input **e**, see Fig. 1.12) that can be stored in the network and recalled later whenever the network settles within each stored pattern's basin of attraction. The first quantitative analysis of storage capacity (Amit, Gutfreund and Sompolinsky 1987) considered a fully connected Hopfield (1982) autoassociator model, in which units are binary elements with an equal probability of being 'on' or 'off' in each pattern, and the number C of inputs per unit is the same as the number N of output units. (Actually it is equal to $N - 1$, since a unit is taken not to connect to itself.) Learning is taken to occur by clamping the desired patterns on the network and using a modified Hebb rule, in which the mean of the presynaptic and postsynaptic firings is subtracted from the firing on any one learning trial (this amounts to a covariance learning rule, and is described more fully in Appendix A4 of Rolls and Treves (1998)). With such fully distributed random patterns, the number of patterns that can be learned is (for C large) $p \approx 0.14C = 0.14N$, hence well below what could be achieved with orthogonal patterns or with an 'orthogonalizing' synaptic matrix. Many variations of this 'standard' autoassociator model have been analyzed subsequently.

Treves and Rolls (1991) have extended this analysis to autoassociation networks that are much more biologically relevant in the following ways:

First, some or many connections between the recurrent collaterals and the dendrites may be missing (this is referred to as diluted connectivity, and results in a non-symmetric synaptic connection matrix in which w_{ij} does not equal w_{ji}, one of the original assumptions made in order to introduce the energy formalism in the Hopfield model).

Second, the neurons need not be restricted to binary threshold neurons, but can have a threshold linear activation function (Rolls 2008d). This enables the neurons to assume real continuously variable firing rates, which are what is found in the brain (Rolls and Tovee 1995, Treves, Panzeri, Rolls, Booth and Wakeman 1999, Rolls 2008d).

Third, the representation need not be fully distributed (with half the neurons 'on', and half 'off'), but instead can have a small proportion of the neurons firing above the spontaneous rate, which is what is found in parts of the brain such as the hippocampus that are involved in memory (see Treves and Rolls (1994), Chapter 6 of Rolls and Treves (1998), and Rolls (2008d)). Such a representation is defined as being sparse, and the sparseness a of the representation can be measured, by extending the binary notion of the proportion of neurons that are firing, as

$$a = \frac{(\sum\limits_{i=1}^{N} y_i/N)^2}{\sum\limits_{i=1}^{N} y_i^2/N} \tag{1.23}$$

where y_i is the firing rate of the ith neuron in the set of N neurons. Treves and Rolls (1991) have shown that such a network does operate efficiently as an autoassociative network, and can store (and recall correctly) a number of different patterns p as follows

$$p \approx \frac{C^{RC}}{a \ln(\frac{1}{a})} k \tag{1.24}$$

where C^{RC} is the number of synapses on the dendrites of each neuron devoted to the recurrent collaterals from other neurons in the network, and k is a factor that depends weakly on the detailed structure of the rate distribution, on the connectivity pattern, etc., but is roughly in the order of 0.2–0.3.

The main factors that determine the maximum number of memories that can be stored in an autoassociative network are thus the number of connections on each neuron devoted to the recurrent collaterals, and the sparseness of the representation. For example, for $C^{RC} = 12,000$ and $a = 0.02$, p is calculated to be approximately $36,000$. This storage capacity can be realized, with little interference between patterns, if the learning rule includes some form of heterosynaptic long-term depression that counterbalances the effects of associative long-term potentiation (Treves and Rolls (1991); see Appendix A4 of Rolls and Treves (1998), and Rolls (2008d)).

It should be noted that the number of neurons N (which is greater than C^{RC}, the number of recurrent collateral inputs received by any neuron in the network from the other neurons in the network) is not a parameter that influences the number of different memories that can be stored in the network. The implication of this is that increasing the number of neurons (without increasing the number of connections per neuron) does not increase the number of different patterns that can be stored (see Rolls and Treves (1998) Appendix A4), although it may enable simpler encoding of the firing patterns, for example more orthogonal encoding, to be used. This latter point may account in part for why there are generally in the brain more neurons in a recurrent network than there are connections per neuron (Rolls 2008d).

The non-linearity inherent in the NMDA receptor-based Hebbian plasticity present in the brain may help to make the stored patterns more sparse than the input patterns, and this may be especially beneficial in increasing the storage capacity of associative networks in the brain by allowing participation in the storage of especially those relatively few neurons with high firing rates in the exponential firing rate distributions typical of neurons in sensory systems (see Section 1.11.3).

1.10.3.7 Context

The environmental context in which learning occurs can be a very important factor that affects retrieval in humans and other animals. Placing the subject back into the same context in which the original learning occurred can greatly facilitate retrieval.

Context effects arise naturally in association networks if some of the activity in the network reflects the context in which the learning occurs. Retrieval is then better when that context is present, for the activity contributed by the context becomes part of the retrieval cue for the memory, increasing the correlation of the current state with what was stored. (A strategy for retrieval arises simply from this property. The strategy is to keep trying to recall as many fragments of the original memory situation, including the context, as possible, as this will provide a better cue for complete retrieval of the memory than just a single fragment.)

The effects that mood has on memory including visual memory retrieval may be accounted for by backprojections from brain regions such as the amygdala and orbitofrontal cortex in which the current mood, providing a context, is represented, to brain regions involved in memory such as the perirhinal cortex, and in visual representations such as the inferior temporal visual cortex (see Rolls and Stringer (2001) and Rolls (2008d)). The very well-known effects of context in the human memory literature could arise in the simple way just described. An implication of the explanation is that context effects will be especially important at late stages of memory or information processing systems in the brain, for there information from a wide range of modalities will be mixed, and some of that information could reflect the context in which the learning takes place. One part of the brain where such effects may be strong is the hippocampus, which is implicated in the memory of recent episodes, and which receives inputs derived from most of the cortical information processing streams, including those involved in space (Rolls 2008d).

1.10.4 Use of autoassociation networks in the brain

Because of its 'one-shot' rapid learning, and ability to complete, this type of network is well suited for episodic memory storage, in which each episode must be stored and recalled later from a fragment, and kept separate from other episodic memories (Rolls 2008e). It does not take a long time (the 'many epochs' of backpropagation networks) to train this network, because it does not have to 'discover the structure' of a problem. Instead, it stores information in the form in which it is presented to the memory, without altering the representation. An autoassociation network may be used for this function in the CA3 region of the hippocampus

P.31

(Rolls 1987, McNaughton and Morris 1987, Rolls 1989a, Rolls 1989b, Treves and Rolls 1994, Rolls and Treves 1998, Rolls 2008d).

An autoassociation memory can also be used as a short-term memory, in which iterative processing round the recurrent collateral loop keeps a representation active until another input cue is received. This may be used to implement many types of short-term memory in the brain (see Section 3.2). For example, it may be used in the perirhinal cortex and adjacent temporal lobe cortex to implement short-term visual object memory (Miyashita and Chang 1988, Amit 1995); in the dorsolateral prefrontal cortex to implement a short-term memory for spatial responses (Goldman-Rakic 1996); and in the prefrontal cortex to implement a short-term memory for where eye movements should be made in space (see Chapter 3).

Such an autoassociation memory in the temporal lobe visual cortical areas may be used to implement the firing that continues for often 300 ms after a very brief (16 ms) presentation of a visual stimulus (Rolls and Tovee 1994), and may be one way in which a short memory trace is implemented to facilitate invariant learning about visual stimuli (Rolls 2008d). In all these cases, the short-term memory may be implemented by the recurrent excitatory collaterals that connect nearby pyramidal cells in the cerebral cortex. The connectivity in this system, that is the probability that a neuron synapses on a nearby neuron, may be in the region of 10% (Braitenberg and Schutz 1991, Abeles 1991).

The recurrent connections between nearby neocortical pyramidal cells may also be important in defining the response properties of cortical cells, which may be triggered by external inputs (from for example the thalamus or a preceding cortical area), but may be considerably dependent on the synaptic connections received from nearby cortical pyramidal cells.

The cortico-cortical backprojection connectivity can be interpreted as a system that allows the forward-projecting neurons in one cortical area to be linked autoassociatively with the backprojecting neurons in the next cortical area (see Section 1.9 and Rolls (2008d)). This would be implemented by associative synaptic modification in for example the backprojections. This particular architecture may be especially important in constraint satisfaction (as well as recall), that is it may allow the networks in the two cortical areas to settle into a mutually consistent state. This would effectively enable information in higher cortical areas, which would include information from more divergent sources, to influence the response properties of neurons in earlier cortical processing stages. This is an important function in cortical information processing of interacting associative networks.

1.11 Noise, and the sparse distributed representations found in the brain

The statistics of the noise received by a neuron in a network is important for understanding the effects of noise on brain function. We show in Section 5.6 that as the size of the network is increased, the statistical fluctuations caused by the stochastic firing of the neurons become smaller, and indeed decrease approximately in proportion to the square root of the size of the network. Given that, as we show in this book, the noise can be advantageous to some brain computations, we need to be able to assess how much noise generated by neuronal spiking is actually present under normal conditions in the brain. If there is a great deal of noise, the system may be very unstable. If there is too little noise, then operations such as probabilistic decision-making with its advantages could be jeopardized. The aim of this section is to consider the neuronal encoding of information, the firing rate distributions of neurons, and the size of networks in the brain, in order to provide a basis for understanding how noise affects the operation of the brain.

When considering the operation of many neuronal networks in the brain, it is found that many useful properties arise if each input to the network (arriving on the axons as **x**) is encoded in the activity of an ensemble or population of the axons or input lines (distributed encoding), and is not signalled by the activity of a single input, which is called local encoding. We start off with some definitions, in Section 1.11.2 summarize how many of the useful properties of neuronal networks depend on distributed encoding (Rolls 2008d), and then in Sections 1.11.3 and 1.11.4 we review evidence on the encoding actually found in visual cortical areas. The distribution of the firing rates of the neurons active when any one stimulus or event is being represented is important to understand, for this distribution affects the noise caused by the neuronal spiking in a cortical area.

1.11.1 Definitions

A *local representation* is one in which all the information that a particular stimulus or event occurred is provided by the activity of one of the neurons. In a famous example, a single neuron might be active only if one's grandmother was being seen, an example used by Jerry Lettvin and discussed by Barlow (1972). An implication is that most neurons in the brain regions where objects or events are represented would fire only very rarely. A problem with this type of encoding is that a new neuron would be needed for every object or event that has to be represented. There are many other disadvantages of this type of encoding, many of which have been described elsewhere (Rolls 2008d). Moreover, there is evidence that objects are represented in the brain by a different type of encoding.

A *fully distributed representation* is one in which all the information that a particular stimulus or event occurred is provided by the activity of the full set of neurons. If the neurons are binary (e.g. either active or not), the most distributed encoding is when half the neurons are active for any one stimulus or event.

A *sparse distributed representation* is a distributed representation in which a small proportion of the neurons is active at any one time to represent any one stimulus or event. In a sparse representation with binary neurons, less than half of the neurons are active for any one stimulus or event. For binary neurons, we can use as a measure of the sparseness the proportion of neurons in the active state. For neurons with real, continuously variable, values of firing rates, the sparseness a^p of the representation provided by the population can be measured, by extending the binary notion of the proportion of neurons that are firing, as

$$a^p = \frac{(\sum\limits_{i=1}^{N} y_i/N)^2}{\sum\limits_{i=1}^{N} y_i^2/N} \qquad (1.25)$$

where y_i is the firing rate of the ith neuron in the set of N neurons (Treves and Rolls 1991). This is referred to as the population sparseness, and measures of sparseness are considered in detail in Section 1.11.3. A low value of the sparseness a^p indicates that few neurons are firing for any one stimulus.

Coarse coding utilizes overlaps of receptive fields, and can compute positions in the input space using differences between the firing levels of coactive cells (e.g. colour-tuned cones in the retina). The representation implied is very distributed. Fine coding (in which, for example, a neuron may be 'tuned' to the exact orientation and position of a stimulus) implies more local coding.

1.11.2 Advantages of different types of coding

One advantage of distributed encoding is that the similarity between two representations can be reflected by the correlation between the two patterns of activity that represent the different stimuli. We have already introduced the idea that the input to a neuron is represented by the activity of its set of input axons x_j, where j indexes the axons, numbered from $j = 1, C$ (see Equation 1.19). Now the set of activities of the input axons is a vector (a vector is an ordered set of numbers; a summary of linear algebra useful in understanding neuronal networks is provided by Rolls (2008d)). We can denote as $\mathbf{x}^1$ the vector of axonal activity that represents stimulus 1, and $\mathbf{x}^2$ the vector that represents stimulus 2. Then the similarity between the two vectors, and thus the two stimuli, is reflected by the correlation between the two vectors. The correlation will be high if the activity of each axon in the two representations is similar; and will become more and more different as the activity of more and more of the axons differs in the two representations. Thus the similarity of two inputs can be represented in a graded or continuous way if (this type of) distributed encoding is used. This enables generalization to similar stimuli, or to incomplete versions of a stimulus (if it is, for example, partly seen or partly remembered), to occur. With a local representation, either one stimulus or another is represented, each by its own neuron firing, and similarities between different stimuli are not encoded.

Another advantage of distributed encoding is that the number of different stimuli that can be represented by a set of C components (e.g. the activity of C axons) can be very large. A simple example is provided by the binary encoding of an 8-element vector. One component can code for which of two stimuli has been seen, 2 components (or bits in a computer byte) for 4 stimuli, 3 components for 8 stimuli, 8 components for 256 stimuli, etc. That is, the number of stimuli increases exponentially with the number of components (or, in this case, axons) in the representation. (In this simple binary illustrative case, the number of stimuli that can be encoded is 2^C.) Put the other way round, even if a neuron has only a limited number of inputs (e.g. a few thousand), it can nevertheless receive a great deal of information about which stimulus was present. This ability of a neuron with a limited number of inputs to receive information about which of potentially very many input events is present is probably one factor that makes computation by the brain possible. With local encoding, the number of stimuli that can be encoded increases only linearly with the number C of axons or components (because a different component is needed to represent each new stimulus). (In our example, only 8 stimuli could be represented by 8 axons.)

In the brain, there is now good evidence that in a number of brain systems, including the high-order visual and olfactory cortices, and the hippocampus, distributed encoding with the earlier described two properties, of representing similarity, and of exponentially increasing encoding capacity as the number of neurons in the representation increases, is found (Rolls and Tovee 1995, Abbott, Rolls and Tovee 1996, Rolls, Treves and Tovee 1997a, Rolls, Treves, Robertson, Georges-Francois and Panzeri 1998, Rolls, Franco, Aggelopoulos and Reece 2003b, Rolls, Aggelopoulos, Franco and Treves 2004, Franco, Rolls, Aggelopoulos and Treves 2004, Aggelopoulos, Franco and Rolls 2005, Rolls, Franco, Aggelopoulos and Jerez 2006a) (see Rolls (2008d) and Sections 1.11.3 and 1.11.4). For example, in the primate inferior temporal visual cortex, the number of faces or objects that can be represented increases approximately exponentially with the number of neurons in the population. If we consider instead the information about which stimulus is seen, we see that this rises approximately linearly with the number of neurons in the representation. This corresponds to an exponential rise in the number of stimuli encoded, because information is a log measure. A similar result has been found for the encoding of position in space by the primate hippocampus (Rolls, Treves, Robertson, Georges-Francois and Panzeri 1998).

It is particularly important that the information can be read from the ensemble of neurons using a simple measure of the similarity of vectors, the correlation (or dot product, see Appendix 1 of Rolls (2008d)) between two vectors. The importance of this is that it is essentially vector similarity operations that characterize the operation of many neuronal networks (see Chapter 2 and Rolls (2008d)). The neurophysiological results show that both the ability to reflect similarity by vector correlation, and the utilization of exponential coding capacity, are properties of real neuronal networks found in the brain.

To emphasize one of the points being made here, although the binary encoding used in the 8-bit vector described earlier has optimal capacity for binary encoding, it is not optimal for vector similarity operations. For example, the two very similar numbers 127 and 128 are represented by 01111111 and 10000000 with binary encoding, yet the correlation or bit overlap of these vectors is 0. The brain, in contrast, uses a code that has the attractive property of exponentially increasing capacity with the number of neurons in the representation, though it is different from the simple binary encoding of numbers used in computers; and at the same time the brain codes stimuli in such a way that the code can be read off with simple dot product or correlation-related decoding, which is what is specified for the elementary neuronal network operation shown in Equation 1.19 (see Section 1.11).

1.11.3 The firing rate distributions of real neurons and sparseness of the representation in the cortex

The firing rate distribution of the neurons active for any one stimulus and the sparseness of the representation are important factors that influence the memory storage capacity, the noise caused by the neuronal firing, and the stability of networks in the brain. Whereas most theoretical studies have been with rectangular firing rate distributions in which a small proportion of neurons representing any one memory is in a high firing rate state and the other neurons are in a low firing rate state, it is found that in the brain when a large number of different stimuli must be represented by a network, the firing rate probability distribution is often approximately exponential, as described next, and the implications of this are considered in Section 1.11.4.

1.11.3.1 Single neuron sparseness a^s

Equation 1.26 defines a measure of the single neuron sparseness, a^s:

$$a^s = \frac{(\sum\limits_{s=1}^{S} y_s/S)^2}{(\sum\limits_{s=1}^{S} y_s^2)/S} \tag{1.26}$$

where y_s is the mean firing rate of the neuron to stimulus s in the set of S stimuli (Rolls and Treves 1998). For a binary representation, a^s is 0.5 for a fully distributed representation, and $1/S$ if a neuron responds to one of a set of S stimuli. Another measure of sparseness is the kurtosis of the distribution, which is the fourth moment of the distribution. It reflects the length of the tail of the distribution. (An actual distribution of the firing rates of a neuron to a set of 65 stimuli is shown in Fig. 1.13. The sparseness a^s for this neuron was 0.69 (see Rolls, Treves, Tovee and Panzeri (1997b).)

It is important to understand and quantify the sparseness of representations in the brain, because many of the useful properties of neuronal networks such as generalization and completion only occur if the representations are not local (see Section 1.10 and Rolls (2008d)), and because the value of the sparseness is an important factor in how many memories can be stored in such neural networks. Relatively sparse representations (low values of a^s) might be expected in memory systems as this will increase the number of different memories that can be stored and retrieved. Less sparse representations might be expected in sensory systems, as this could allow more information to be represented (see Rolls (2008d)).

Barlow (1972) proposed a single neuron doctrine for perceptual psychology. He proposed that sensory systems are organized to achieve as complete a representation as possible with the minimum number of active neurons. He suggested that at progressively higher levels of sensory processing, fewer and fewer cells are active, and that each represents a more and more specific happening in the sensory environment. He suggested that 1,000 active neurons (which he called cardinal cells) might represent the whole of a visual scene. An important principle involved in forming such a representation was the reduction of redundancy. The implication of Barlow's (1972) approach was that when an object is being recognized, there are, towards the end of the visual system, a small number of neurons (the cardinal cells) that are so specifically tuned that the activity of these neurons encodes the information that one particular object is being seen. (He thought that an active neuron conveys something of the order of complexity of a word.) The encoding of information in such a system is described as local, in that knowing the activity of just one neuron provides evidence that a particular stimulus (or, more exactly, a given 'trigger feature') is present. Barlow (1972) eschewed 'combinatorial rules of usage of nerve cells', and believed that the subtlety and sensitivity of perception results from the mechanisms determining when a single cell becomes active. In contrast, with distributed or ensemble encoding, the activity of several or many neurons must be known in order to identify which stimulus is present, that is, to read the code. It is the relative firing of the different neurons in the ensemble that provides the information about which object is present.

At the time Barlow (1972) wrote, there was little actual evidence on the activity of neurons in the higher parts of the visual and other sensory systems. There is now considerable evidence, which is now described.

First, it has been shown that the representation of which particular object (face) is present is actually rather distributed. Baylis, Rolls and Leonard (1985) showed this with the responses of temporal visual cortical neurons that typically responded to several members of a set of five faces, with each neuron having a different profile of responses to each face (see further Rolls (2008d) and Rolls (2008b)). It would be difficult for most of these single cells to tell which of even five faces, let alone which of hundreds of faces, had been seen. (At the same time, the neurons discriminated between the faces reliably, as shown by the values of d', taken, in the case of the neurons, to be the number of standard deviations of the neuronal responses that separated the response to the best face in the set from that to the least effective face in the set. The values of d' were typically in the range 1–3.)

Second, the distributed nature of the representation can be further understood by the finding that the firing rate probability distribution of single neurons, when a wide range of natural visual stimuli are being viewed, is approximately exponential, with rather few stimuli producing high firing rates, and increasingly large numbers of stimuli producing lower and lower firing rates, as illustrated in Fig. 1.14a (Rolls and Tovee 1995, Baddeley, Abbott, Booth, Sengpiel, Freeman, Wakeman and Rolls 1997, Treves, Panzeri, Rolls, Booth and Wakeman 1999, Franco, Rolls, Aggelopoulos and Jerez 2007).

For example, the responses of a set of temporal cortical neurons to 23 faces and 42 non-face natural images were measured, and a distributed representation was found (Rolls and Tovee 1995). The tuning was typically graded, with a range of different firing rates to the set of faces, and very little response to the non-face stimuli (see example in Fig. 1.13). The spontaneous firing rate of the neuron in Fig. 1.13 was 20 spikes/s, and the histogram bars indicate the change of firing rate from the spontaneous value produced by each stimulus.

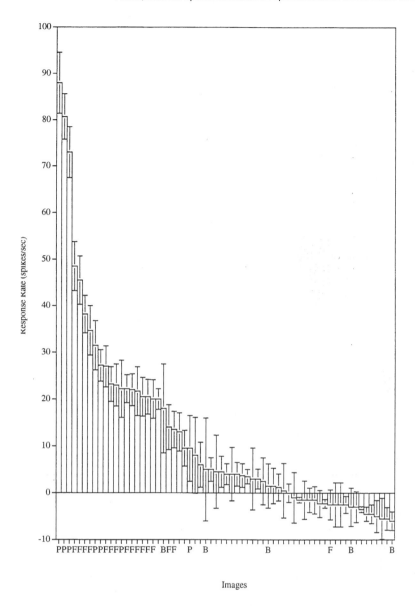

Fig. 1.13 Firing rate distribution of a single neuron in the temporal visual cortex to a set of 23 face (F) and 45 non-face images of natural scenes. The firing rate (mean ± standard deviation) to each of the 68 stimuli is shown. P–face profile; B–body part, such as a hand. The neuron does not respond to just one of the 68 stimuli. Instead, it responds to a small proportion of stimuli with high rates, to more stimuli with intermediate rates, and to many stimuli with almost no change of firing. This is typical of the distributed representations found in temporal cortical visual areas. (After Rolls, Treves, Tovee and Panzeri 1997b.)

Stimuli that are faces are marked F, or P if they are in profile. B refers to images of scenes that included either a small face within the scene, sometimes as part of an image that included a whole person, or other body parts, such as hands (H) or legs. The non-face stimuli are unlabelled. The neuron responded best to three of the faces (profile views), had some response to some of the other faces, and had little or no response, and sometimes had a small decrease

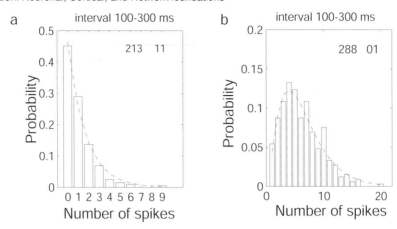

Fig. 1.14 Firing rate probability distributions for two neurons in the inferior temporal visual cortex tested with a set of 20 face and non-face stimuli. (a) A neuron with a good fit to an exponential probability distribution (dashed line). (b) A neuron that did not fit an exponential firing rate distribution (but which could be fitted by a gamma distribution, dashed line). The firing rates were measured in an interval 100–300 ms after the onset of the visual stimuli, and similar distributions are obtained in other intervals. (After Franco, Rolls, Aggelopoulos and Jerez 2007.)

of firing rate below the spontaneous firing rate, to the non-face stimuli. The sparseness value a^s for this cell across all 68 stimuli was 0.69, and the response sparseness a_r^s (based on the evoked responses minus the spontaneous firing of the neuron) was 0.19. It was found that the sparseness of the representation of the 68 stimuli by each neuron had an average across all neurons of 0.65 (Rolls and Tovee 1995). This indicates a rather distributed representation. (If neurons had a continuum of firing rates equally distributed between zero and the maximum rate, a^s would be 0.75, while if the probability of each response decreased linearly, to reach zero at the maximum rate, a^s would be 0.67.) If the spontaneous firing rate was subtracted from the firing rate of the neuron to each stimulus, so that the changes of firing rate, that is the active responses of the neurons, were used in the sparseness calculation, then the 'response sparseness' a_r^s had a lower value, with a mean of 0.33 for the population of neurons, or 0.60 if calculated over the set of faces rather than over all the face and non-face stimuli. Thus the representation was rather distributed. (It is, of course, important to remember the relative nature of sparseness measures, which (like the information measures to be discussed later) depend strongly on the stimulus set used.) Thus we can reject a cardinal cell representation. As shown earlier, the readout of information from these cells is actually much better in any case than would be obtained from a local representation, and this makes it unlikely that there is a further population of neurons with very specific tuning that use local encoding.

These data provide a clear answer to whether these neurons are grandmother cells: they are not, in the sense that each neuron has a graded set of responses to the different members of a set of stimuli, with the prototypical distribution similar to that of the neuron illustrated in Fig. 1.13. On the other hand, each neuron does respond very much more to some stimuli than to many others, and in this sense is tuned to some stimuli.

Figure 1.14 shows data of the type shown in Fig. 1.13 as firing rate probability density functions, that is as the probability that the neuron will be firing with particular rates. These data were from inferior temporal cortex neurons, and show when tested with a set of 20 face and non-face stimuli how fast the neuron will be firing in a period 100–300 ms after the visual stimulus appears (Franco, Rolls, Aggelopoulos and Jerez 2007). Figure 1.14a shows an example of a neuron where the data fit an exponential firing rate probability distribution, with

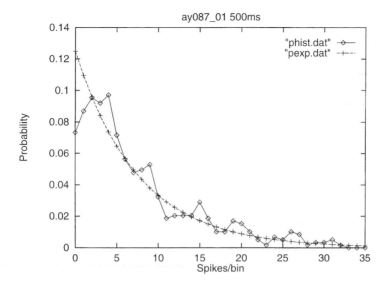

Fig. 1.15 The probability of different firing rates measured in short (e.g. 100 ms or 500 ms) time windows of a temporal cortex neuron calculated over a 5 min period in which the macaque watched a video showing natural scenes, including faces. An exponential fit (+) to the data (diamonds) is shown. (After Baddeley, Abbott, Booth, Sengpiel, Freeman, Wakeman and Rolls 1997.)

many occasions on which the neuron was firing with a very low firing rate, and decreasingly few occasions on which it fired at higher rates. This shows that the neuron can have high firing rates, but only to a few stimuli. Figure 1.14b shows an example of a neuron where the data do not fit an exponential firing rate probability distribution, with insufficiently few very low rates. Of the 41 responsive neurons in this data set, 15 had a good fit to an exponential firing rate probability distribution; the other 26 neurons did not fit an exponential but did fit a gamma distribution in the way illustrated in Fig. 1.14b. For the neurons with an exponential distribution, the mean firing rate across the stimulus set was 5.7 spikes/s, and for the neurons with a gamma distribution was 21.1 spikes/s (t=4.5, df=25, p< 0.001). It may be that neurons with high mean rates to a stimulus set tend to have few low rates ever, and this accounts for their poor fit to an exponential firing rate probability distribution, which fits when there are many low firing rate values in the distribution as in Fig. 1.14a.

The large set of 68 stimuli used by Rolls and Tovee (1995) was chosen to produce an approximation to a set of stimuli that might be found to natural stimuli in a natural environment, and thus to provide evidence about the firing rate distribution of neurons to natural stimuli. Another approach to the same fundamental question was taken by Baddeley, Abbott, Booth, Sengpiel, Freeman, Wakeman, and Rolls (1997) who measured the firing rates over short periods of individual inferior temporal cortex neurons while monkeys watched continuous videos of natural scenes. They found that the firing rates of the neurons were again approximately exponentially distributed (see Fig. 1.15), providing further evidence that this type of representation is characteristic of inferior temporal cortex (and indeed also V1) neurons.

The actual distribution of the firing rates to a wide set of natural stimuli is of interest, because it has a rather stereotypical shape, typically following a graded unimodal distribution with a long tail extending to high rates (see for example Figs. 1.14a and 1.15). The mode of the distribution is close to the spontaneous firing rate, and sometimes it is at zero firing. If the number of spikes recorded in a fixed time window is taken to be constrained by a fixed maximum rate, one can try to interpret the distribution observed in terms of optimal information transmission (Shannon 1948), by making the additional assumption that the coding is noiseless. An exponential distribution, which maximizes entropy (and hence information transmission for noiseless codes) is the most efficient in terms of energy consumption if its mean takes an optimal value that is a decreasing function of the relative metabolic cost of emitting a spike (Levy and Baxter 1996). This argument would favour sparser coding schemes the more energy expensive neuronal firing is (relative to rest). Although the tail of actual firing rate distributions is often approximately exponential (see for example Figs. 1.14a and 1.15; Baddeley, Abbott, Booth, Sengpiel, Freeman, Wakeman and Rolls (1997); Rolls, Treves, Tovee and Panzeri (1997b); and Franco, Rolls, Aggelopoulos and Jerez (2007)), the maximum entropy argument cannot apply as such, because noise is present and the noise level varies as a function of the rate, which makes entropy maximization different from information maximization. Moreover, a mode at low but non-zero rate, which is often observed (see e.g. Fig. 1.14b), is inconsistent with the energy efficiency hypothesis.

A simpler explanation for the characteristic firing rate distribution arises by appreciating that the value of the activation of a neuron across stimuli, reflecting a multitude of contributing factors, will typically have a Gaussian distribution; and by considering a physiological input–output transform (i.e. activation function), and realistic noise levels. In fact, an input–output transform that is supralinear in a range above threshold results from a fundamentally linear transform and fluctuations in the activation, and produces a variance in the output rate, across repeated trials, that increases with the rate itself, consistent with common observations. At the same time, such a supralinear transform tends to convert the Gaussian tail of the activation distribution into an approximately exponential tail, without implying a fully exponential distribution with the mode at zero. Such basic assumptions yield excellent fits with observed distributions (Treves, Panzeri, Rolls, Booth and Wakeman 1999), which often differ from exponential in that there are too few very low rates observed, and too many low rates (Rolls, Treves, Tovee and Panzeri 1997b, Franco, Rolls, Aggelopoulos and Jerez 2007).

This peak at low but non-zero rates may be related to the low firing rate spontaneous activity that is typical of many cortical neurons. Keeping the neurons close to threshold in this way may maximize the speed with which a network can respond to new inputs (because time is not required to bring the neurons from a strongly hyperpolarized state up to threshold). The advantage of having low spontaneous firing rates may be a further reason why a curve such as an exponential cannot sometimes be exactly fitted to the experimental data.

A conclusion of this analysis was that the firing rate distribution may arise from the threshold non-linearity of neurons combined with short-term variability in the responses of neurons (Treves, Panzeri, Rolls, Booth and Wakeman 1999).

However, given that the firing rate distribution for some neurons is approximately exponential, some properties of this type of representation are worth elucidation. The sparseness of such an exponential distribution of firing rates is 0.5. This has interesting implications, for to the extent that the firing rates are exponentially distributed, this fixes an important parameter of cortical neuronal encoding to be close to 0.5. Indeed, only one parameter specifies the shape of the exponential distribution, and the fact that the exponential distribution is at least a close approximation to the firing rate distribution of some real cortical neurons implies that the sparseness of the cortical representation of stimuli is kept under precise control. The utility of this may be to ensure that any neuron receiving from this representation can perform a dot product operation between its inputs and its synaptic weights that produces similarly distributed outputs; and that the information being represented by a population of cortical neurons is kept high. It is interesting to realize that the representation that is stored in an associative network may be more sparse than the 0.5 value for an exponential firing rate

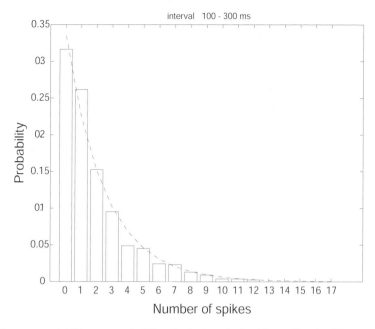

Fig. 1.16 An exponential firing rate probability distribution obtained by pooling the firing rates of a population of 41 inferior temporal cortex neurons tested to a set of 20 face and non-face stimuli. The firing rate probability distribution for the 100–300 ms interval following stimulus onset was formed by adding the spike counts from all 41 neurons, and across all stimuli. The fit to the exponential distribution (dashed line) was high. (After Franco, Rolls, Aggelopoulos and Jerez 2007.)

distribution, because the non-linearity of learning introduced by the voltage dependence of the NMDA receptors effectively means that synaptic modification in, for example, an auto-associative network will occur only for the neurons with relatively high firing rates, i.e. for those that are strongly depolarized (Rolls 2008d).

The single neuron selectivity reflects response distributions of individual neurons across time to different stimuli. As we have seen, part of the interest of measuring the firing rate probability distributions of individual neurons is that one form of the probability distribution, the exponential, maximizes the entropy of the neuronal responses for a given mean firing rate, which could be used to maximize information transmission consistent with keeping the firing rate on average low, in order to minimize metabolic expenditure (Levy and Baxter 1996, Baddeley, Abbott, Booth, Sengpiel, Freeman, Wakeman and Rolls 1997). Franco, Rolls, Aggelopoulos and Jerez (2007) showed that while the firing rates of some single inferior temporal cortex neurons (tested in a visual fixation task to a set of 20 face and non-face stimuli) do fit an exponential distribution, and others with higher spontaneous firing rates do not, as described earlier, it turns out that there is a very close fit to an exponential distribution of firing rates if all spikes from all the neurons are considered together. This interesting result is shown in Fig. 1.16.

One implication of the result shown in Fig. 1.16 is that a neuron with inputs from the inferior temporal visual cortex will receive an exponential distribution of firing rates on its afferents, and this is therefore the type of input that needs to be considered in theoretical models of neuronal network function in the brain (see Section 1.10 and Rolls (2008d)). The second implication is that at the level of single neurons, an exponential probability density function is consistent with minimizing energy utilization, and maximizing information transmission, for a given mean firing rate (Levy and Baxter 1996, Baddeley, Abbott, Booth, Sengpiel, Freeman,

Wakeman and Rolls 1997).

1.11.3.2 Population sparseness a^p

If instead we consider the responses of a population of neurons taken at any one time (to one stimulus), we might also expect a sparse graded distribution, with few neurons firing fast to a particular stimulus. It is important to measure the population sparseness, for this is a key parameter that influences the number of different stimuli that can be stored and retrieved in networks such as those found in the cortex with recurrent collateral connections between the excitatory neurons, which can form autoassociation or attractor networks if the synapses are associatively modifiable (Hopfield 1982, Treves and Rolls 1991, Rolls and Treves 1998, Rolls and Deco 2002) (see Section 1.10). Further, in physics, if one can predict the distribution of the responses of the system at any one time (the population level) from the distribution of the responses of a component of the system across time, the system is described as ergodic, and a necessary condition for this is that the components are uncorrelated (Lehky, Sejnowski and Desimone 2005). Considering this in neuronal terms, the average sparseness of a population of neurons over multiple stimulus inputs must equal the average selectivity to the stimuli of the single neurons within the population provided that the responses of the neurons are uncorrelated (Foldiak 2003).

The sparseness a^p of the population code may be quantified (for any one stimulus) as

$$a^p = \frac{(\sum\limits_{n=1}^{N} y_n/N)^2}{(\sum\limits_{n=1}^{N} y_n^2)/N} \tag{1.27}$$

where y_n is the mean firing rate of neuron n in the set of N neurons.

This measure, a^p, of the sparseness of the representation of a stimulus by a population of neurons has a number of advantages. One is that it is the same measure of sparseness that has proved to be useful and tractable in formal analyses of the capacity of associative neural networks and the interference between stimuli that use an approach derived from theoretical physics (Rolls and Treves 1990, Treves 1990, Treves and Rolls 1991, Rolls and Treves 1998, Rolls 2008d). We note that high values of a^p indicate broad tuning of the population, and that low values of a^p indicate sparse population encoding.

Franco, Rolls, Aggelopoulos and Jerez (2007) measured the population sparseness of a set of 29 inferior temporal cortex neurons to a set of 20 visual stimuli that included faces and objects. Figure 1.17a shows, for any one stimulus picked at random, the normalized firing rates of the population of neurons. The rates are ranked with the neuron with the highest rate on the left. For different stimuli, the shape of this distribution is on average the same, though with the neurons in a different order. (The rates of each neuron were normalized to a mean of 10 spikes/s before this graph was made, so that the neurons can be combined in the same graph, and so that the population sparseness has a defined value, as described by Franco, Rolls, Aggelopoulos and Jerez (2007).) The population sparseness a^p of this normalized (i.e. scaled) set of firing rates is 0.77.

Figure 1.17b shows the probability distribution of the normalized firing rates of the population of (29) neurons to any stimulus from the set. This was calculated by taking the probability distribution of the data shown in Fig. 1.17a. This distribution is not exponential because of the normalization of the firing rates of each neuron, but becomes exponential as shown in Fig. 1.16 without the normalization step.

A very interesting finding of Franco, Rolls, Aggelopoulos and Jerez (2007) was that when the single cell sparseness a^s and the population sparseness a^p were measured from the same

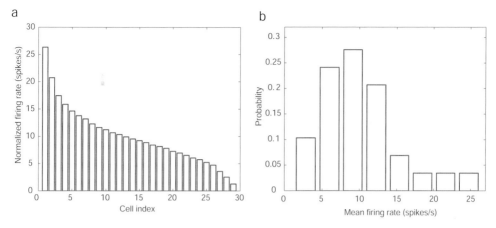

Fig. 1.17 Population sparseness. (a) The firing rates of a population of inferior temporal cortex neurons to any one stimulus from a set of 20 face and non-face stimuli. The rates of each neuron were normalized to the same average value of 10 spikes/s, then for each stimulus, the cell firing rates were placed in rank order, and then the mean firing rates of the first ranked cell, second ranked cell, etc. were taken. The graph thus shows how, for any one stimulus picked at random, the expected normalized firing rates of the population of neurons. (b) The population normalized firing rate probability distributions for any one stimulus. This was computed effectively by taking the probability density function of the data shown in (a). (After Franco, Rolls, Aggelopoulos and Jerez 2007.)

set of neurons in the same experiment, the values were very close, in this case 0.77. (This was found for a range of measurement intervals after stimulus onset, and also for a larger population of 41 neurons.)

The single cell sparseness a^s and the population sparseness a^p can take the same value if the response profiles of the neurons are uncorrelated, that is each neuron is independently tuned to the set of stimuli (Lehky et al. 2005). Franco, Rolls, Aggelopoulos and Jerez (2007) tested whether the response profiles of the neurons to the set of stimuli were uncorrelated in two ways. In a first test, they found that the mean (Pearson) correlation between the response profiles computed over the 406 neuron pairs was low, 0.049 ± 0.013 (sem). In a second test, they computed how the multiple cell information available from these neurons about which stimulus was shown increased as the number of neurons in the sample was increased, and showed that the information increased approximately linearly with the number of neurons in the ensemble. The implication is that the neurons convey independent (non-redundant) information, and this would be expected to occur if the response profiles of the neurons to the stimuli are uncorrelated (see further Rolls (2008d)).

We now consider the concept of ergodicity. The single neuron selectivity, a^s, reflects response distributions of individual neurons across time and therefore stimuli in the world (and has sometimes been termed 'lifetime sparseness'). The population sparseness a^p reflects response distributions across all neurons in a population measured simultaneously (to for example one stimulus). The similarity of the average values of a^s and a^p (both 0.77 for inferior temporal cortex neurons (Franco, Rolls, Aggelopoulos and Jerez 2007)) indicates, we believe for the first time experimentally, that the representation (at least in the inferior temporal cortex) is ergodic. The representation is ergodic in the sense of statistical physics, where the average of a single component (in this context a single neuron) across time is compared with the average of an ensemble of components at one time (cf. Masuda and Aihara (2003) and Lehky, Sejnowski and Desimone (2005)). This is described further next.

In comparing the neuronal selectivities a^s and population sparsenesses a^p, we formed a table in which the columns represent different neurons, and the stimuli different rows (Foldiak 2003). We are interested in the probability distribution functions (and not just their summary values a^s, and a^p), of the columns (which represent the individual neuron selectivities) and the rows (which represent the population tuning to any one stimulus). We could call the system strongly ergodic (cf. Lehky et al. (2005)) if the selectivity (probability density or distribution function) of each individual neuron is the same as the average population sparseness (probability density function). (Each neuron would be tuned to different stimuli, but have the same shape of the probability density function.) We have seen that this is not the case, in that the firing rate probability distribution functions of different neurons are different, with some fitting an exponential function, and some a gamma function (see Fig. 1.14). We can call the system weakly ergodic if individual neurons have different selectivities (i.e. different response probability density functions), but the average selectivity (measured in our case by $<a^s>$) is the same as the average population sparseness (measured by $<a^p>$), where $< >$ indicates the ensemble average. We have seen that for inferior temporal cortex neurons the neuron selectivity probability density functions are different (see Fig. 1.14), but that their average $<a^s>$ is the same as the average (across stimuli) $<a^p>$ of the population sparseness, 0.77, and thus conclude that the representation in the inferior temporal visual cortex of objects and faces is weakly ergodic (Franco, Rolls, Aggelopoulos and Jerez 2007).

We note that weak ergodicity necessarily occurs if $<a^s>$ and $<a^p>$ are the same and the neurons are uncorrelated, that is each neuron is independently tuned to the set of stimuli (Lehky et al. 2005). The fact that both hold for the inferior temporal visual cortex neurons studied by Franco, Rolls, Aggelopoulos and Jerez (2007) thus indicates that their responses are uncorrelated, and this is potentially an important conclusion about the encoding of stimuli by these neurons. This conclusion is confirmed by the linear increase in the information with the number of neurons which is the case not only for this set of neurons (Franco, Rolls, Aggelopoulos and Jerez 2007), but also in other data sets for the inferior temporal visual cortex (Rolls, Treves and Tovee 1997a, Booth and Rolls 1998). Both types of evidence thus indicate that the encoding provided by at least small subsets (up to e.g. 20 neurons) of inferior temporal cortex neurons is approximately independent (non-redundant), which is an important principle of cortical encoding.

1.11.3.3 Comparisons of sparseness between areas: the hippocampus, insula, orbitofrontal cortex, and amygdala

In the study of Franco, Rolls, Aggelopoulos and Jerez (2007) on inferior temporal visual cortex neurons, the selectivity of individual cells for the set of stimuli, or single cell sparseness a^s, had a mean value of 0.77. This is close to a previously measured estimate, 0.65, which was obtained with a larger stimulus set of 68 stimuli (Rolls and Tovee 1995). Thus the single neuron probability density functions in these areas do not produce very sparse representations. Therefore the goal of the computations in the inferior temporal visual cortex may not be to produce sparse representations (as has been proposed for V1 (Field 1994, Olshausen and Field 1997, Vinje and Gallant 2000, Olshausen and Field 2004)). Instead one of the goals of the computations in the inferior temporal visual cortex may be to compute invariant representations of objects and faces (Rolls 2000a, Rolls and Deco 2002, Rolls 2007d, Rolls and Stringer 2006, Rolls 2008d), and to produce not very sparse distributed representations in order to maximize the information represented (Rolls 2008d). In this context, it is very interesting that the representations of different stimuli provided by a population of inferior temporal cortex neurons are decorrelated, as shown by the finding that the mean (Pearson) correlation between the response profiles to a set of 20 stimuli computed over 406 neuron pairs was low, 0.049 ± 0.013 (sem) (Franco, Rolls, Aggelopoulos and Jerez 2007). The implication is that decorrelation is being achieved in the inferior temporal visual cortex, but not by forming a sparse code. It will be interesting to investigate the mechanisms for this.

In contrast, the representation in some memory systems may be more sparse. For example, in the hippocampus in which spatial view cells are found in macaques, further analysis of data described by Rolls, Treves, Robertson, Georges-Francois and Panzeri (1998) shows that for the representation of 64 locations around the walls of the room, the mean single cell sparseness $<a^s>$ was 0.34 ± 0.13 (sd), and the mean population sparseness a^p was 0.33 ± 0.11. The more sparse representation is consistent with the view that the hippocampus is involved in storing memories, and that for this, more sparse representations than in perceptual areas are relevant. These sparseness values are for spatial view neurons, but it is possible that when neurons respond to combinations of spatial view and object (Rolls, Xiang and Franco 2005), or of spatial view and reward (Rolls and Xiang 2005), the representations are more sparse. It is of interest that the mean firing rate of these spatial view neurons across all spatial views was 1.77 spikes/s (Rolls, Treves, Robertson, Georges-Francois and Panzeri 1998). (The mean spontaneous firing rate of the neurons was 0.1 spikes/s, and the average across neurons of the firing rate for the most effective spatial view was 13.2 spikes/s.) It is also notable that weak ergodicity is implied for this brain region too (given the similar values of $<a^s>$ and $<a^p>$), and the underlying basis for this is that the response profiles of the different hippocampal neurons to the spatial views are uncorrelated. Further support for these conclusions is that the information about spatial view increases linearly with the number of hippocampal spatial view neurons (Rolls, Treves, Robertson, Georges-Francois and Panzeri 1998), again providing evidence that the response profiles of the different neurons are uncorrelated.

Further evidence is now available on ergodicity in three further brain areas, the macaque insular primary taste cortex, the orbitofrontal cortex, and the amygdala. In all these brain areas sets of neurons were tested with an identical set of 24 oral taste, temperature, and texture stimuli. (The stimuli were: taste - 0.1 M NaCl (salt), 1 M glucose (sweet), 0.01 M HCl (sour), 0.001 M quinine HCl (bitter), 0.1 M monosodium glutamate (umami), and water; temperature - 10°C, 37°C and 42°C; flavour - blackcurrant juice; viscosity - carboxymethyl-cellulose 10 cPoise, 100 cPoise, 1000 cPoise and 10000 cPoise; fatty / oily - single cream, vegetable oil, mineral oil, silicone oil (100 cPoise), coconut oil, and safflower oil; fatty acids - linoleic acid and lauric acid; capsaicin; and gritty texture.) Further analysis of data described by Verhagen, Kadohisa and Rolls (2004) showed that in the primary taste cortex the mean value of a^s across 58 neurons was 0.745 and of a^p (normalized) was 0.708. Further analysis of data described by Rolls, Verhagen and Kadohisa (2003c), Verhagen, Rolls and Kadohisa (2003), Kadohisa, Rolls and Verhagen (2004) and Kadohisa, Rolls and Verhagen (2005a) showed that in the orbitofrontal cortex the mean value of a^s across 30 neurons was 0.625 and of a^p was 0.611. Further analysis of data described by Kadohisa, Rolls and Verhagen (2005b) showed that in the amygdala the mean value of a^s across 38 neurons was 0.811 and of a^p was 0.813. Thus in all these cases, the mean value of a^s is close to that of a^p, and weak ergodicity is implied. The values of a^s and a^p are also relatively high, implying the importance of representing large amounts of information in these brain areas about this set of stimuli by using a very distributed code, and also perhaps about the stimulus set, some members of which may be rather similar to each other.

Part of the impact of this evidence for understanding noise in the brain is that the spiking noise will be influenced by the firing rate distribution that is present at any one time, and by the sparseness of the representation. When the size of autoassociation or attractor neuronal networks is scaled up, the noise decreases in proportion to the square root of the number of neurons (Mattia and Del Giudice 2002, Mattia and Del Giudice 2004), or in the diluted connectivity case, to the number of connections per neuron. However, the actual amount of noise present, and how it affects the operation of the networks as they are scaled up, has not been investigated yet with biologically realistic firing rate distributions, overlap between the representations provided by different neurons, and sparseness, and it will be interesting to perform the appropriate simulations in future to answer these questions.

In fact, the noise will be related to the firing rate distribution of all the afferent neurons, so that the mean spiking rate of all the afferent neurons, including those with low rates at or below the spontaneous rate, need to be taken into account in considering the noise.

Other factors related to the neuronal firing that affect the way that noise operates in a network of neurons in the brain include the following.

1.11.3.4 Noise and the firing rates of the neurons

The mean firing rates of the neurons when they are responding is also a factor that influences the noise. As shown in Section A.5, in the extended mean-field framework there are two sources of noise in the spiking networks considered: the randomly arriving external Poissonian spike trains, and the noise due to statistical fluctuations arising from the spiking of the finite number of N neurons in the network. In this extended mean-field framework, an estimate of the firing rate $\nu(t)$ of neurons in a population is a stochastic process $\nu_N(t)$ due to this finite-size noise, and $\nu_N(t) \simeq \nu(t) + \sqrt{\nu(t)/N}\gamma(t)$, where $\gamma(t)$ is Gaussian white noise with zero mean and unit variance, and $\nu(t)$ is the probability of emitting a spike per unit time in the infinite network. Thus the noise effects in a finite-sized system with approximately Poisson firing of the neurons increase according to the $\sqrt{\nu}$. In some cortical areas, such as the inferior temporal visual cortex, the mean firing rates to the most effective stimulus tend to be high, up to 100 spikes/s. In other cortical areas, such as those involved in taste processing, the firing rates to the most effective stimulus tend to be lower, typically 20 spikes/s (Rolls 2008d). These differences may reflect the need in some areas, such as the multiple stage cortical visual system, for rapid processing and settling in each cortical area (Panzeri, Rolls, Battaglia and Lavis 2001), whereas for taste and olfactory processing, the stimuli in the environment vary less rapidly, and there are fewer cortical processing stages, so the need for high firing rates and thus rapid information exchange between neurons may be smaller. The point here is that one source of noise in the system is the finite-size noise, and that this increases with the firing rate. Other factors of course may affect the noise. For example, with low firing rates the firing may be variable due to relatively little activation of NMDA receptors with their long, stabilizing, time constant (see Section 1.11.3.6).

1.11.3.5 The Poisson-related, noisy, firing of the neurons

The spiking of typical cortical neurons is close to that of a Poisson process, that is, for a given mean firing rate, the emission times of the spikes are approximately random and independent of each other. There are of course departures from pure Poisson statistics, caused for example by the refractory periods of neurons, which prevent an action potential of a neuron following a previous action potential with a time of less than 1–2 ms. However, this is a minor effect, given that the mean firing rate of cortical neurons to their most effective stimulus is rarely more than 100 spikes/s. As described in Sections 1.10.3.4 and by Rolls (2008d), these approximately Poisson statistics of neurons are convenient, for they mean that in a network there are always some neurons that are close to the firing threshold, and these neurons will respond rapidly to a new input, which facilitates the rapid setting of autoassociation networks into a high firing rate (low energy) attractor basin (see Section 1.10.3.4).

Not only are the spikes of a single neuron relatively random in time, but the spike times of different cortical neurons are relatively independent of each other, as shown by the finding that strong cross-correlations between the spikes of different neurons are not a strong feature of neocortical neurons (Rolls 2008d). Although stimulus-dependent cross-correlations have been proposed as a method used by the cortex to encode information, the evidence is that when the cortex is working normally in the awake behaving macaque processing natural visual scenes to which attention is being paid, then stimulus-dependent cross-correlations contribute very little to information encoding, as described elsewhere (Rolls, Aggelopoulos, Franco and Treves 2004, Franco, Rolls, Aggelopoulos and Treves 2004, Aggelopoulos, Franco and Rolls 2005, Rolls 2008d). Even stimulus-independent cross-correlations between the spike trains of simultaneously recorded neocortical neurons are typically rare and, if present, are weak under normal conditions when for example a cortical visual area such as the inferior temporal visual cortex is performing its normal functions (Rolls, Aggelopoulos, Franco and Treves 2004). Of course if the system is tested under less physiological conditions, for example during drowsiness or under anaesthesia, then synchronized neuronal activity can occur, and the electrocorticogram will show synchronized activity evident as slow waves in the frequency range 2–20 Hz or more.

1.11.3.6 Noise and the time constants of the synapses, neurons, and network

The time constants of the synapses are an important factor in determining how noise including that from the stochastic firing of other neurons in the network affects the operation of the network (see Rolls (2008d)). Whereas the AMPA and GABA receptors operate with relatively short time constants (in the order of 5 ms), the NMDA receptors have a much longer time constant (in the order of 100 ms) which dampens the effect of high frequency noise and helps to promote stability and dampen oscillations in a cortical network (Wang 2002, Wang 1999, Compte, Constantinidis, Tegner, Raghavachari, Chafee, Goldman-Rakic and Wang 2003). The effective time constant of the whole network (or more generally its behaviour as a function of the frequency of the inputs) will depend on the relative contribution of the NMDA and other receptors, with NMDA receptors perhaps increasing in importance when strong activation is present in the network. In the decision-making network described in Chapter 5, the NMDA receptors accounted for approximately 90% of the excitatory synaptic currents in the recurrent collateral connections compared to 10% for AMPA, when the system was in an attractor. (In this particular network, the recurrent collateral connections contribute approximately 33% to the total synaptic current in the neurons, as do the external excitatory inputs, and the GABA receptor mediated currents.)

The neuronal membrane and dendritic time constants are in the order of 20 ms, as determined by the membrane capacitance and the leak current (see Equation 2.1), but are likely to decrease when a neuron is strongly activated and other synaptically activated ion channels are open. Their effect on the temporal dynamics of the network is less than that of the synapses (Treves 1993, Battaglia and Treves 1998).

The overall temporal properties of the network will be influenced by the synaptic and membrane time constants. For example, if the time constant of the excitatory effects on the neurons is relatively short with respect to the inhibitory time constant, the network will tend to oscillate (Brunel and Hakim 1999). If the excitatory time constant is longer, an asynchronous state (i.e. without oscillations) will occur. Given that the time constant of GABA synapses is in the order of 10–20 ms, of AMPA receptors is 2–5 ms, and of NMDA receptors is approximately 100 ms, oscillations are less likely to occur when the system is strongly activated so that the neurons are firing fast, and the voltage-dependent NMDA receptors are contributing significantly to the currents in the system (Buehlmann and Deco 2008).

1.11.3.7 Oscillations

The evidence described in Section 1.11.4 indicates that any stimulus-dependent synchronization or cross-correlation between neurons is not quantitatively an important way in which information is encoded by neuronal activity, and nor would it be useful for encoding syntactic relations such as that a feature encoded by neuron 1 is above (as compared to the left of) the feature encoded by neuron 2 (Rolls 2008d). Is the much simpler property of oscillation useful computationally in neural dynamics (Buzsáki 2006, Lisman and Buzsaki 2008)?

The phase of the hippocampal theta cycle at which place neurons fire as a rat traverses a place field provides some additional evidence about where the rat is in the place field (O'Keefe and Recce 1993), but whether this is used by the system is not known. More generally, a possible benefit of neuronal oscillations is that if the spikes of different neurons are bunched together this might provide a sufficiently large signal to push an attractor system into its non-linear positive feedback mode to help somewhat more rapid falling into an attractor (Buehlmann and Deco 2008). The possible cost is that there is an advantage to having the neurons firing asynchronously, for then some of them will always be close to threshold and therefore ready to fire very soon when an input is received, and this helps rapid memory retrieval in large attractor networks for the recall process starts through the modified recurrent collateral synapses within 1–2 ms of the retrieval cue being applied (Treves 1993, Battaglia and Treves 1998, Panzeri, Rolls, Battaglia and Lavis 2001, Rolls 2008d), rather than having to wait 10 ms or more for the right phase of any oscillation that might be present to arrive and enable the neurons to respond.

The factors that influence the temporal dynamics of cortical neurons are complex and multiple, with not only the relative contributions of AMPA, NMDA, and GABA receptor activated ion channels important to the stability of the system (Brunel and Hakim 1999, Buehlmann and Deco 2008), but also their absolute values important too as when the ion channels opened by these receptors are open much electrical shunting and a reduction of the overall time constant of the system will occur. Given the complexity of these factors that influence the temporal dynamics, instabilities revealed as rhythmic activity including oscillations are difficult to control, and are expected to occur unless the regulation is very precise. For example, when animals sleep, alpha waves at approximately 10 Hz appear in the neocortex (and are a major indicator of slow wave sleep), and these may be related to the fact that with most cortical neurons firing at low rates in slow wave sleep, there is little activation of the long time constant NMDA receptors, as well as little driving by sensory inputs to activate some neurons strongly and inhibit others, so that in this situation slow waves appear. The parsimonious explanation is that in this case, the rhythmic activity is a consequence of this particular combination of factors affecting the different components each with their own time constant in the cortical circuitry, and that the waves are not there to perform any useful function. It may be that this parsimonious explanation holds in many other cases of cortical rhythmic activity, and this would appear to be the appropriate view until clear evidence for a function other than the bunching of spikes referred to earlier (Buehlmann and Deco 2008), a phenomenon that we might term 'synchronous resonance' because of some computational similarities with the effects produced by stochastic resonance described in Section 7.4 on page 217. Oscillations are also considered in Section 8.18.

1.11.3.8 The parameter values during the transient dynamical situation in which the network is falling into an attractor

The values of the sparseness of the representation, firing rate distribution, overall synaptic time constants etc that apply during the transient dynamical phase in the settling of a decision-making network or memory retrieval network when the firing rates are still relatively low (see e.g. Fig. 5.5 in the period 0–300 ms) are of particular relevance to the stochastic behaviour of the network. More analyses of the parameter values when in this transient state are of interest, though the parameter values when in the high firing rate attractor state are also important in determining the stability of the network. More detailed investigations are needed of the transient phase, and part of the utility of integrate-and-fire network simulations is that the effects of these parameters can be explored in the transient settling stage, when the firing rates are still low and the NMDA voltage-dependent receptor activated synaptic currents may not be contributing as much as with high firing rates.

1.11.4 Information theoretic understanding of neuronal representations

The approaches just described in Section 1.11.3 on the sparseness of representations are complemented by quantitative analyses of neuronal encoding using information theory, as described in this section.

Because single neurons are the computing elements of the brain and send the results of their processing by spiking activity to other neurons, we can understand brain processing by understanding what is encoded by the neuronal firing at each stage of the brain (e.g. each cortical area), and determining how what is encoded changes from stage to stage. Each neuron responds differently to a set of stimuli (with each neuron tuned differently to the members of the set of stimuli), and it is this that allows different stimuli to be represented. We can only address the richness of the representation therefore by understanding the differences in the responses of different neurons, and the impact that this has on the amount of information that is encoded. These issues can only be adequately and directly addressed at the level of the activity of single neurons and of populations of single neurons, and understanding at this neuronal level (rather than at the level of thousands or millions of neurons as revealed by functional neuroimaging (Rolls, Grabenhorst and Franco 2009b)) is essential for understanding brain computation.

Information theory provides the means for quantifying how much neurons communicate to other neurons, and thus provides a quantitative approach to fundamental questions about information processing in the brain. To investigate what in neuronal activity carries information, one must compare the amounts of information carried by different codes, that is different descriptions of the same activity. To investigate the speed of information transmission, one must define and measure information rates from neuronal responses. To investigate to what extent the information provided by different cells is redundant or instead independent, again one must measure amounts of information in order to provide quantitative evidence. To compare the information carried by the number of spikes, by the timing of the spikes within the response of a single neuron, and by the relative time of firing of different neurons reflecting for example stimulus-dependent neuronal synchronization, information theory again provides a quantitative and well-founded basis for the necessary comparisons. To compare the information carried by a single neuron or a group of neurons with that reflected in the behaviour of the human or animal, one must again use information theory, as it provides a single measure which can be applied to the measurement of the performance of all these different cases. In all these situations, there is no quantitative and well-founded alternative to information theory.

Rolls and Treves (1998) and Rolls (2008d) introduce the fundamental elements of information theory relevant to understanding neuronal representations. More general treatments can be found in many books on the subject (e.g. Abramson (1963), Hamming (1990), Cover and Thomas (1991) and Rieke, Warland, de Ruyter van Steveninck and Bialek (1996)). A short glossary of some of the relevant terms is provided in Section 1.11.4.3. The actual measurement of information from pairs of stimuli s and the neuronal responses r obtained on each trial (where r might be the responses of a whole population of neurons on that trial) proceeds in principle using Equation 1.33 on page 63, though in practice is much more complicated, with details of the methods used described by Rolls (2008d).

1.11.4.1 The information from single neurons

p. 45

Examples of the responses of single neurons (in this case in the inferior temporal visual cortex) to sets of objects and/or faces are shown in Fig. 1.13 and elsewhere (Rolls 2008d, Rolls 2008b). We now summarize how much information these types of neuronal response convey about the set of stimuli S, and about each stimulus s in the set. The mutual information $I(S, R)$ that the set of responses R encode about the set of stimuli S is calculated with Equation 1.33 and corrected for the limited sampling using the analytic bias correction procedure described by Panzeri and Treves (1996) as described in detail by Rolls, Treves, Tovee and Panzeri (1997b). Some of the findings are as follows:

1. Much information is available about the stimulus presented in the number of spikes emitted by single neurons in a fixed time period, the firing rate. If we consider neurons in the macaque inferior temporal visual cortex that encode information about faces and objects, we found that they reflected in their firing rates for the post-stimulus period 100 to 600 ms on average 0.36 bits of mutual information about which of 20 face stimuli was presented (Rolls, Treves, Tovee and Panzeri 1997b). Similar values have been found in other experiments (Tovee, Rolls, Treves and Bellis 1993, Tovee and Rolls 1995, Rolls, Tovee and Panzeri 1999, Rolls, Franco, Aggelopoulos and Jerez 2006a, Booth and Rolls 1998). The analyses just described for neurons with visual responses are general, in that they apply in a similar way to olfactory and taste neurons recorded in the macaque orbitofrontal cortex (Rolls, Critchley and Treves 1996a, Rolls 2008d, Rolls, Critchley, Verhagen and Kadohisa 2009a), to neurons that represent the reward value of visual stimuli in the orbitofrontal cortex (Rolls, Grabenhorst and Franco 2009b), to spatial view cells recorded in the macaque hippocampus (Rolls, Treves, Robertson, Georges-Francois and Panzeri 1998), and to head direction cells recorded in the primate presubiculum (Robertson, Rolls, Georges-Francois and Panzeri 1999).

The finding that typically 0.3–0.6 bits of mutual information between the stimuli and the responses of a single neuron (with stimulus set sizes of typically 2–20 stimuli) is encoded in all these parts of the brain of course attests to the fact that the brain is noisy, and helps to quantify the amount of noise. If the neuronal firing rates to a set of stimuli are graded, as they always are (see examples in Figs. 1.13 and 1.14), but were noiseless, with a different firing rate for each stimulus, then the mutual information available with a set of 2 stimuli would be 1 bit, with 8 stimuli 3 bits, and with 16 stimuli 4 bits. To further clarify this quantitative evidence on the amount of noise in neuronal firing from trial to trial, the standard deviation of the mean response (measured in a 500 ms period) to each stimulus is shown in Fig. 1.13, which shows that there is considerable variability from trial to trial. Another way to quantify this variability is by the coefficient of variation, which is the standard deviation divided by the mean. We see for example in Fig. 1.13 (which is typical) that even for the high firing rates to the most effective stimuli producing mean rates of 90–100 spikes/s, the trial by trial coefficient of variation was in the order of 0.1, and higher for lower firing rate responses. Another simple measure of the variability of neuronal responses is the Fano factor, which is the ratio of the variance to the mean. A Poisson process has a Fano factor of 1.0, and we see that when firing at 90–100 spikes/s the neuron illustrated in Fig. 1.13 has a Fano factor of approximately 1.0. Moreover, the noise is not just a general increase or decrease for all neurons in the population from trial to trial, for even with the multiple cell information analysis described next when using decoding procedures that are insensitive to a similar change of rate in all neurons from trial to trial, the information does not rise to more than 0.3–0.4 bits per neuron. The overall implication is that there is considerable noise in the brain, and the noise is typically largely independent for different neurons in the awake behaving situation while a task is being performed.

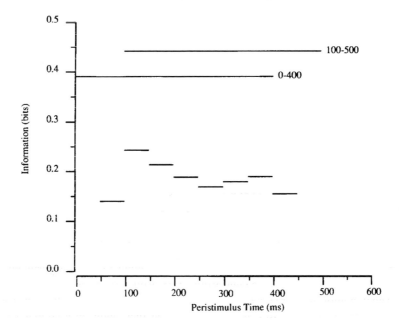

Fig. 1.18 The average information $I(S, R)$ available in short temporal epochs (50 ms as compared to 400 ms) of the spike trains of single inferior temporal cortex neurons about which face had been shown. (From Tovee and Rolls 1995.)

2. Much of this firing rate information is available in short periods, with a considerable proportion available in as little as 20 ms. For example, Tovee, Rolls, Treves and Bellis (1993) and Tovee and Rolls (1995) measured the information available in short epochs of the firing of single inferior temporal cortex neurons to faces and objects, and found that a considerable proportion of the information available in a long time period of 400 ms was available in time periods as short as 20 ms and 50 ms. For example, in periods of 20 ms, 30% of the information present in 400 ms using temporal encoding with the first three principal components was available. Moreover, the exact time when the epoch was taken was not crucial, with the main effect being that rather more information was available if information was measured near the start of the spike train, when the firing rate of the neuron tended to be highest (see Figs. 1.18 and 1.19). The conclusion was that much information was available when temporal encoding could not be used easily, that is in very short time epochs of 20 or 50 ms.

It is also useful to note from Figs. 1.18 and 1.19 the typical time course of the responses of many temporal cortex visual neurons in the awake behaving primate (see further Rolls (2008d)). Although the firing rate and availability of information is highest in the first 50–100 ms of the neuronal response, the firing is overall well sustained in the 500 ms stimulus presentation period. Cortical neurons in the primate temporal lobe visual system (Rolls 2008d, Rolls 2008b), in the taste cortex (Rolls, Yaxley and Sienkiewicz 1990), and in the olfactory cortex (Rolls, Critchley and Treves 1996a), do not in general have rapidly adapting neuronal responses to sensory stimuli. This may be important for associative learning: the outputs of these sensory systems can be maintained for sufficiently long while the stimuli are present for synaptic modification to occur. Although rapid synaptic adaptation within a spike train is seen in some experiments in brain slices (Markram and Tsodyks 1996, Abbott, Varela, Sen and Nelson 1997), it is not a very marked effect in at least some brain systems *in vivo*,

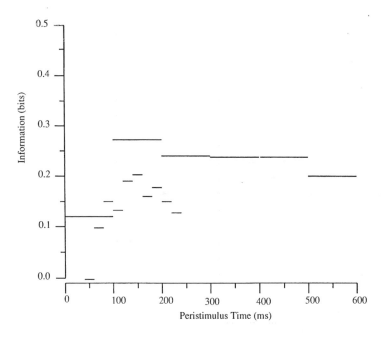

Fig. 1.19 The average information $I(S, R)$ about which face had been shown available in short temporal epochs (20 ms and 100 ms) of the spike trains of single inferior temporal cortex neurons. (From Tovee and Rolls 1995.)

when they operate in normal physiological conditions with normal levels of acetylcholine, etc.

This rapid availability of information enables the next stage of processing to read the information quickly, and thus for multistage processing to operate rapidly (Panzeri, Rolls, Battaglia and Lavis 2001). This time is the order of time over which a receiving neuron might be able to utilize the information, given its synaptic and membrane time constants. In this time, a sending neuron is most likely to emit 0, 1, or 2 spikes.

3. This rapid availability of information is confirmed by population analyses, which indicate that across a population of neurons, much information is available in short time periods (Rolls, Franco, Aggelopoulos and Jerez 2006a).

4. More information is available using this rate code in a short period (of e.g. 20 ms) than from just the first spike (Rolls, Franco, Aggelopoulos and Jerez 2006a).

5. Little information is available by time variations within the spike train of individual neurons for static visual stimuli (in periods of several hundred milliseconds), apart from a small amount of information from the onset latency of the neuronal response (Tovee, Rolls, Treves and Bellis 1993). The same holds for the relative time of arrival of spikes from different neurons (Rolls, Franco, Aggelopoulos and Jerez 2006a). A static stimulus encompasses what might be seen in a single visual fixation, what might be tasted with a stimulus in the mouth, what might be smelled in a single breath, etc. For a time-varying stimulus, clearly the firing rate will vary as a function of time.

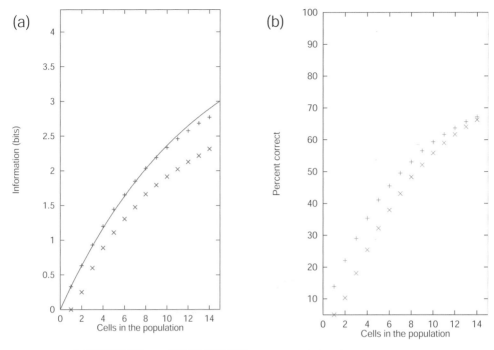

Fig. 1.20 (a) The values for the average information available in the responses of different numbers of inferior temporal cortex neurons on each trial, about which of a set of 20 face stimuli has been shown. The decoding method was Dot Product (DP, $\times$) or Probability Estimation (PE, $+$). The full line indicates the amount of information expected from populations of increasing size, when assuming random correlations within the constraint given by the ceiling (the information in the stimulus set, I = 4.32 bits). (b) The per cent correct for the corresponding data to those shown in (a). The measurement period was 500 ms. (After Rolls, Treves and Tovee 1997b.)

1.11.4.2 The information from multiple cells

1. Across a population of neurons, the firing rate information provided by each neuron tends to be independent; that is, the information increases approximately linearly with the number of neurons. This has been found in many investigations, including for face and object representations in the inferior temporal visual cortex (Rolls, Treves and Tovee 1997a, Abbott, Rolls and Tovee 1996, Booth and Rolls 1998, Rolls, Franco, Aggelopoulos and Reece 2003b, Aggelopoulos, Franco and Rolls 2005, Rolls, Franco, Aggelopoulos and Jerez 2006a), for neurons responding to spatial view in the hippocampus (Rolls, Treves, Robertson, Georges-Francois and Panzeri 1998), for neurons responding to head direction in the primate presubiculum (Robertson, Rolls, Georges-Francois and Panzeri 1999), and for neurons responding to olfactory and to taste stimuli in the orbitofrontal cortex (Rolls 2008d, Rolls, Critchley, Verhagen and Kadohisa 2009a). An example is shown in Fig. 1.20. This applies of course only when there is a large amount of information to be encoded, that is with a large number of stimuli. The outcome is that the number of stimuli that can be encoded rises exponentially in the number of neurons in the ensemble. (For a small stimulus set, the information saturates gradually as the amount of information available from the neuronal population approaches that required to code for the stimulus set (Rolls, Treves and Tovee 1997a).) This applies up to the number of neurons tested and the stimulus set sizes used, but as the number of neurons becomes very large, this is likely to hold less well. An implication of the independence is that the response profiles to a set of stimuli of different neurons are uncorrelated, and, consistent

with this, the mean (Pearson) correlation between the response profiles computed over a set of 406 inferior temporal cortex neuron pairs to a set of 20 face and object stimuli was low, 0.049 ± 0.013 (sem) (Franco, Rolls, Aggelopoulos and Jerez 2007).

2. The information in the firing rate across a population of neurons can be read moderately efficiently by a decoding procedure as simple as a dot product (Rolls, Treves and Tovee 1997a, Abbott, Rolls and Tovee 1996, Booth and Rolls 1998). This is the simplest type of processing that might be performed by a neuron, as it involves taking a dot product of the incoming firing rates with the receiving synaptic weights to obtain the activation (e.g. depolarization) of the neuron. This type of information encoding ensures that the simple emergent properties of associative neuronal networks such as generalization, completion, and graceful degradation can be realized very naturally and simply.

3. There is little additional information to the great deal available in the firing rates from any stimulus-dependent cross-correlations or synchronization that may be present. This has been tested by simultaneous recording from inferior temporal cortex neurons that provide information about which face or object has been shown (Rolls, Franco, Aggelopoulos and Reece 2003b, Rolls, Aggelopoulos, Franco and Treves 2004, Franco, Rolls, Aggelopoulos and Treves 2004), and is found under natural scene conditions, when a macaque is paying attention and searching in a complex natural scene for one of two objects (Aggelopoulos, Franco and Rolls 2005), a situation in which feature binding and segmentation must be performed (Rolls 2008d). Stimulus-dependent synchronization might in any case only be useful for grouping different neuronal populations (Singer 1999), and would not easily provide a solution to the binding problem in vision (Rolls 2008d). Instead, the binding problem in vision may be solved by the presence of neurons that respond to combinations of features in a given spatial position with respect to each other (Aggelopoulos and Rolls 2005, Elliffe, Rolls and Stringer 2002, Rolls 2008d).

4. There is little additional information to the great deal available in the firing rates from the order on which spikes arrive from different neurons (Rolls, Franco, Aggelopoulos and Jerez 2006a).

1.11.4.3 Information theory terms: a short glossary

1. The **amount of information**, or **surprise**, in the occurrence of an event (or symbol) s_i of probability $P(s_i)$ is

$$I(s_i) = \log_2 1/P(s_i) = -\log_2 P(s_i).\qquad(1.28)$$

(The measure is in bits if logs to the base 2 are used.) This is also the amount of **uncertainty** removed by the occurrence of the event.

2. The average amount of information per source symbol over the whole alphabet (S) of symbols s_i is the **entropy**,

$$H(S) = -\sum_i P(s_i) \log_2 P(s_i)\qquad(1.29)$$

(or *a priori* entropy).

3. The probability of the pair of symbols s and s' is denoted $P(s, s')$, and is $P(s)P(s')$ only when the two symbols are **independent**.

4. Bayes' theorem (given the output s', what was the input s ?) states that

$$P(s|s') = \frac{P(s'|s)P(s)}{P(s')} \tag{1.30}$$

where $P(s'|s)$ is the **forward** conditional probability (given the input s, what will be the output s' ?), and $P(s|s')$ is the **backward** (or posterior) conditional probability (given the output s', what was the input s ?). The prior probability is $P(s)$.

5. **Mutual information**. Prior to reception of s', the probability of the input symbol s was $P(s)$. This is the *a priori* probability of s. After reception of s', the probability that the input symbol was s becomes $P(s|s')$, the conditional probability that s was sent given that s' was received. This is the *a posteriori* probability of s. The difference between the *a priori* and *a posteriori* uncertainties measures the gain of information due to the reception of s'. Once averaged across the values of both symbols s and s', this is the **mutual information**, or **transinformation**

$$I(S, S') = \sum_{s,s'} P(s, s')\{\log_2[1/P(s)] - \log_2[1/P(s|s')]\} \tag{1.31}$$

$$= \sum_{s,s'} P(s, s') \log_2[P(s|s')/P(s)].$$

Alternatively,

$$I(S, S') = H(S) - H(S|S'). \tag{1.32}$$

$H(S|S')$ is sometimes called the **equivocation** (of S with respect to S').

6. **The mutual information carried by a neuronal response and its averages**. Where we have pairs of events s and r (where r might be a neuronal response to a stimulus s given on an individual trial), denoting with $P(s, r)$ the *joint probability* of the pair of events s and r, and using Bayes' theorem, we can express the **mutual information** $I(S, R)$ between the whole set of stimuli S and responses R

$$I(S, R) = \sum_{s,r} P(s, r) \log_2 \frac{P(s, r)}{P(s)P(r)}. \tag{1.33}$$

With this overview, it is now time to consider how networks in the brain are influenced by noise.

2 Stochastic neurodynamics

2.1 Introduction

We have provided an introduction to attractor networks in Section 1.10. In this Chapter we describe neural networks with integrate-and-fire spiking dynamics in Section 2.2. The spiking of the neurons in the brain is almost random in time for a given mean rate, i.e. the spiking is approximately Poissonian, and this randomness introduces noise into the system, which makes the system behave stochastically. The integrate-and-fire approach is a computationally realizable way to simulate the operation of neural networks when operating in this stochastic way, and to investigate how networks operate when subject to this noise. The approach incorporates the synaptic currents produced by activation of synaptic receptors, and each of the neurons shows spiking that is very much like that of neurons recorded in the brain. From such integrate-and-fire models it is possible to make predictions not only about synaptic and neuronal activity, but also about the signals recorded in functional neuroimaging experiments, and the behaviour of the system, in terms for example of its probabilistic choices and its reaction times. This makes a close link with human behaviour possible (Rolls, Grabenhorst and Deco 2010b, Rolls, Grabenhorst and Deco 2010c).

We then provide underlying reasons for why the system operates in this noisy way (Section 2.4), and in this Chapter, and elsewhere in this book, describe some of the advantages of this stochastic operation.

Although integrate-and-fire networks provide very useful models of neuronal activity in the brain, it is helpful to complement them with mean-field approaches, which allow formal mathematical analysis. We therefore later in the Chapter (in Sections 2.6 and 2.7) introduce such mean-field approaches, and in particular describe an approach in which it is possible to have a mean-field model that has the same parameters as an integrate-and-fire model, providing the advantages of both approaches (see further Appendix A). In Section A.5 we introduce some more advanced mean-field approaches, whereby mean-field approaches can be extended to incorporate some of the effects of the noise that is generated by neuronal spiking.

2.2 Network dynamics: the integrate-and-fire approach

The attractor networks described in Section 1.10, and analyzed in Appendix 4 of Rolls and Treves (1998), were described in terms of the steady-state activation of networks of neuron-like units. Those may be referred to as 'static' properties, in the sense that they do not involve the time dimension. In order to address 'dynamical' questions, the time dimension has to be reintroduced into the formal models used.

Consider for example a real network whose operation has been described by an auto-associative formal model that acquires, with learning, a given attractor structure. How does the state of the network approach, in real time during a retrieval operation, one of those attractors? How long does it take? How does the noise in the system produced by the spiking of the neurons influence the state that is finally reached, and the time that this takes? This is crucial for understanding decision-making, memory recall, and the stability of short-term memory and attention. How does the amount of information that can be read off the network's activity evolve with time? Also, which of the potential steady states is indeed a stable state that can be reached asymptotically by the net? How is the stability of different states modulated by external agents? These are examples of dynamical properties, which to be studied require the use of models endowed with some dynamics.

2.2.1 From discrete to continuous time

Already at the level of simple models in which each unit is described by an input–output relation, one may introduce equally simple 'dynamical' rules, in order both to fully specify the model, and to simulate it on computers. These rules are generally formulated in terms of 'updatings': time is considered to be discrete, a succession of time steps, and at each time step the output of one or more of the units is set, or updated, to the value corresponding to its input variable. The input variable may reflect the outputs of other units in the net as updated at the previous time step or, if delays are considered, the outputs as they were at a prescribed number of time steps in the past. If all units in the net are updated together, the dynamics is referred to as parallel; if instead only one unit is updated at each time step, the dynamics is sequential. (One main difference between the Hopfield (1982) model of an autoassociator and a similar model considered earlier by Little (1974) is that the latter was based on parallel rather than sequential dynamics.) Many intermediate possibilities obviously exist, involving the updating of groups of units at a time. The order in which sequential updatings are performed may for instance be chosen at random at the beginning and then left the same in successive cycles across all units in the net; or it may be chosen anew at each cycle; yet a third alternative is to select at each time step a unit, at random, with the possibility that a particular unit may be selected several times before some of the other ones are ever updated. The updating may also be made probabilistic, with the output being set to its new value only with a certain probability, and otherwise remaining at the current value.

Variants of these dynamical rules have been used for decades in the analysis and computer simulation of physical systems in statistical mechanics (and field theory). They can reproduce in simple but effective ways the stochastic nature of transitions among discrete quantum states, and they have been subsequently considered appropriate also in the simulation of neural network models in which units have outputs that take discrete values, implying that a change from one value to another can only occur in a sudden jump. To some extent, different rules are equivalent, in that they lead, in the evolution of the activity of the net along successive steps and cycles, to the same set of possible steady states. For example, it is easy to realize that when no delays are introduced, states that are stable under parallel updating are also stable under sequential updating. The reverse is not necessarily true, but on the other hand states that are stable when updating one unit at a time are stable irrespective of the updating order. Therefore, static properties, which can be deduced from an analysis of stable states, are to some extent robust against differences in the details of the dynamics assigned to the model. (This is a reason for using these dynamical rules in the study of the thermodynamics of physical systems.) Such rules, however, bear no relation to the actual dynamical processes by which the activity of real neurons evolves in time, and are therefore inadequate for the discussion of dynamical issues in neural networks.

A first step towards realism in the dynamics is the substitution of discrete time with continuous time. This somewhat parallels the substitution of the discrete output variables of the most rudimentary models with continuous variables representing firing rates. Although continuous output variables may evolve also in discrete time, and as far as static properties are concerned differences are minimal, with the move from discrete to continuous outputs the main raison d'etre for a dynamics in terms of sudden updatings ceases to exist, since continuous variables can change continuously in continuous time. A paradox arises immediately, however, if a continuous time dynamics is assigned to firing rates. The paradox is that firing rates, although in principle continuous if computed with a generic time-kernel, tend to vary in jumps as new spikes – essentially discrete events – come to be included in the kernel. To avoid this paradox, a continuous time dynamics can be assigned, instead, to instantaneous continuous variables such as membrane potentials. Hopfield (1984), among others, has introduced a model of an autoassociator in which the output variables represent membrane potentials and evolve continuously in time, and has suggested that under certain conditions the stable states attainable by such a network are essentially the same as for a network of binary units evolving in discrete time. If neurons in the central nervous system communicated with each other via the transmission of graded membrane potentials, as they do in some peripheral and invertebrate neural systems, this model could be an excellent starting point. The fact that, centrally, transmission is primarily via the emission of discrete spikes makes a model based on membrane potentials as output variables inadequate to correctly represent spiking dynamics.

2.2.2 Continuous dynamics with discontinuities: integrate-and-fire neuronal networks

In principle, a solution would be to keep the membrane potential as the basic dynamical variable, evolving in continuous time, and to use as the output variable the spike emission times, as determined by the rapid variation in membrane potential corresponding to each spike. A point-like neuron can generate spikes by altering the membrane potential V according to continuous equations of the Hodgkin–Huxley type, as described in Section 1.6.

From the point of view of formal models of neural networks, this level of description is too complicated to be the basis for an analytic (formal mathematical) understanding of the operation of networks, and it must be simplified. The most widely used simplification is the so-called integrate-and-fire model (see for example MacGregor (1987) and Brunel and Wang (2001)), which is legitimized by the observation that (sodium) action potentials are typically brief and self-similar events. If, in particular, the only relevant variable associated with the spike is its time of emission (at the soma, or axon hillock), which essentially coincides with the time the potential V reaches a certain threshold level V_{thr}, then the conductance changes underlying the rest of the spike can be omitted from the description, and substituted with the ad hoc prescription that (i) a spike is emitted, with its effect on receiving units and on the unit itself; and (ii) after a brief time corresponding to the duration of the spike plus a refractory period, the membrane potential is reset and resumes its integration of the input current I. After a spike the membrane potential is taken to be reset to a value V_{reset}. This type of simplified dynamics of the membrane potentials is thus in continuous time with added discontinuities: continuous in between spikes, with discontinuities occurring at different times for each neuron in a population, every time a neuron emits a spike.

A leaky integrate-and-fire neuron along the lines just introduced can be modelled as follows. The model describes the depolarization of the membrane potential V (which typically is dynamically changing as a result of the synaptic effects described below between approximately –70 and –50 mV) until threshold V_{thr} (typically –50 mV) is reached when a spike is emitted and the potential is reset to V_{reset} (typically –55 mV). The membrane time constant τ_m is set by the membrane capacitance C_m and the membrane leakage conductance g_m where $\tau_m = C_m/g_m$. V_L denotes the resting potential of the cell, typically –70 mV. Changes in the membrane potential are defined by the following equation

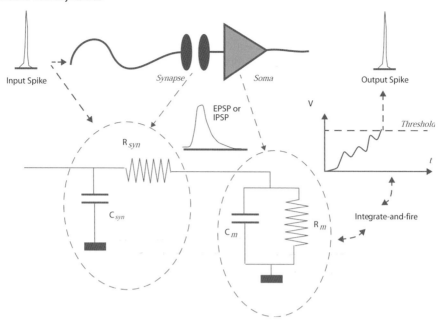

Fig. 2.1 Integrate-and-fire neuron. The basic circuit of an integrate-and-fire model consists of the neuron's membrane capacitance C_m in parallel with the membrane's resistance R_m (the reciprocal of the membrane conductance g_m) driven by a synaptic current with a conductance and time constant determined by the synaptic resistance R_{syn} (the reciprocal of the synaptic conductance g_j) and capacitance C_{syn} shown in the Figure. These effects produce excitatory or inhibitory postsynaptic potentials, EPSPs or IPSPs. These potentials are integrated by the cell, and if a threshold V_{thr} is reached a δ-pulse (spike) is fired and transmitted to other neurons, and the membrane potential is reset. (After Deco and Rolls 2003.)

$$C_{\mathrm{m}} \frac{\mathrm{d}V(t)}{\mathrm{d}t} = g_{\mathrm{m}}\left(V_{\mathrm{L}} - V(t)\right) + \sum_j g_j \left(V_{\mathrm{AMPA}} - V(t)\right) + \sum_j g_j \left(V_{\mathrm{GABA}} - V(t)\right). \quad (2.1)$$

The first term on the right of the equation describes how the membrane potential decays back towards the resting potential of the cell depending on how far the cell potential V is from the resting potential V_{L}, and the membrane leak conductance g_{m}. The second term on the right represents the excitatory synaptic current that could be injected through AMPA receptors. This is a sum over all AMPA synapses indexed by j on the cell. At each synapse, the current is driven into the cell by the difference between the membrane potential V and the reversal potential V_{AMPA} of the channels opened by AMPA receptors, weighted by the synaptic conductance g_j. This synaptic conductance changes dynamically as a function of the time since a spike reached the synapse, as shown below. Due to their reversal potential V_{AMPA} of typically 0 mV, these currents will tend to depolarize the cell, that is to move the membrane potential towards the firing threshold. The third term on the right represents the inhibitory synaptic current that could be injected through GABA receptors. Due to their reversal potential V_{GABA} of typically –70 mV, these currents will tend to hyperpolarize the cell, that is to move the membrane potential away from the firing threshold. There may be other types of receptor on a cell, for example NMDA receptors, and these operate analogously though with interesting differences, as described in Section 2.2.3.

The opening of each synaptic conductance g_j is driven by the arrival of spikes at the presynaptic terminal j, and its closing can often be described as a simple exponential process. A simplified equation for the dynamics of $g_j(t)$ is then

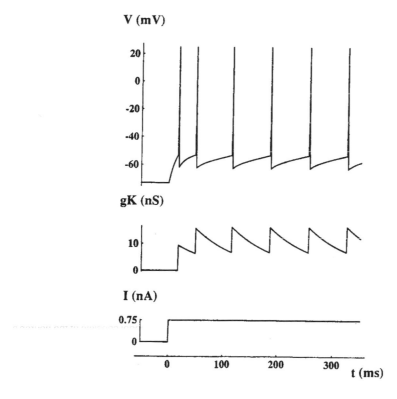

Fig. 2.2 Model behaviour of an integrate-and-fire neuron: the membrane potential and adaptation-producing potassium conductance in response to a step of injected current. The spikes were added to the graph by hand, as they do not emerge from the simplified voltage equation. (From Treves 1993.)

$$\frac{dg_j(t)}{dt} = -\frac{g_j(t)}{\tau} + \Delta g_j \sum_l \delta(t - \Delta t - t_l). \tag{2.2}$$

According to Equation 2.2, each synaptic conductance opens instantaneously by a fixed amount Δg_j a time Δt after the emission of the presynaptic spike at t_l. Δt summarizes delays (axonal, synaptic, and dendritic), and each opening superimposes linearly, without saturating, on previous openings. The value of τ in this equation will be different for AMPA receptors (typically 2–10 ms), NMDA receptors (typically 100 ms), and GABA receptors (typically 10 ms).

In order to model the phenomenon of adaptation in the firing rate, prominent especially with pyramidal cells, it is possible to include a time-varying intrinsic (potassium-like) conductance in the cell membrane (Brown, Gahwiler, Griffith and Halliwell 1990) (shown as g_K in Equations 2.3 and 2.4). This can be done by specifying that this conductance, which if open tends to shunt the membrane and thus to prevent firing, opens by a fixed amount with the potential excursion associated with each spike, and then relaxes exponentially to its closed state. In this manner sustained firing driven by a constant input current occurs at lower rates after the first few spikes, in a way similar, if the relevant parameters are set appropriately, to the behaviour observed *in vitro* of many pyramidal cells (for example, Lanthorn, Storn and Andersen (1984), Mason and Larkman (1990)).

The equations for the dynamics of each neuron with adaptation are then

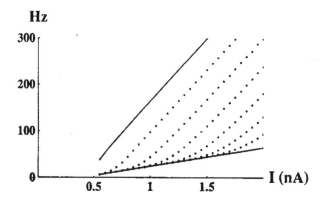

Fig. 2.3 Current-to-frequency transduction in a pyramidal cell modelled as an integrate-and-fire neuron. The top solid curve is the firing frequency in the absence of adaptation, $\Delta g_{\mathrm{K}} = 0$. The dotted curves are the instantaneous frequencies computed as the inverse of the ith interspike interval (top to bottom, $i = 1, ..., 6$). The bottom solid curve is the adapted firing curve ($i \to \infty$). With or without adaptation, the input–output transform is close to threshold–linear. (From Treves 1993.)

$$C_{\mathrm{m}}\frac{\mathrm{d}V(t)}{\mathrm{d}t} = g_{\mathrm{m}}\left(V_{\mathrm{L}}-V(t)\right)+\sum_{j} g_{j}\left(V_{\mathrm{AMPA}}-V(t)\right)+\sum_{j} g_{j}\left(V_{\mathrm{GABA}}-V(t)\right)+g_{\mathrm{K}}(t)(V_{\mathrm{K}}-V(t))$$

$$(2.3)$$

and

$$\frac{\mathrm{d}g_{\mathrm{K}}(t)}{\mathrm{d}t} = -\frac{g_{\mathrm{K}}(t)}{\tau_{\mathrm{K}}}+\sum_{k}\Delta g_{\mathrm{K}}\delta(t-t_{k}) \qquad (2.4)$$

supplemented by the prescription that when at time $t = t_{k+1}$ the potential reaches the level V_{thr}, a spike is emitted, and hence included also in the sum of Equation 2.4, and the potential resumes its evolution according to Equation 2.3 from the reset level V_{reset}. The resulting behaviour is exemplified in Fig. 2.2, while Fig. 2.3 shows the input–output transform (current to frequency transduction) operated by an integrate-and-fire unit of this type with firing rate adaptation. One should compare this with the transduction operated by real cells, as exemplified for example in Fig. 2.4. Implementations of spike frequency adaptation are described further by Treves (1993), Liu and Wang (2001), Deco and Rolls (2005c), Deco and Rolls (2005d), and in Section 8.13.4.

It should be noted that synaptic conductance dynamics is not always included in integrate-and-fire models: sometimes it is substituted with current dynamics, which essentially amounts to neglecting non-linearities due to the appearance of the membrane potential in the driving force for synaptic action (see for example, Amit and Tsodyks (1991), Gerstner (1995)); and sometimes it is simplified altogether by assuming that the membrane potential undergoes small sudden jumps when it receives instantaneous pulses of synaptic current (see the review in Gerstner (1995)). The latter simplification is quite drastic and changes the character of the dynamics markedly; whereas the former can be a reasonable simplification in some circumstances, but it produces serious distortions in the description of inhibitory GABA$_{\mathrm{A}}$ currents, which, having an equilibrium (Cl^{-}) synaptic potential close to the operating range of the membrane potential, are quite sensitive to the instantaneous value of the membrane potential itself.

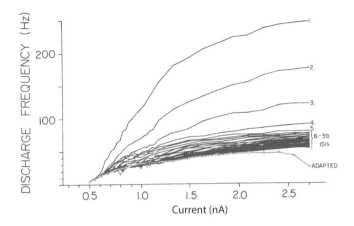

Fig. 2.4 Frequency–current plot (the closest experimental analogue of the activation function) for a CA1 pyramidal cell. The firing frequency (in Hz) in response to the injection of 1.5 s long, rectangular depolarizing current pulses has been plotted against the strength of the current pulses (in nA) (abscissa). The first 39 interspike intervals (ISIs) are plotted as instantaneous frequency (1 / ISI), together with the average frequency of the adapted firing during the last part of the current injection (circles and broken line). The plot indicates a current threshold at approximately 0.5 nA, a linear range with a tendency to saturate, for the initial instantaneous rate, above approximately 200 Hz, and the phenomenon of adaptation, which is not reproduced in simple non-dynamical models (see further Appendix A5 of Rolls and Treves 1998). (Reprinted with permission from Lanthorn, Storm and Andersen 1984.)

2.2.3 An integrate-and-fire implementation with NMDA receptors and dynamical synapses

In this subsection the mathematical equations that describe the spiking activity and synapse dynamics in some of the integrate-and-fire attractor network simulations performed by Deco and Rolls (Deco and Rolls 2003, Deco, Rolls and Horwitz 2004, Deco and Rolls 2005d, Deco and Rolls 2005c, Deco and Rolls 2005b, Deco and Rolls 2006, Loh, Rolls and Deco 2007a, Rolls, Loh, Deco and Winterer 2008b, Rolls, Loh and Deco 2008a, Rolls, Grabenhorst and Deco 2010b, Rolls, Grabenhorst and Deco 2010c) are set out, in order to show in more detail how an integrate-and-fire simulation is implemented which includes in addition NMDA receptors. The architecture of such a network is illustrated in Fig. 5.3. The simulation is that of Deco, Rolls and Horwitz (2004), and follows in general the formulation described by Brunel and Wang (2001).

Each neuron is described by an integrate-and-fire model. The subthreshold membrane potential $V(t)$ of each neuron evolves according to the following equation:

$$C_m \frac{dV(t)}{dt} = -g_m(V(t) - V_L) - I_{syn}(t) \tag{2.5}$$

where $I_{syn}(t)$ is the total synaptic current flow into the cell, V_L is the resting potential of –70 mV , C_m is the membrane capacitance taken to be 0.5 nF for excitatory neurons and 0.2 nF for inhibitory neurons, and g_m is the membrane leak conductance taken to be 25 nS for excitatory neurons and 20 nS for inhibitory neurons. When the membrane potential $V(t)$ reaches the firing threshold $V_{thr} = -50$ mV a spike is generated, and the membrane potential is reset to $V_{reset} = -55$ mV (see McCormick, Connors, Lighthall and Prince (1985)). The neuron is unable to spike during the first τ_{ref} which is the absolute refractory period.

The total synaptic current is given by the sum of glutamatergic excitatory components (NMDA and AMPA) and inhibitory components (GABA). The external excitatory contributions (ext) from outside the network are produced through AMPA receptors ($I_{\text{AMPA,ext}}$), while the excitatory recurrent synapses (rec) within the attractor network act through AMPA and NMDA receptors ($I_{\text{AMPA,rec}}$ and $I_{\text{NMDA,rec}}$). The total synaptic current is therefore given by:

$$I_{\text{syn}}(t) = I_{\text{AMPA,ext}}(t) + I_{\text{AMPA,rec}}(t) + I_{\text{NMDA,rec}}(t) + I_{\text{GABA}}(t) \tag{2.6}$$

where

$$I_{\text{AMPA,ext}}(t) = g_{\text{AMPA,ext}}(V(t) - V_{\text{E}}) \sum_{j=1}^{N_{\text{ext}}} s_j^{\text{AMPA,ext}}(t) \tag{2.7}$$

$$I_{\text{AMPA,rec}}(t) = g_{\text{AMPA,rec}}(V(t) - V_{\text{E}}) \sum_{j=1}^{N_E} w_j s_j^{\text{AMPA,rec}}(t) \tag{2.8}$$

$$I_{\text{NMDA,rec}}(t) = \frac{g_{\text{NMDA,rec}}(V(t) - V_{\text{E}})}{(1 + [\text{Mg}^{++}] \exp(-0.062V(t))/3.57)} \sum_{j=1}^{N_E} w_j s_j^{\text{NMDA,rec}}(t) \tag{2.9}$$

$$I_{\text{GABA}}(t) = g_{\text{GABA}}(V(t) - V_{\text{I}}) \sum_{j=1}^{N_I} s_j^{\text{GABA}}(t) \tag{2.10}$$

where $V_{\text{E}} = 0$ mV, $V_{\text{I}} = -70$ mV, w_j are the synaptic weights, each receptor has its own fraction s_j of open channels, and its own synaptic conductance g. Typical values of the conductances for pyramidal neurons are: $g_{\text{AMPA,ext}}$=2.08, $g_{\text{AMPA,rec}}$=0.052, $g_{\text{NMDA,rec}}$=0.164, and g_{GABA}=0.67 nS; and for interneurons: $g_{\text{AMPA,ext}}$=1.62, $g_{\text{AMPA,rec}}$=0.0405, $g_{\text{NMDA,rec}}$=0.129 and g_{GABA}=0.49 nS. In the work of Brunel and Wang (2001) the conductances were set so that in an unstructured network the excitatory neurons had a spontaneous spiking rate of 3 Hz and the inhibitory neurons a spontaneous rate of 9 Hz, and we have followed a similar procedure.

In the preceding equations the reversal potential of the excitatory synaptic currents $V_{\text{E}} = 0$ mV and of the inhibitory synaptic currents $V_{\text{I}} = -70$ mV. The different form for the NMDA receptor-activated channels implements the voltage-dependence of NMDA receptors, and the voltage-dependence is controlled by the extracellular magnesium concentration (Jahr and Stevens 1990). This voltage-dependency, and the long time constant of the NMDA receptors, are important in the effects produced through NMDA receptors (Brunel and Wang 2001, Wang 1999). The synaptic strengths w_j are specified in our papers (Deco and Rolls 2003, Deco, Rolls and Horwitz 2004, Deco and Rolls 2005d, Deco and Rolls 2005c, Deco and Rolls 2005b, Deco and Rolls 2006, Rolls, Grabenhorst and Deco 2010b, Rolls, Grabenhorst and Deco 2010c, Insabato, Pannunzi, Rolls and Deco 2010), and depend on the architecture being simulated.

The synaptic currents also evolve dynamically in time, and are modelled as follows. The fractions of open channels s are given by:

$$\frac{ds_j^{\text{AMPA,ext}}(t)}{dt} = -\frac{s_j^{\text{AMPA,ext}}(t)}{\tau_{\text{AMPA}}} + \sum_k \delta(t - t_j^k) \tag{2.11}$$

$$\frac{ds_j^{\text{AMPA,rec}}(t)}{dt} = -\frac{s_j^{\text{AMPA,rec}}(t)}{\tau_{\text{AMPA}}} + \sum_k \delta(t - t_j^k) \tag{2.12}$$

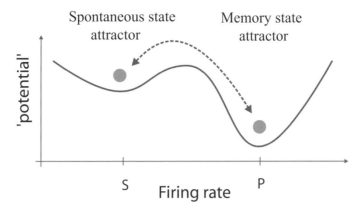

Fig. 2.5 Energy landscape. The noise influences when the system will jump out of the spontaneous firing stable (low energy) state S, and whether it jumps into the high firing rate state labelled P (with persistent or continuing firing in a state which is even more stable with even lower energy), which might correspond to a short-term memory, or to a decision.

$$\frac{ds_j^{\text{NMDA,rec}}(t)}{dt} = -\frac{s_j^{\text{NMDA,rec}}(t)}{\tau_{\text{NMDA,decay}}} + \alpha x_j(t)(1 - s_j^{\text{NMDA,rec}}(t)) \qquad (2.13)$$

$$\frac{dx_j(t)}{dt} = -\frac{x_j(t)}{\tau_{\text{NMDA,rise}}} + \sum_k \delta(t - t_j^k) \qquad (2.14)$$

$$\frac{ds_j^{\text{GABA}}(t)}{dt} = -\frac{s_j^{\text{GABA}}(t)}{\tau_{\text{GABA}}} + \sum_k \delta(t - t_j^k) \qquad (2.15)$$

where the sums over k represent a sum over spikes formulated as δ-Peaks ($\delta(t)$) emitted by presynaptic neuron j at time t_j^k. The rise time constant for NMDA synapses is $\tau_{\text{NMDA,rise}} = 2$ ms (Spruston, Jonas and Sakmann 1995, Hestrin, Sah and Nicoll 1990), the rise time constants for AMPA and GABA are neglected because they are smaller than 1 ms, and $\alpha = 0.5$ ms^{-1}. All synapses have a delay of 0.5 ms. The decay time constant for the AMPA synapses is $\tau_{\text{AMPA}} = 2$ ms (Spruston et al. 1995, Hestrin et al. 1990), for NMDA synapses is $\tau_{\text{NMDA,decay}} = 100$ ms (Spruston et al. 1995, Hestrin et al. 1990), and for GABA synapses $\tau_{\text{GABA}} = 10$ ms (Salin and Prince 1996, Xiang, Huguenard and Prince 1998).

Further details of the values of the parameters used for example in a model of decision-making (Deco and Rolls 2006) are given in Section 2.7.3.

2.3 Attractor networks, energy landscapes, and stochastic dynamics

Autoassociation attractor systems (described in Section 1.10) have two types of stable fixed points: a spontaneous state with a low firing rate, and one or more attractor states with high firing rates in which the positive feedback implemented by the recurrent collateral connections maintains a high firing rate. We sometimes refer to this latter state as the persistent state, because the high firing normally persists to maintain a set of neurons active, which might implement a short-term memory.

The stable points of the system can be visualized in an energy landscape (see Fig. 2.5). The area in the energy landscape within which the system will move to a stable attractor state

p. 30

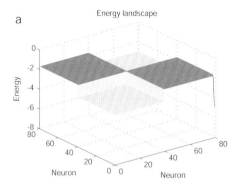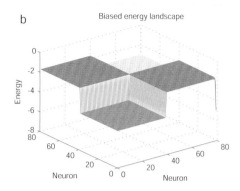

a   Energy landscape   b   Biased energy landscape

Fig. 2.6 (a) Energy landscape without any differential bias applied. The landscape is for a network with neurons 1–40 connected by strengthened synapses so that they form attractor 1, and neurons 41–80 connected by synapses strengthened (by the same amount) so that they form attractor 2. The energy basins in the two-dimensional landscape are calculated by Equation 2.17. In this two-dimensional landscape, there are two stable attractors, and each will be reached equally probably under the influence of noise. This scenario might correspond to decision-making where the input λ_1 to attractor 1 has the same value as the input λ_2 to attractor 2, and the network is equally likely under the influence of the noise to fall into attractor 1 representing decision 1 as into attractor 2 representing decision 2. (b) Energy landscape with bias applied to neurons 1–40. This make the basin of attraction deeper for attractor 1, as calculated with Equation 2.17. Thus, under the influence of noise caused by the randomness in the firing of the neurons, the network will reach attractor 1 more probably than it will reach attractor 2. This scenario might correspond to decision-making, where the evidence for decision 1 is stronger than for decision 2, so that a higher firing rate is applied as λ_1 to neurons 1–40. The scenario might also correspond to memory recall, in which memory 1 might be probabilistically more likely to be recalled than memory 2 if the evidence for memory 1 is stronger. Nevertheless, memory 2 will be recalled sometimes in what operates as a non-deterministic system. (See colour plates Appendix B.)

is called its basin of attraction. The attractor dynamics can be pictured by energy landscapes, which indicate the basins of attraction by valleys, and the attractor states or fixed points by the bottom of the valleys (see Fig. 2.5).

The stability of an attractor is characterized by the average time in which the system stays in the basin of attraction under the influence of noise. The noise provokes transitions to other attractor states. One source of noise results from the interplay between the Poissonian character of the spikes and the finite-size effect due to the limited number of neurons in the network, as analyzed in Chapter 5.

Two factors determine the stability. First, if the depths of the attractors are shallow (as in the left compared to the right valley in Figure 2.5), then less force is needed to move a ball from one valley to the next. Second, high noise will make it more likely that the system will jump over an energy boundary from one state to another. We envision that the brain as a dynamical system has characteristics of such an attractor system including statistical fluctuations (see Section A.5, where the effects of noise are defined quantitatively). The noise could arise not only from the probabilistic spiking of the neurons which has significant effects in finite size integrate-and-fire networks (Deco and Rolls 2006), but also from any other source of noise in the brain or the environment (Faisal, Selen and Wolpert 2008), including the effects of distracting stimuli.

In an attractor network in which a retrieval cue is provided to initiate recall but then removed, a landscape can be defined in terms of the synaptic weights. An example is shown in

Fig. 2.6a. The basins in the landscape can be defined by the strengths of the synaptic weights which describe the stable operating points of the system, where the depth of the basins can be defined in terms of the synaptic weight space, in terms defined by an associative rule operating on the firing rates of pairs of neurons during the learning as follows

$$w_{ij} = y_i y_j \tag{2.16}$$

where y_i is the firing rate of the postsynaptic neuron, y_j is the firing rate of the presynaptic neuron, and w_{ij} is the strength of the synapses connecting these neurons.

Hopfield (1982) showed how many stable states a simple attractor system might contain, and this is the capacity of the network described in Section 1.10.3.6. He showed that the recall process in his attractor network can be conceptualized as movement towards basins of attraction, and his equation defines the energy at a given point in time as being a function of the synaptic weights and the current firing rates as follows

$$E = -\frac{1}{2} \sum_{i,j} w_{ij} (y_i - <y>)(y_j - <y>). \tag{2.17}$$

where y_i is the firing rate of the postsynaptic neuron, y_j is the firing rate of the presynaptic neuron, w_{ij} is the strength of the synapse connecting them, and $<y>$ is the mean firing rate of the neurons. We note that the system defined by Hopfield had an energy function, in that the neurons were connected by symmetric synaptic weights (produced for example by associative synaptic modification of the recurrent collateral synapses) and there was no self-coupling (Hertz, Krogh and Palmer 1991, Moreno-Botel, Rinzel and Rubin 2007, Hopfield and Herz 1995).

The situation is more complicated in an attractor network if it does not have a formal energy function. One such condition is when the connectivity is randomly diluted, for then the synaptic weights between pairs of neurons will not be symmetric. Indeed, in general, neuronal systems do not admit such an energy function. (This is the case in that it is not in general possible to define the flow in terms of the gradient of an energy function. Hopfield defined first an energy function, and from there derived dynamics.) However, such diluted connectivity systems can still operate as attractor systems (Treves 1993, Treves 1991a, Treves 1991b, Treves and Rolls 1991, Treves, Rolls and Simmen 1997, Rolls and Treves 1998, Battaglia and Treves 1998), and the concept of an energy function and landscape is useful for discussion purposes. In practice, a Lyapunov function can be used to prove analytically that there is a stable fixed point such as an attractor basin (Khalil 1996), and even in systems where this can not be proved analytically, it may still be possible to show numerically that there are stable fixed points, to measure the flow towards those fixed points which describes the depth of the attractor basin as we have done for this type of network (Loh, Rolls and Deco 2007a), and to use the concept of energy or potential landscapes to help visualize the properties of the system.

If an external input remains on during the retrieval process, this will influence the energy function of such a network, and its stable points, as implied by Equation 2.17, and as illustrated in Fig. 2.6b. In this situation, the external inputs bias the stable points of the system. Indeed, in this situation, a landscape, though not necessarily formally an energy landscape, can be specified by a combination of the synaptic weights and external inputs that bias the firing rates. The noise introduced into the network by for example the random neuronal spiking can be conceptualized as influencing the way that the system flows across this fixed landscape shaped by the synaptic weights, and by the external inputs if they remain on during operation of the network, in what is referred to as a 'clamped' condition, the normal condition that applies during decision-making (see Chapter 5).

In more detail, the flow, which is the time derivative of the neuronal activity, specifies the landscape in an attractor system. The flow is defined in the mean field analysis described in Appendix A in terms of the effects of the synaptic weights between the neurons and the external inputs (Loh, Rolls and Deco 2007a). The flow is the force that drives the system towards the attractor given a parameter value in phase space, i.e. the firing rates of the pools. This is measured by fixing the value of the firing rate of the selective pool and letting the other values converge to their fixed point. The flow can then be computed with this configuration (Mascaro and Amit 1999). This landscape is thus fixed by the synaptic and the external inputs. The noise, produced for example by the almost Poissonian spiking of the neurons, can be conceptualized as influencing the way that the system flows across this fixed landscape. Moreover, the noise can enable the system to jump over a barrier in this fixed landscape, as illustrated in Figs. 2.5 and 6.1 on page 169.

In Fig. 6.1 (and in Fig. 2.6a) the decision basins of attraction are equally deep, because the inputs λ_1 and λ_2 to the decision-making network are equal, that is, ΔI the difference between them is zero. If λ_1 is greater than λ_2, the basin will be deeper for λ_1. The shape of the landscape is thus a function of the synaptic weights and the biassing inputs to the system. This is illustrated in Fig. 2.6b. Noise can be thought of as provoking movements across the 'effective energy landscape' conceptualized in this way.

The way in which we conceptualise the operation of an attractor network used for noise-driven stochastic decision-making, stimulus detection, etc, is as follows. The noise in the system (caused for example by statistical fluctuations produced by the Poisson-like neuronal firing in a finite-sized system as described in Section 2.4) produces changes in neuronal firing. These changes may accumulate stochastically, and eventually may become sufficiently large that the firing is sufficient to produce energy to cause the system to jump over an energy barrier (see Fig. 2.7). Opposing this noise-driven fluctuation will be the flow being caused by the shape and depth of the fixed energy landscape defined by the synaptic weights and the applied external input bias or biases. The noisy statistical fluctuation is a diffusion-like process. If the spontaneous firing rate state is stable with the decision cues applied (see Fig. 5.14 middle on page 165), eventually the noise may provoke a transition over the energy barrier in an escaping time, and the system will drop, again noisily, down the valley on the other side of the hill. The rate of change of the firing rate is again measured by the flow, and is influenced by the synaptic weights and applied biases, and by the statistical fluctuations. In this scenario, the reaction times will depend on the amount of noise, influenced by the size of the network, and by the fixed 'effective energy landscape' as determined by the synaptic weights and applied biasing inputs λ, which will produce an escaping time as defined further in Appendix A.

If the spontaneous state is not stable (see Fig. 5.14 right), the reaction times will be influenced primarily by the flow as influenced by the gradient of the energy landscape, and by the noise caused by the random neuronal firings. An escaping time from a stable spontaneous state attractor will not in this situation contribute to the reaction times.

The noise-driven escaping time from the stable spontaneous state is important in understanding long and variable reaction times, and such reaction times are present primarily in the scenario when the parameters make the spontaneous state stable, as described further by Marti, Deco, Mattia, Gigante and Del Giudice (2008). While in a spontaneous stable state the system may be thought of as being driven by the noise, and it is primarily when the system has reached a ridge at the edge of the spontaneous valley and the system is close to a bifurcation point into a high firing rate close to the ridge that the attractor system can be thought of as accumulating evidence from the input stimuli (Deco, Scarano and Soto-Faraco 2007b). While in the spontaneous state valley (see Fig. 5.14 middle), the inputs can be thought of as biasing the 'effective energy landscape' across which the noise is driving the system stochastically.

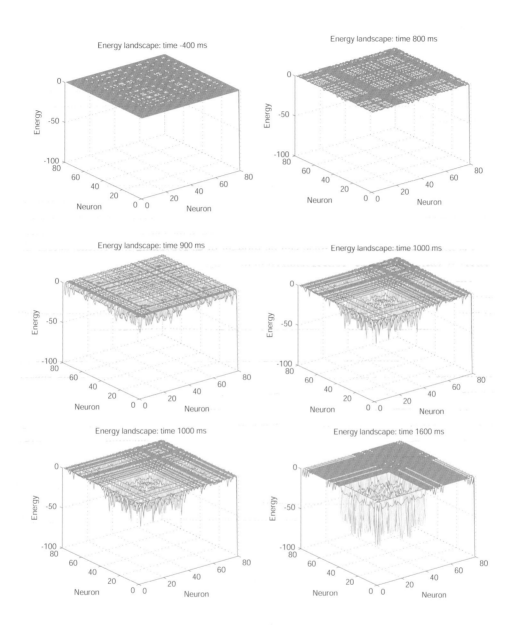

Fig. 2.7 Energy states shown at different times after the decision cues were applied. Neurons 1–40 are in attractor 1 and are connected by strong weights with each other; and neurons 41–80 are in attractor 2 and are connected by strong weights with each other. The energy state is defined in Equation 2.17, and the energy between any pair of neurons is a product of the firing rates of each neuron and the synaptic weight that connects them. These are the energy states for the trial shown in Fig. 6.2b and c on page 172. Time 0 is the time when the decision stimuli were applied (and this corresponds to time 2 s in Fig. 6.2).

2.4 Reasons why the brain is inherently noisy and stochastic

We may raise the conceptually important issue of why the operation of what is effectively memory retrieval is probabilistic. Part of the answer is shown in Fig. 5.13 on page 158, in which it is seen that even when a fully connected recurrent attractor network has 4,000 neurons, the operation of the network is still probabilistic (Deco and Rolls 2006). Under these conditions, the probabilistic spiking of the excitatory (pyramidal) cells in the recurrent collateral firing, rather than variability in the external inputs to the network, is what makes the major contribution to the noise in the network (Deco and Rolls 2006). Thus, once the firing in the recurrent collaterals is spike implemented by integrate-and-fire neurons, the probabilistic behaviour seems inevitable, even up to quite large attractor network sizes.

We may then ask why the spiking activity of any neuron is probabilistic, and what the advantages are that this may confer. The answer suggested (Rolls 2008d) is that the spiking activity is approximately Poisson-like (as if generated by a random process with a given mean rate, both in the brain and in the integrate-and-fire simulations we describe), because the neurons are held close to (just slightly below) their firing threshold, so that any incoming input can rapidly cause sufficient further depolarization to produce a spike. It is this ability to respond rapidly to an input, rather than having to charge up the cell membrane from the resting potential to the threshold, a slow process determined by the time constant of the neuron and influenced by that of the synapses, that enables neuronal networks in the brain, including attractor networks, to operate and retrieve information so rapidly (Treves 1993, Rolls and Treves 1998, Battaglia and Treves 1998, Rolls 2008d, Panzeri, Rolls, Battaglia and Lavis 2001). The spike trains are essentially Poisson-like because the cell potential hovers noisily close to the threshold for firing (Dayan and Abbott 2001), the noise being generated in part by the Poisson-like firing of the other neurons in the network (Jackson 2004). The noise and spontaneous firing help to ensure that when a stimulus arrives, there are always some neurons very close to threshold that respond rapidly, and then communicate their firing to other neurons through the modified synaptic weights, so that an attractor process can take place very rapidly (Rolls 2008d).

The implication of these concepts is that the operation of networks in the brain is inherently noisy because of the Poisson-like timing of the spikes of the neurons, which itself is related to the mechanisms that enable neurons to respond rapidly to their inputs (Rolls 2008d). However, the consequence of the Poisson-like firing is that, even with quite large attractor networks of thousands of neurons with hundreds of neurons representing each pattern or memory, the network inevitably settles probabilistically to a given attractor state. This results, *inter alia*, in decision-making being probabilistic. Factors that influence the probabilistic behaviour of the network include the strength of the inputs (with the difference in the inputs / the magnitude of the inputs being relevant to decision-making and Weber's Law as described in Chapter 5); the depth and position of the basins of attraction, which if shallow or correlated with other basins will tend to slow the network; noise in the synapses due to probabilistic release of transmitters (Koch 1999, Abbott and Regehr 2004); and, perhaps, the mean firing rates of the neurons during the decision-making itself, and the firing rate distribution (see Section 1.11). In terms of synaptic noise, if the probability of release, p, is constant, then the only effect is that the releases in response to a Poisson presynaptic train at rate r are described by a Poisson process with rate pr, rather than r. However, if the synapse has short-term depression or facilitation, the statistics change (Fuhrmann, Segev, Markram and Tsodyks 2002b, Abbott and Regehr 2004). A combination of these factors contributes to the probabilistic behaviour of the network becoming poorer and slower as the difference between the inputs divided by the base frequency is decreased, as shown in Figs. 5.7 and 5.12 (Deco and Rolls 2006). Other factors that may contribute to noise in the brain are reviewed by Faisal, Selen and Wolpert (2008).

In signal-detection theory, behavioural indeterminacy is accounted for in terms of noisy input (Green and Swets 1966). In the present approach, indeterminacy is accounted for in part in terms of internal noise inherent to the operation of spiking neuronal circuitry, as is made evident in this book (see also Wang (2008) and Faisal et al. (2008)).

In fact, there are many sources of noise, both external to the brain and internal (Faisal, Selen and Wolpert 2008). A simple measure of the variability of neuronal responses is the Fano factor, which is the ratio of the variance to the mean. A Poisson process has a Fano factor of 1, and neurons often have Fano factors close to one, though in some systems and for some neurons they can be lower (less variable than a Poisson process), or higher (Faisal et al. 2008). The sources of noise include the following, with citations to original findings provided by Faisal et al. (2008).

First, there is sensory noise, arising external to the brain. Thermodynamic noise is present for example in chemical sensing (including smell and gustation) because molecules arrive at the receptor at random rates owing to diffusion. Quantum noise can also be a factor, with photons arriving at the photoreceptor at a rate governed by a Poisson process. The peripheral nervous system (e.g. the retina, and in the olfactory bulb the glomeruli) often sums inputs over many receptors in order to limit this type of noise. In addition to these processes inherent in the transduction and subsequent amplification of the signals, there may of course be variation in the signal itself arriving from the world.

Second, there is cellular or neuronal noise. arising even in the absence of variation in synaptic input. This cellular noise arises from the stochastic opening and closing of voltage or ligand-gated ion channels, the fusing of synaptic vesicles, and the diffusion and binding of signalling molecules to the receptors. The channel noise becomes appreciable in small axons and cell bodies (where the number of channels and molecules involved becomes relatively small). This channel noise can affect not only the membrane potential of the cell body and the initiation of spikes, but also spike propagation in thin axons. There may also be some cross-talk between neurons, caused by processes such as ephaptic coupling, changes in extracellular ion concentration after electrical signalling, and spillover of neurotransmitters between adjacent unrelated synapses.

Third, there is synaptic noise evident for example as spontaneous miniature postsynaptic currents (mPSCs) that are generated by the quantal nature of released neurotransmitter vesicles (Fatt and Katz 1950, Manwani and Koch 2001). This synaptic noise can be produced by processes that include spontaneous opening of intracellular Ca^{2+} stores, synaptic Ca^{2+}-channel noise, spontaneous triggering of the vesicle-release pathway, and spontaneous fusion of a vesicle with the membrane. In addition, at the synapses there is variation in the number of transmitter molecules in a released vesicle, and there are random diffusion processes of the molecules (Franks, Stevens and Sejnowski 2003). In addition, there is randomness in whether an action potential releases a vesicle, with the probability sometimes being low, and being influenced by factors such as synaptic plasticity and synaptic adaptation (Faisal et al. 2008, Abbott and Regehr 2004, Manwani and Koch 2001).

The sources of noise in the brain arise from the effects described earlier that make the firing properties of neurons very often close to Poisson-like. If the system were infinitely large (as the mean field approach described in the following sections assumes) then the Poisson firing would average out to produce a steady average firing rate averaged across all the neurons. However, if the system has a finite number of neurons, then the Poisson or semirandom effects will not average out, and the average firing rate of the population will fluctuate statistically. These effects are referred to as coherent fluctuations. The magnitude of these fluctuations decreases with the square root of the size of the network (the number of neurons in the population in a fully connected network, and the number of connections per neuron in a network with diluted connectivity) (Mattia and Del Giudice 2002). The magnitude of these fluctuations is described by the standard deviation of the firing rates divided by the mean rate, and this can be used to measure the noise in the network. These effects are described in detail in Appendix A, for example in Section A.5 on page 267, where the noise due to these finite size fluctuations is defined (see also Section 1.11.3.4).

The signal-to-noise ratio can be measured in a network by the average firing rate change produced by the signal in the neuronal population divided by the standard deviation of the firing rates. Non-linearities in the system will affect how the signal-to-noise ratio described in this way is reflected in the output of the system. Another type of measure that reflects the signal-to-noise ratio is the stability of the system as described by its trial to trial variability, as described in Section 3.8.3.3 and by Rolls, Loh, Deco and Winterer (2008b).

Although many approaches have seen noise as a problem in the brain that needs to be averaged out or resolved (Faisal, Selen and Wolpert 2008), and this is certainly important in sensory systems, the thrust of this book is to show that in fact noise inherent in brain activity has a number of advantages by making the dynamics stochastic, which allows for many remarkable features of the brain, including creativity, probabilistic decision-making, stochastic resonance, unpredictability, conflict resolution, symmetry breaking, allocation to discrete categories, and many other important properties described in Chapter 8 and throughout this book.

2.5 Brain dynamics with and without stochasticity: an introduction to mean-field theory

The challenge to unravel the primary mechanisms underlying brain functions demands explicit description of the computation performed by the neuronal and synaptic substrate (Rolls and Deco 2002, Rolls 2008d). Computational neuroscience aims to understand that computation by the construction and simulation of 'microscopic' (a term used in physics) models based on local networks with large numbers of neurons and synapses that lead to the desired global behaviour of the whole system. A suitable level of description of the microscopic level is captured by the spiking and synaptic dynamics of one-compartment, point-like models of neurons, such as Integrate-and-Fire-Models, as described earlier (Brunel and Amit 1997). These dynamics allow the use of realistic biophysical constants (like conductances, delays, etc.) in a thorough study of the actual time scales and firing rates involved in the evolution of the neural activity underlying cognitive processes, for comparison with experimental data.

Nevertheless, an in-depth analytical study of these detailed microscopic models is not possible, and therefore a reduction of the hypothesized models is necessary in order to establish a systematic relation between structure (parameters), dynamics, and functional behaviour (i.e. to solve the 'inverse' problem). A simulation only of phenomena of a complex system, such as the brain, is in general not useful, because there are usually no explicit underlying first principles.

In order to overcome this problem, statistical physics methods have been introduced, to mathematically analyze a reduced version of the system. In particular, **mean-field** techniques (Brunel and Wang 2001, Brunel and Amit 1997) are used for simplifying the detailed spiking dynamics at least for the stationary conditions, i.e. for periods after the dynamical transients. The stationary dynamics of a population of neurons can be described by a reduced rate-equation describing the asymptotic stationary states of the associated average population rate. The set of stationary, self-reproducing rates for the different populations in the network can be found by solving a set of coupled self-consistency equations. These reduced dynamics allow a thorough analysis of the parameter space enabling a selection of the parameter region that shows the emergent behaviour of interest (e.g. short-term memory, attention). After that, with this set of parameters, one can perform the simulations described by the full integrate-and-fire scheme, which include the effects of statistical fluctuations caused by the spiking-related noise, and are able to describe the computation performed by the transients.

We stress the fact that statistical fluctuations (noise) are not just a concomitant of neural processing, but can play an important and unique role in cortical function. We show how computations can be performed through stochastic dynamical effects, including the role of noise in enabling probabilistic jumping across barriers in the energy landscape describing the flow of the dynamics in attractor networks. We show how these effects help to understand short-term memory, attention, long-term memory, and decision-making in the brain, and to establish a link between neuronal firing variability and probabilistic behaviour.

2.6   Network dynamics: the mean-field approach

Units whose potential and conductances follow the integrate-and-fire equations in Section 2.2 can be assembled together in a model network of any composition and architecture. It is convenient to imagine that units are grouped into classes, such that the parameters quantifying the electrophysiological properties of the units (in this case the neurons) are uniform, or nearly uniform, within each class, while the parameters assigned to synaptic connections are uniform or nearly uniform for all connections from a given presynaptic class to another given postsynaptic class. The parameters that have to be set in a model at this level of description are quite numerous, as listed in Tables 2.1 and 2.2.

In the limit in which the parameters are uniform within each class or pair of classes, a mean-field treatment can be applied to analyze a model network, by summing equations that describe the dynamics of individual neurons to obtain a more limited number of equations that describe the dynamical behaviour of groups of neurons (Frolov and Medvedev 1986). The treatment is exact in the further limit in which very many neurons belong to each class, and is an approximation if each class includes just a few neurons. Suppose that N_C is the number of classes defined. Summing Equations 2.3 and 2.4 (on page 70) across neurons of the same class results in N_C functional equations describing the evolution in time of the fraction of neurons of a particular class that at a given instant have a given membrane potential. In other words, from a treatment in which the evolution of the variables associated with each neuron is followed separately, one moves to a treatment based on density functions, in which the common behaviour of neurons of the same class is followed together, keeping track solely of the portion of neurons at any given value of the membrane potential. Summing Equation 2.2 across connections with the same class of origin and destination, results in $N_C \times N_C$ equations describing the dynamics of the overall summed conductance opened on the membrane of a neuron of a particular class by all the neurons of another given class. A more explicit description of mean-field equations is given in Section 2.7, and a derivation of the equations in Appendix A.

The system of mean-field equations can have many types of asymptotic solutions for long times, including chaotic, periodic, and stationary ones. The stationary solutions are stationary in the sense of the mean fields, but in fact correspond to the neurons of each class firing tonically at a certain rate. They are of particular interest as the dynamical equivalent of the steady states analyzed by using non-dynamical model networks. In fact, since the neuronal current-to-frequency transfer function resulting from the dynamical equations is rather similar to a threshold linear function (see Fig. 2.3), and since each synaptic conductance is constant in

Table 2.1 Cellular parameters (chosen according to the class of each neuron)

V_{rest}	Resting potential
V_{thr}	Threshold potential
V_{ahp}	Reset potential
V_K	Potassium conductance equilibrium potential
C	Membrane capacitance
τ_K	Potassium conductance time constant
g_0	Leak conductance
Δg_K	Extra potassium conductance following a spike
Δt	Overall transmission delay

Table 2.2 Synaptic parameters (chosen according to the classes of presynaptic and postsynaptic neurons)

V_α	Synapse equilibrium potential
τ_α	Synaptic conductance time constant
Δg_α	Conductance opened by one presynaptic spike
Δt_α	Delay of the connection

time, the stationary solutions are essentially the same as the states described using model networks made up of threshold linear, non-dynamical neurons. Thus the dynamical formulation reduces to the simpler formulation in terms of steady-state rates when applied to asymptotic stationary solutions; but, among simple rate models, it is equivalent only to those that allow description of the continuous nature of neuronal output, and not to those, for example based on binary units, that do not reproduce this fundamental aspect. The advantages of the dynamical formulation are that (i) it enables one to describe the character and prevalence of other types of asymptotic solutions, and (ii) it enables one to understand how the network reaches, in time, the asymptotic behaviour.

A summary of this mean-field approach, and how it can be developed, are given next, with a fuller description in Appendix A. Applications of this mean-field approach to models of attention, memory, and decision-making are given in the next Chapters of this book, and elsewhere (Rolls and Deco 2002, Rolls 2008d).

2.7 Mean-field based theory

Starting from the mathematical models of biologically realistic single neurons (i.e. spiking neurons), one can derive models that describe the joint activity of pools (or populations) of equivalent neurons. As described in the previous Section, it is convenient in this simplified approach to neural dynamics to consider all neurons of the same type in a small cortical volume as a computational unit of a neural network. This computational unit is called a neuronal pool, population, or assembly. The mathematical description of the dynamical evolution of neuronal pool activity in multimodular networks, associated with different cortical areas, establishes the roots of the dynamical approach that are used in Chapters 4 and 5, and in Rolls and Deco (2002) chapters 9–11. In this Section (2.7), we introduce the mathematical fundamentals utilized for a neurodynamical description of pool activity. Beginning at the microscopic level and using single spiking neurons to form the pools of a network, we derive the mathematical formulation of the neurodynamics of cell assemblies. Further, we introduce the basic architecture of neuronal pool networks that fulfil the basic mechanisms consistent with the biased competition hypothesis (Rolls and Deco 2002). Each of these networks corresponds to cortical areas that also communicate with each other. We describe the dynamical interaction between different modules or networks, which provides a basis for the implementation of attentional top-down bias (Rolls and Deco 2002).

In Section 2.7.2, in order to be explicit, we summarize the mean-field model used in the model of decision-making described in Chapter 5. A derivation of the mean-field approach that is consistent with the spiking dynamics is described in Appendix A. In Sections A.5 and A.6 we show how noise can be introduced into a mean-field approach.

2.7.1 Population activity

We now introduce more thoroughly the concept of a neuronal pool and the differential equations representing the neurodynamics of pool activity.

Starting from individual spiking neurons one can derive a differential equation that describes the dynamical evolution of the averaged activity of a pool of extensively many equivalent neurons. Several areas of the brain contain groups of neurons that are organized in populations of units with (somewhat) similar properties (though in practice the neurons convey independent information, as described in Section 1.11.4). These groups for mean-field modelling purposes are usually called pools of neurons and are constituted by a large and similar population of identical spiking neurons that receive similar external inputs and are mutually coupled by synapses of similar strength. Assemblies of motor neurons (Kandel, Schwartz and Jessell 2000) and the columnar organization in the visual and somatosensory cortex (Hubel and Wiesel 1962, Mountcastle 1957) are examples of these pools. Each single cell in a pool can be described by a spiking model. We take the activity level of each pool of neurons as the relevant dependent variable rather than the spiking activity of individual neurons. We therefore derive a dynamical model for the mean activity of a neural population. In a population of M neurons, the mean activity $A(t)$ is determined by the proportion of active neurons by counting the number of spikes $n_{\text{spikes}}(t, t + \Delta t)$ in a small time interval Δt and dividing by M and by Δt (Gerstner 2000, Gerstner and Kistler 2002), i.e. formally

$$A(t) = \lim_{\Delta t \to 0} \frac{n_{\text{spikes}}(t, t + \Delta t)}{M \Delta t} . \tag{2.18}$$

As indicated by Gerstner (2000), and as depicted in Fig. 2.8, the concept of pool activity is quite different from the definition of the average firing rate of a single neuron. Contrary to the concept of temporal averaging over many spikes of a single cell, which requires that the input is slowly varying compared with the size of the temporal averaging window, a coding scheme based on pool activity allows rapid adaptation to real-world situations with quickly changing inputs. It is possible to derive dynamical equations for pool activity levels by utilizing the mean-field approximation (Wilson and Cowan 1972, Abbott 1991, Amit and Tsodyks 1991, Wilson 1999). The mean-field approximation consists of replacing the temporally averaged discharge rate of a cell with an equivalent momentary activity of a neural population (ensemble average), which corresponds to the assumption of ergodicity. (Ergodicity refers to a situation in which the statistical properties of a single unit measured over a long time are similar to the statistical properties of many units measured over a short time, and applies if the units are independent of each other (Lehky, Sejnowski and Desimone 2005, Franco, Rolls, Aggelopoulos and Jerez 2007). This applies to neurons recorded in the cerebral cortex involved in the representation of objects and faces, and relates

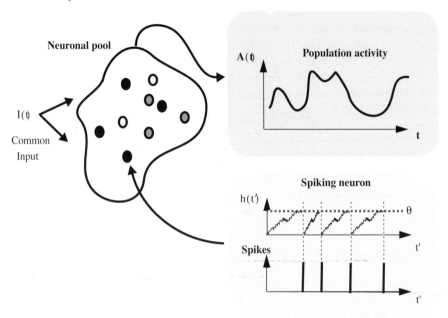

Fig. 2.8 Population averaged rate of a neuronal pool of spiking neurons (top) and the action potential generating mechanism of single neurons (bottom). In a neuronal pool, the mean activity $A(t)$ is determined by the proportion of active neurons by counting the number of spikes in a small time interval Δt and dividing by the number of neurons in the pool and by Δt. Spikes are generated by a threshold process. The firing time t' is given by the condition that the membrane potential $h(t')$ reaches the firing threshold θ. The membrane potential $h(t')$ is given by the integration of the input signal weighted by a given kernel (see text for details). (After Rolls and Deco 2002.)

to the finding that the tuning profiles of the neurons to a set of visual stimuli are uncorrelated with each other (Franco, Rolls, Aggelopoulos and Jerez 2007), as described in Section 1.11.3.2.)

According to this, the mean-field approach can be used to mathematically analyze a reduced version of the system. At the spiking level, when stimuli and tasks are imposed, the neural system will undergo transient change to eventually reach a steady state (an attractor). To make this precise, the 'whole system' behaviour that we may witness in many experiments represents the steady state of the underlying neurodynamical system. *Mean-field* techniques (Brunel and Wang 2001, Brunel and Amit 1997) allow us to express the steady state of a population of neurons by a reduced rate-equation describing the asymptotic steady states of the associated average population rate. This steady state solution for each population can be described by the population transfer function (Ricciardi and Sacerdote 1979, Renart, Brunel and Wang 2003), which provides the average spiking rate at a given time across the population (the population rate) as a function of the average input current.

For more than one population, the network is partitioned into populations of neurons whose input currents share the same statistical properties and fire spikes independently at the same rate. The set of stationary, self-reproducing rates ν_x for different populations x in the network can be found by solving a set of coupled self-consistency equations (one equation for each population of neurons x), given by:

$$\nu_x = \phi(\mu_x, \sigma_x). \tag{2.19}$$

where $\phi(\mu, \sigma)$ is the population transfer function defined in equation A.28 on page 266, and

μ_x and σ_x are the mean and standard deviation of the input currents to pool x defined in Equation A.18 on page 264.

Section 2.7.2 provides as an example the set of equations for the mean-field analysis used in the model of decision-making described in Chapter 5. The architecture of the network is shown in Figs. 5.1 and 5.3, and the application of the mean field analysis to define the different operating points of the network is described in Sections 5.5 and 5.8. This mean-field analysis is consistent with the integrate-and-fire spiking simulation described in Section 2.2.3, that is, the same parameters used in the mean-field analysis can then be used in the integrate-and-fire simulations. Part of the value of the mean-field analysis is that it provides a way of determining the parameters that will lead to the specified steady state behaviour (in the absence of noise), and these parameters can then be used in a well-defined system for the integrate-and-fire simulations to investigate the full dynamics of the system in the presence of the noise generated by the random spike timings of the neurons.

2.7.2 The mean-field approach used in the model of decision-making

The mean-field approximation used by Deco and Rolls (2006) was derived by Brunel and Wang (2001), assuming that the network of integrate-and-fire neurons is in a stationary state. This mean-field formulation includes synaptic dynamics for AMPA, NMDA and GABA activated ion channels (Brunel and Wang 2001). In this formulation the potential of a neuron is calculated as:

$$\tau_x \frac{dV(t)}{dt} = -V(t) + \mu_x + \sigma_x \sqrt{\tau_x} \eta(t) \tag{2.20}$$

where $V(t)$ is the membrane potential, x labels the populations, τ_x is the effective membrane time constant, μ_x is the mean value the membrane potential would have in the absence of spiking and fluctuations, σ_x measures the magnitude of the fluctuations, and η is a Gaussian process with absolute exponentially decaying correlation function with time constant τ_{AMPA}. The quantities μ_x and σ_x^2 are given by:

$$\mu_x = \frac{(T_{ext}\nu_{ext} + T_{AMPA}n_x^{AMPA} + \rho_1 n_x^{NMDA})V_E + \rho_2 n_x^{NMDA}\langle V \rangle + T_I n_x^{GABA}V_I + V_L}{S_x} \tag{2.21}$$

$$\sigma_x^2 = \frac{g_{AMPA,ext}^2(\langle V \rangle - V_E)^2 N_{ext}\nu_{ext}\tau_{AMPA}^2 \tau_x}{g_m^2 \tau_m^2}. \tag{2.22}$$

where ν_{ext} Hz is the external incoming spiking rate, $\tau_m = C_m/g_m$ with the values for the excitatory or inhibitory neurons depending on the population considered, and the other quantities are given by:

$$S_x = 1 + T_{\text{ext}} \nu_{\text{ext}} + T_{\text{AMPA}} n_x^{\text{AMPA}} + (\rho_1 + \rho_2) n_x^{\text{NMDA}} + T_1 n_x^{\text{GABA}} \tag{2.23}$$

$$\tau_x = \frac{C_m}{g_m S_x} \tag{2.24}$$

$$n_x^{\text{AMPA}} = \sum_{j=1}^{p} r_j w_{jx}^{\text{AMPA}} \nu_j \tag{2.25}$$

$$n_x^{\text{NMDA}} = \sum_{j=1}^{p} r_j w_{jx}^{\text{NMDA}} \psi(\nu_j) \tag{2.26}$$

$$n_x^{\text{GABA}} = \sum_{j=1}^{p} r_j w_{jx}^{\text{GABA}} \nu_j \tag{2.27}$$

$$\psi(\nu) = \frac{\nu \tau_{\text{NMDA}}}{1 + \nu \tau_{\text{NMDA}}} \left(1 + \frac{1}{1 + \nu \tau_{\text{NMDA}}} \sum_{n=1}^{\infty} \frac{(-\alpha \tau_{\text{NMDA,rise}})^n T_n(\nu)}{(n+1)!} \right) \tag{2.28}$$

$$T_n(\nu) = \sum_{k=0}^{n} (-1)^k \binom{n}{k} \frac{\tau_{\text{NMDA,rise}} (1 + \nu \tau_{\text{NMDA}})}{\tau_{\text{NMDA,rise}} (1 + \nu \tau_{\text{NMDA}}) + k \tau_{\text{NMDA,decay}}} \tag{2.29}$$

$$\tau_{\text{NMDA}} = \alpha \tau_{\text{NMDA,rise}} \tau_{\text{NMDA,decay}} \tag{2.30}$$

$$T_{\text{ext}} = \frac{g_{\text{AMPA,ext}} \tau_{\text{AMPA}}}{g_m} \tag{2.31}$$

$$T_{\text{AMPA}} = \frac{g_{\text{AMPA,rec}} N_E \tau_{\text{AMPA}}}{g_m} \tag{2.32}$$

$$\rho_1 = \frac{g_{\text{NMDA}} N_E}{g_m J} \tag{2.33}$$

$$\rho_2 = \beta \frac{g_{\text{NMDA}} N_E (\langle V_x \rangle - V_E)(J - 1)}{g_m J^2} \tag{2.34}$$

$$J = 1 + \gamma \exp(-\beta \langle V_x \rangle) \tag{2.35}$$

$$T_1 = \frac{g_{\text{GABA}} N_I \tau_{\text{GABA}}}{g_m} \tag{2.36}$$

$$\langle V_x \rangle = \mu_x - (V_{\text{thr}} - V_{\text{reset}}) \nu_x \tau_x, \tag{2.37}$$

where p is the number of excitatory populations, r_x is the fraction of neurons in the excitatory population x, w_{jx} the weight of the connections from population x to population j, ν_x is the spiking rate of the population x, $\gamma = [\text{Mg}^{++}]/3.57$, $\beta = 0.062$ and the average membrane potential $\langle V_x \rangle$ has a value between -55 mV and -50 mV.

The spiking rate of a population as a function of the defined quantities is then given by:

$$\nu_x = \phi(\mu_x, \sigma_x), \tag{2.38}$$

where

$$\phi(\mu_x, \sigma_x) = \left(\tau_{\text{rp}} + \tau_x \int_{\beta(\mu_x, \sigma_x)}^{\alpha(\mu_x, \sigma_x)} du \sqrt{\pi} \exp(u^2)[1 + \text{erf}(u)] \right)^{-1} \tag{2.39}$$

$$\alpha(\mu_x, \sigma_x) = \frac{(V_{\text{thr}} - \mu_x)}{\sigma_x} \left(1 + 0.5 \frac{\tau_{\text{AMPA}}}{\tau_x} \right) + 1.03 \sqrt{\frac{\tau_{\text{AMPA}}}{\tau_x}} - 0.5 \frac{\tau_{\text{AMPA}}}{\tau_x} \tag{2.40}$$

$$\beta(\mu_x, \sigma_x) = \frac{(V_{\text{reset}} - \mu_x)}{\sigma_x} \tag{2.41}$$

with $\mathrm{erf}(u)$ the error function and τ_{rp} the refractory period which is considered to be 2 ms for excitatory neurons and 1 ms for inhibitory neurons. To solve the equations defined by (2.38) for all xs we integrate numerically (2.37) and the differential Equation 2.42, which has fixed point solutions corresponding to Equation 2.38:

$$\tau_x \frac{d\nu_x}{dt} = -\nu_x + \phi(\mu_x, \sigma_x).$$

(2.42)

The values of the parameters used in a mean-field analysis of decision-making are provided in Section 2.7.3, with further details of the parameters in Chapter 5.

2.7.3 The model parameters used in the mean-field analyses of decision-making

The fixed parameters of the model are shown in Table 2.3, and not only provide information about the values of the parameters used in the simulations, but also enable them to be compared to experimentally measured values.

Table 2.3 Parameters used in the integrate-and-fire simulations

N_E	800
N_I	200
r	0.1
w_+	2.2
w_I	1.015
N_{ext}	800
ν_{ext}	2.4 kHz
C_m (excitatory)	0.5 nF
C_m (inhibitory)	0.2 nF
g_m (excitatory)	25 nS
g_m (inhibitory)	20 nS
V_L	−70 mV
V_{thr}	−50 mV
V_{reset}	−55 mV
V_E	0 mV
V_I	−70 mV
$g_{AMPA,ext}$ (excitatory)	2.08 nS
$g_{AMPA,rec}$ (excitatory)	0.104 nS
g_{NMDA} (excitatory)	0.327 nS
g_{GABA} (excitatory)	1.25 nS
$g_{AMPA,ext}$ (inhibitory)	1.62 nS
$g_{AMPA,rec}$ (inhibitory)	0.081 nS
g_{NMDA} (inhibitory)	0.258 nS
g_{GABA} (inhibitory)	0.973 nS
$\tau_{NMDA,decay}$	100 ms
$\tau_{NMDA,rise}$	2 ms
τ_{AMPA}	2 ms
τ_{GABA}	10 ms
α	0.5 ms^{-1}

2.7.4 Mean-field neurodynamics used to analyze competition and cooperation between networks

The computation underlying brain functions emerges from the collective dynamics of spiking neuronal networks. A spiking neuron transforms a large set of incoming input spike trains, coming from different neurons, into an output spike train. Thus, at the microscopic level, neuronal circuits of the brain encode and process information by spatiotemporal spike patterns. In a network of interconnected neurons, stable patterns (called 'attractors') can be formed and activated by external inputs. Furthermore, a pattern activated by an input could eventually then stably be maintained by the system even after input offset. These patterns could correspond to memories, perceptual representations, or thoughts. It has been hypothesized that each cortical area represents a set of alternative hypotheses, encoded in the activities of cell assemblies (Hebb 1949). Representations of conflicting hypotheses compete with each other; however, each area represents only a part of the environment or internal state. In order to arrive at a coherent global representation, different cortical areas bias each others' internal representations by communicating their current states to other areas, thereby favouring certain sets of local hypotheses over others. By recurrently biasing each other's competitive internal dynamics, the neocortical system arrives at a global representation in which each area's state is maximally consistent with those of the other areas. This view has been referred to as the biased-competition hypothesis.

In addition to this competition-centered view, a cooperation-centered picture of brain dynamics, where global representations find their neural correlate in assemblies of co-activated neurons, has been formulated (Hebb 1949, Deco and Rolls 2005a). Co-activation is achieved by increased connectivity among the members of each assembly. Reverberatory communication between the members of the assembly then leads to persistent activation to provide temporally extended representations (Lorente de No 1933, Goldman-Rakic 1995).

The mean-field approach has been applied to biased-competition and cooperation networks and has been used to model single neuronal responses, fMRI activation patterns, psychophysical measurements, effects of pharmacological agents, and effects of local cortical lesions (Rolls and Deco 2002, Deco and Lee 2002, Corchs and Deco 2002, Deco, Pollatos and Zihl 2002, Corchs and Deco 2004, Deco and Rolls 2002, Deco and Rolls 2003, Deco and Rolls 2004, Deco, Rolls and Horwitz 2004, Szabo, Almeida, Deco and Stetter 2004, Deco and Rolls 2005b, Rolls 2008d). Such a mean-field approach operates without noise, and is deterministic, but can provide a useful model of the typical or average dynamics of a neural system.

2.7.5 A consistent mean-field and integrate-and-fire approach

In the mean field non-linear transfer function, all the biophysical details contained in the original leaky integrate-and-fire (LIF) model are maintained because the transfer function is derived consistently from the original spiking equations, as shown in Appendix A. The steady state of the network, solved through the mean field approach, can be matched to a behavioural steady state, observed in a given experiment. In finding the parameters that generate this behavioural steady state, we solve the inverse problem. We emphasize the importance of having a direct 'vertical' transfer of parameters from the spiking level to the mean field level, so that we have a transparent means of linking the different levels of description. This is critical for making another inverse inference, concerning how the real neural system must operate based on the parameters needed to capture behavioural data in the model. We provide a derivation of a mean-field approach that is consistent with an integrate-and-fire approach in Appendix A.

The classical mean-field approximation (Amit and Brunel 1997, Brunel and Wang 2001) corresponds to the case where the fluctuations can be neglected (i.e. with an infinite number of neurons in each population). In this case, a standard bifurcation analysis of the fixed points can be carried out. In the absence of noise the fixed points of Equation A.31 on page 267 can be determined by setting the time derivative equal to zero and solving for ν_i.

The problem here is that some of the theoretically most informative aspects of experiments reflect the transient states that are passed through to obtain any final steady state and the probabilistic character of these transitions. The solution derived at the mean field level holds only for the steady state of the network. It does not describe the transient states that the network goes through to reach the steady state. Furthermore, the mean field level of description is classically deterministic: there is no noise. In contrast, the spiking level has inherent noise (given the Poisson distribution of spikes, at least for a system with a finite number of neurons). For this reason too, it is important to have the proposed direct link from the mean field to the spiking level. The critical point about the spiking level modeling is that it enables high-level psychological descriptions (e.g. concerning diffusion processes and perceptual memory in decision making, Weber's law and so forth, see Chapter 5) to be linked directly to underlying neurophysiological processes. In doing this, the spiking level integrate-and-fire simulations emphasize particular aspects of those processes, such as the dynamic changes in stochastic states over time, and the functional role of noise in cognition. In particular, we will show in the following chapters how these effects help to understand decision-making and perception in the brain, and to establish a link between neuronal firing variability and probabilistic behaviour.

A full characterization of the dynamics, and specially of its probabilistic behaviour, including the stochastic non-stationary regime of the system, can only be obtained through computer simulations of the spiking, integrate-and-fire, network model. Moreover, these simulations enable comparisons between the model and neurophysiological data. Usually, the simulations of the spiking dynamics of the network are integrated numerically, and the non-stationary evolution of spiking activity is averaged over many trials initialized with different random seeds.

Nevertheless, theoretical work has been able to extend the mean-field approach (described in Appendix A) in such a way that both non-stationary behaviour, and spiking noise fluctuations, are taken into account (Renart et al. 2003, La Camera, Rauch, Luescher, Senn and Fusi 2004, Mattia and Del Giudice 2004, Deco et al. 2007b), and these developments are described in Sections A.5 and A.6.

3 Short-term memory and stochastic dynamics

3.1 Introduction

In this Chapter, we describe evidence that attractor networks implement short-term memory in the brain, and show how noise generated by the spiking of neurons can contribute to the instability of short-term memory.

3.2 Cortical short-term memory systems and attractor networks

There are a number of different short-term memory systems, each implemented in a different cortical area. The particular systems considered here each implement short-term memory by subpopulations of neurons that show maintained activity in a delay period, while a stimulus or event is being remembered. These memories may operate as autoassociative attractor networks, the operation of which is described in Section 1.10. The actual autoassociation could be implemented by associatively modifiable synapses between connected pyramidal cells within an area, or by the forward and backward connections between adjacent cortical areas in a hierarchy.

p. 30

One short-term memory system is in the dorso-lateral prefrontal cortex, area 46. This is involved in remembering the locations of spatial responses, in for example delayed spatial response tasks (Goldman-Rakic 1996, Fuster 2000). In such a task performed by a monkey, a light beside one of two response keys illuminates briefly, there is then a delay of several seconds, and then the monkey must touch the appropriate key in order to obtain a food reward. The monkey must not initiate the response until the end of the delay period, and must hold a central key continuously in the delay period. Lesions of the prefrontal cortex in the region of the principal sulcus impair the performance of this task if there is a delay, but not if there is no delay. Some neurons in this region fire in the delay period, while the response is being remembered (Fuster 1973, Fuster 1989, Goldman-Rakic 1996, Fuster 2000). Different neurons fire for the two different responses or spatial locations.

There is an analogous system in a more dorsal and posterior part of the prefrontal cortex involved in remembering the position in visual space to which an eye movement (a saccade) should be made (Funahashi, Bruce and Goldman-Rakic 1989, Goldman-Rakic 1996). In this case, the monkey may be asked to remember which of eight lights appeared, and after the delay to move his eyes to the light that was briefly illuminated. The short-term memory function is topographically organized, in that lesions in small parts of the system impair remembered eye movements only to that eye position. Moreover, neurons in the appropriate part of the topographic map respond to eye movements in one but not in other directions (see Fig. 3.1). Such a memory system could be easily implemented in such a topographically organized system by having local cortical connections between nearby pyramidal cells which implement an attractor network (Fig. 3.2). Then triggering activity in one part of the topographically

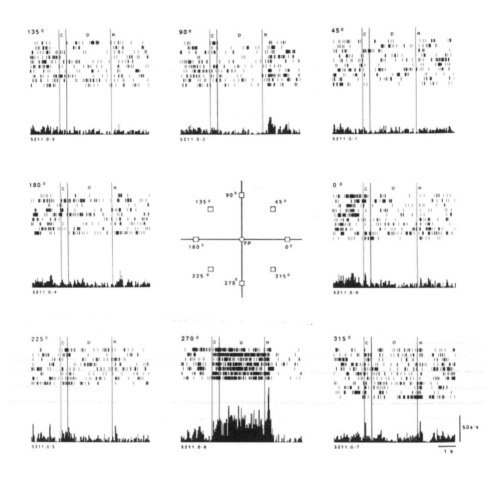

Fig. 3.1 The activity of a single neuron in the dorsolateral prefrontal cortical area involved in remembered saccades. Each row is a single trial, with each spike shown by a vertical line. A cue is shown in the cue (C) period, there is then a delay (D) period without the cue in which the cue position must be remembered, then there is a response (R) period. The monkey fixates the central fixation point (FP) during the cue and delay periods, and saccades to the position where the cue was shown, in one of the eight positions indicated, in the response period. The neuron increased its activity primarily for saccades to position 270°. The increase of activity was in the cue, delay, and response period while the response was made. The time calibration is 1 s. (Reproduced with permission from Funahashi, Bruce and Goldman-Rakic 1989.)

organized system would lead to sustained activity in that part of the map, thus implementing a short-term or working memory for eye movements to that position in space.

Another short-term memory system is implemented in the inferior temporal visual cortex, especially more ventrally towards the perirhinal cortex (Rolls 2008d). This memory is for whether a particular visual stimulus (such as a face) has been seen recently. This is implemented in two ways. One way is that some neurons respond more to a novel than to a familiar visual stimulus in such tasks, or in other cases respond to the familiar, selected, stimulus

Local autoassociation networks in the prefrontal cortex
for delayed spatial responses, eg. saccades

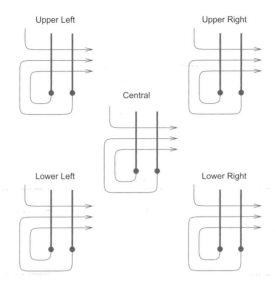

Fig. 3.2 A possible cortical model of a topographically organized set of attractor networks in the prefrontal cortex that could be used to remember the position to which saccades should be made. Excitatory local recurrent collateral Hebb-modifiable connections would enable a set of separate attractors to operate. The input that would trigger one of the attractors into continuing activity in a memory delay period would come from the parietal cortex, and topology of the inputs would result in separate attractors for remembering different positions in space. (For example inputs for the Upper Left of space would trigger an attractor that would remember the upper left of space.) Neurons in different parts of this cortical area would have activity related to remembering one part of space; and damage to a part of this cortical area concerned with one part of space would result in impairments in remembering targets to which to saccade for only one part of space.

(Baylis and Rolls 1987, Miller and Desimone 1994). The other way is that some neurons, especially more ventrally, continue to fire in the delay period of a delayed match to sample task (Fahy, Riches and Brown 1993, Miyashita 1993), and fire for several hundred milliseconds after a 16 ms visual stimulus has been presented (Rolls and Tovee 1994). These neurons can be considered to reflect the implementation of an attractor network between the pyramidal cells in this region (Amit 1995). A cortical syndrome that may reflect loss of such a short-term visual memory system is simultanagnosia, in which more than one visual stimulus cannot be remembered for more than a few seconds (Warrington and Weiskrantz 1973, Kolb and Whishaw 2008).

Continuing firing of neurons in short-term memory tasks in the delay period is also found in other cortical areas. For example, it is found in an object short-term memory task (delayed match to sample described in Section 3.3) in the inferior frontal convexity cortex, in a region connected to the ventral temporal cortex (Fuster 1989, Wilson, O'Sclaidhe and Goldman-Rakic 1993). However, whether this network is distinct from the network in the dorsolateral prefrontal cortex involved in spatial short-term memory is not clear, as some neurons in these regions may be involved in both spatial and object short-term memory tasks (Rao, Rainer and Miller 1997). Heterogeneous populations of neurons, some with more spatial input, others with more object input, and top-down attentional modulation of the different subpopulations depending on the task, start to provide an explanation for these findings (see Section 3.6).

Delay-related neuronal firing is also found in the parietal cortex when monkeys are remembering a target to which a saccade should be made (Andersen 1995); and in the motor cortex when a monkey is remembering a direction in which to reach with the arm (Georgopoulos 1995).

Another short-term memory system is human auditory–verbal short-term memory, which appears to be implemented in the left hemisphere at the junction of the temporal, parietal, and occipital lobes. Patients with damage to this system are described clinically as showing conduction aphasia, in that they cannot repeat a heard string of words (cannot conduct the input to the output) (Warrington and Weiskrantz 1973, Kolb and Whishaw 2008).

3.3 Prefrontal cortex short-term memory networks, and their relation to temporal and parietal perceptual networks

As described in Section 3.2, a common way that the brain uses to implement a short-term memory is to maintain the firing of neurons during a short memory period after the end of a stimulus (see Rolls and Treves (1998) and Fuster (2000)). In the inferior temporal cortex this firing may be maintained for a few hundred milliseconds even when the monkey is not performing a memory task (Rolls and Tovee 1994, Rolls, Tovee, Purcell, Stewart and Azzopardi 1994b, Rolls, Tovee and Panzeri 1999, Desimone 1996). In more ventral temporal cortical areas such as the entorhinal cortex the firing may be maintained for longer periods in delayed match to sample tasks (Suzuki, Miller and Desimone 1997), and in the prefrontal cortex for even tens of seconds (Fuster 1997, Fuster 2000). In the dorsolateral and inferior convexity prefrontal cortex the firing of the neurons may be related to the memory of spatial responses or objects (Goldman-Rakic 1996, Wilson, O'Sclaidhe and Goldman-Rakic 1993) or both (Rao, Rainer and Miller 1997), and in the principal sulcus / arcuate sulcus region to the memory of places for eye movements (Funahashi, Bruce and Goldman-Rakic 1989) (see Figs. 1.5 and 1.7). The firing may be maintained by the operation of associatively modified recurrent collateral connections between nearby pyramidal cells producing attractor states in autoassociative networks (see Section 1.10).

For the short-term memory to be maintained during periods in which new stimuli are to be perceived, there must be separate networks for the perceptual and short-term memory functions, and indeed two coupled networks, one in the inferior temporal visual cortex for perceptual functions, and another in the prefrontal cortex for maintaining the short-term memory during intervening stimuli, provide a precise model of the interaction of perceptual and short-term memory systems (Renart, Parga and Rolls 2000, Renart, Moreno, Rocha, Parga and Rolls 2001) (see Fig. 3.3). In particular, this model shows how a prefrontal cortex attractor (autoassociation) network could be triggered by a sample visual stimulus represented in the inferior temporal visual cortex in a delayed match to sample task, and could keep this attractor active during a memory interval in which intervening stimuli are shown. Then when the sample stimulus reappears in the task as a match stimulus, the inferior temporal cortex module shows a large response to the match stimulus, because it is activated both by the visual incoming match stimulus, and by the consistent backprojected memory of the sample stimulus still being represented in the prefrontal cortex memory module (see Fig. 3.3).

This computational model makes it clear that in order for ongoing perception to occur unhindered implemented by posterior cortex (parietal and temporal lobe) networks, there must be a separate set of modules that is capable of maintaining a representation over intervening stimuli. This is the fundamental understanding offered for the evolution and functions of the

Inferior temporal cortex (IT)   Prefrontal cortex (PF)

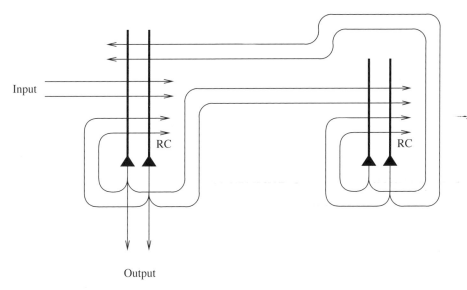

Fig. 3.3 A short-term memory autoassociation network in the prefrontal cortex could hold active a working memory representation by maintaining its firing in an attractor state. The prefrontal module would be loaded with the to-be-remembered stimulus by the posterior module (in the temporal or parietal cortex) in which the incoming stimuli are represented. Backprojections from the prefrontal short-term memory module to the posterior module would enable the working memory to be unloaded, to for example influence ongoing perception (see text). RC, recurrent collateral connections.

dorsolateral prefrontal cortex, and it is this ability to provide multiple separate short-term attractor memories that provides I suggest the basis for its functions in planning.

Renart, Parga and Rolls (2000) and Renart, Moreno, Rocha, Parga and Rolls (2001) performed analyses and simulations which showed that for working memory to be implemented in this way, the connections between the perceptual and the short-term memory modules (see Fig. 3.3) must be relatively weak. As a starting point, they used the neurophysiological data showing that in delayed match to sample tasks with intervening stimuli, the neuronal activity in the inferior temporal visual cortex (IT) is driven by each new incoming visual stimulus (Miller, Li and Desimone 1993b, Miller and Desimone 1994), whereas in the prefrontal cortex, neurons start to fire when the sample stimulus is shown, and continue the firing that represents the sample stimulus even when the potential match stimuli are being shown (Miller, Erickson and Desimone 1996).

The architecture studied by Renart, Parga and Rolls (2000) was as shown in Fig. 3.3, with both the intramodular (recurrent collateral) and the intermodular (forward IT to PF, and backward PF to IT) connections trained on the set of patterns with an associative synaptic modification rule. A crucial parameter is the strength of the intermodular connections, g, which indicates the relative strength of the intermodular to the intramodular connections. (This parameter measures effectively the relative strengths of the currents injected into the neurons by the inter-modular relative to the intra-modular connections, and the importance of setting this parameter to relatively weak values for useful interactions between coupled attractor networks was highlighted by Renart, Parga and Rolls (1999b) and Renart, Parga and Rolls (1999a) (see Rolls (2008d).) The patterns themselves were sets of random numbers, and the simulation utilized a dynamical approach with neurons with continuous (hyper-

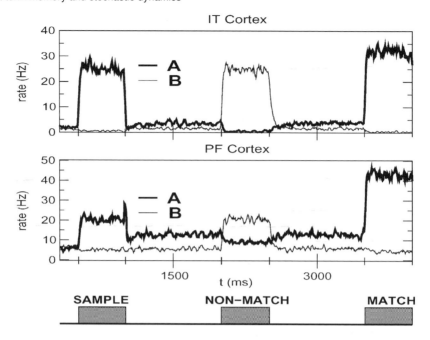

Fig. 3.4 Interaction between the prefrontal cortex (PF) and the inferior temporal cortex (IT) in a delayed match to sample task with intervening stimuli with the architecture illustrated in Fig. 3.3. Above: activity in the IT attractor module. Below: activity in the PF attractor module. The thick lines show the firing rates of the set of neurons with activity selective for the sample stimulus (which is also shown as the match stimulus, and is labelled **A**), and the thin lines the activity of the neurons with activity selective for the non-match stimulus, which is shown as an intervening stimulus between the sample and match stimulus and is labelled **B**. A trial is illustrated in which **A** is the sample (and match) stimulus. The prefrontal cortex module is pushed into an attractor state for the sample stimulus by the IT activity induced by the sample stimulus. Because of the weak coupling to the PF module from the IT module, the PF module remains in this sample-related attractor state during the delay periods, and even while the IT module is responding to the non-match stimulus. The PF module remains in its sample-related state even during the non-match stimulus because once a module is in an attractor state, it is relatively stable. When the sample stimulus reappears as the match stimulus, the PF module shows higher sample stimulus-related firing, because the incoming input from IT is now adding to the activity in the PF attractor network. This in turn also produces a match enhancement effect in the IT neurons with sample stimulus-related selectivity, because the backprojected activity from the PF module matches the incoming activity to the IT module. (After Renart, Parga and Rolls 2000; and Renart, Moreno, Rocha, Parga and Rolls 2001.)

bolic tangent) activation functions (Shiino and Fukai 1990, Kuhn 1990, Kuhn, Bos and van Hemmen 1991, Amit and Tsodyks 1991). The external current injected into IT by the incoming visual stimuli was sufficiently strong to trigger the IT module into a state representing the incoming stimulus. When the sample was shown, the initially silent PF module was triggered into activity by the weak ($g > 0.002$) intermodular connections. The PF module remained firing to the sample stimulus even when IT was responding to potential match stimuli later in the trial, provided that g was less than 0.024, because then the intramodular recurrent connections could dominate the firing (see Fig. 3.4). If g was higher than this, then the PF module was pushed out of the attractor state produced by the sample stimulus. The IT module responded to each incoming potentially matching stimulus provided that g was not greater than approximately 0.024. Moreover, this value of g was sufficiently large that a larger response of the IT module was found when the stimulus matched the sample stimulus (the match enhancement

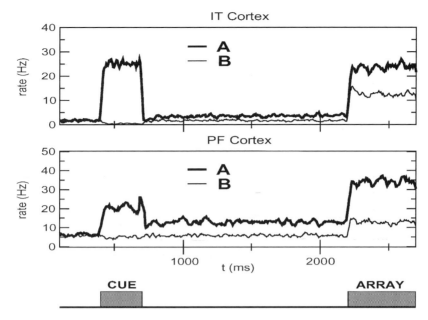

Fig. 3.5 Interaction between the prefrontal cortex (PF) and the inferior temporal cortex (IT) in a visual search task with the architecture illustrated in Fig. 3.3. Above: activity in the IT attractor module. Below: activity in the PF attractor module. The thick lines show the firing rates of the set of neurons with activity selective for search stimulus **A**, and the thin lines the activity of the neurons with activity selective for stimulus **B**. During the cue period either **A** or **B** is shown, to indicate to the monkey which stimulus to select when an array containing both **A** and **B** is shown after a delay period. The trial shown is for the case when **A** is the cue stimulus. When stimulus **A** is shown as a cue, then via the IT module, the PF module is pushed into an attractor state **A**, and the PF module remembers this state during the delay period. When the array **A** + **B** is shown later, there is more activity in the PF module for the neurons selective for **A**, because they have inputs both from the continuing attractor state held in the PF module and from the forward activity from the IT module which now contains both **A** and **B**. This PF firing to **A** in turn also produces greater firing of the population of IT neurons selective for **A** than in the IT neurons selective for **B**, because the IT neurons selective for **A** are receiving both **A**-related visual inputs, and **A**-related backprojected inputs from the PF module. (After Renart, Parga and Rolls 2000; and Renart, Moreno, Rocha, Parga and Rolls 2001.)

effect found neurophysiologically, and a mechanism by which the matching stimulus can be identified). This simple model thus shows that the operation of the prefrontal cortex in short-term memory tasks such as delayed match to sample with intervening stimuli, and its relation to posterior perceptual networks, can be understood by the interaction of two weakly coupled attractor networks, as shown in Figs. 3.3 and 3.4.

The same network can also be used to illustrate the interaction between the prefrontal cortex short-term memory system and the posterior (IT or PP) perceptual regions in visual search tasks, as illustrated in Fig. 3.5. Details of these analyses and simulations are provided by Renart, Parga and Rolls (2000) and Renart, Moreno, Rocha, Parga and Rolls (2001).

3.4 Computational necessity for a separate, prefrontal cortex, short-term memory system

This approach emphasizes that in order to provide a useful brain lesion test of prefrontal cortex short-term memory functions, the task set should require a short-term memory for stimuli over an interval in which other stimuli are being processed, because otherwise the posterior cortex perceptual modules could implement the short-term memory function by their own recurrent collateral connections. This approach also emphasizes that there are many at least partially independent modules for short-term memory functions in the prefrontal cortex (e.g. several modules for delayed saccades; one or more for delayed spatial (body) responses in the dorsolateral prefrontal cortex; one or more for remembering visual stimuli in the more ventral prefrontal cortex; and at least one in the left prefrontal cortex used for remembering the words produced in a verbal fluency task—see Section 3.2; and Section 10.3 of Rolls and Treves (1998)).

This computational approach thus provides a clear understanding of why a separate (prefrontal) mechanism is needed for working memory functions, as elaborated in Section 3.3. It may also be commented that if a prefrontal cortex module is to control behaviour in a working memory task, then it must be capable of assuming some type of executive control. There may be no need to have a single central executive additional to the control that must be capable of being exerted by every short-term memory module. This is in contrast to what has traditionally been assumed (Shallice and Burgess 1996) for the prefrontal cortex (see Rolls (2008d)).

3.5 Synaptic modification is needed to set up but not to reuse short-term memory systems

To set up a new short-term memory attractor, synaptic modification is needed to form the new stable attractor. Once the attractor is set up, it may be used repeatedly when triggered by an appropriate cue to hold the short-term memory state active by continued neuronal firing even without any further synaptic modification (see Section 1.10 and Kesner and Rolls (2001)). Thus manipulations that impair the long term potentiation of synapses (LTP) may impair the formation of new short-term memory states, but not the use of previously learned short-term memory states. Kesner and Rolls (2001) analyzed many studies of the effects of blockade of LTP in the hippocampus on spatial working memory tasks, and found evidence consistent with this prediction. Interestingly, it was found that if there was a large change in the delay interval over which the spatial information had to be remembered, then the task became susceptible, during the transition to the new delay interval, to the effects of blockade of LTP. The implication is that some new learning is required when a rat must learn the strategy of retaining information for longer periods when the retention interval is changed.

In another approach to the implementation of short-term memory, in a recurrent network without long-term synaptic modification to a particular set of neurons, short-term synaptic facilitation of synapses activated in a high firing rate neuronal representation that lasts for 1–2 s can enable that set of neurons to be reactivated by non-specific inputs (Mongillo, Barak and Tsodyks 2008), and this may be useful in a model of synaptic decision-making (Deco, Rolls and Romo 2010).

3.6 What, where, and object–place combination short-term memory in the prefrontal cortex

Given that inputs from the parietal cortex, involved in spatial computation, project more to the dorsolateral prefrontal cortex, for example to the cortex in the principal sulcus, and that inputs from the temporal lobe visual cortex, involved in computations about objects, project ventro-laterally to the prefrontal cortex (see Figs. 1.5, 1.7, and 1.6), it could be that the dorsolateral prefrontal cortex is especially involved in spatial short-term memory, and the inferior con-vexity prefrontal cortex more in object short-term memory (Goldman-Rakic 1996, Goldman-Rakic 1987, Fuster, Bauer and Jervey 1982). This organization-by-stimulus-domain hypothe-sis (Miller 2000) holds that spatial ('where') working memory is supported by the dorsolateral PFC in the neighbourhood of the principal sulcus (Brodmann's area (BA) 46/9 in the middle frontal gyrus (MFG)); and object ('what') working memory is supported by the ventrolateral PFC on the lateral/inferior convexity (BA 45 in the inferior frontal gyrus (IFG)). Consistent with this, lesions of the cortex in the principal sulcus impair delayed spatial response memory, and lesions of the inferior convexity can impair delayed (object) matching to sample short-term memory (Goldman-Rakic 1996) (especially when there are intervening stimuli between the sample and the matching object, as described earlier). Further, Wilson, O'Sclaidhe and Goldman-Rakic (1993) found that neurons dorsolaterally in the prefrontal cortex were more likely to be related to spatial working memory, and in the inferior convexity to object work-ing memory. Some fMRI studies in humans support this topographical organization (Leung, Gore and Goldman-Rakic 2002), and some do not (Postle and D'Esposito 2000, Postle and D'Esposito 1999). Rao, Rainer and Miller (1997) found in a task that had a spatial short-term memory component and an object short-term memory component that some prefrontal cor-tex neurons are involved in both, and thus questioned whether there is segregation of spatial and object short-term memory in the prefrontal cortex. Some segregation does seem likely, in that for example the neurons with activity related to delayed oculomotor saccades are in the frontal eye field (Funahashi, Bruce and Goldman-Rakic 1989) (see Fig. 3.1), and this area has not been implicated in object working memory.

Part of the conceptual resolution of the discussion about segregation of functions in the lateral prefrontal cortex is that sometimes combinations of objects and places must be held in short-term memory. For example, in one such task, a *conditional object–response (assoc-iative) task* with a delay, a monkey was shown one of two stimulus objects (O1 or O2), and after a delay had to make either a rightward or leftward oculomotor saccade response depend-ing on which stimulus was shown. In another such task, a *delayed spatial response task*, the same stimuli were used, but the rule required was different, namely to respond after the delay towards the left or right location where the stimulus object had been shown (Asaad, Rainer and Miller 2000). The tasks could be run on some trial blocks with the mappings reversed. Thus the monkey had to remember in any trial block the particular object–place mapping that was required in the delay period. In this task, neurons were found that responded to combina-tions of the object and the spatial position (Asaad et al. 2000). This bringing together of the spatial and object components of the task is important in how the short-term memory task was solved, as described by Deco and Rolls (2003). However, this makes the conceptual point that at least some mixing of the spatial and object representations in the lateral prefrontal cortex is needed in order for some short-term memory tasks to be solved, and thus total segregation of 'what' and 'where' processing in the lateral prefrontal cortex should not be expected.

Another factor in interpreting fMRI imaging studies of short-term memory is that differ-ences of the timecourse of the activations from the lateral prefrontal cortex have been found in 'what' vs 'where' short-term memory components of a task (Postle and D'Esposito 2000). Deco, Rolls and Horwitz (2004) showed that differences in the fMRI BOLD signal from the dorsal as compared to the ventral prefrontal cortex in working memory tasks may reflect a higher level of inhibition in the dorsolateral prefrontal cortex. They demonstrated this by simulating a 'what' vs 'where' short-term memory task with integrate-and-fire neurons, and then integrated over the synaptic currents and convolved with the haemodynamic response function, as described in Section 6.5, to predict the BOLD signal change. However, the fMRI data were most easily modelled by hypothesizing that the differences between these two prefrontal regions resulted from assigning a greater amount of inhibition to the dorsolateral portion of the prefrontal cortex. Both brain areas may show short-term memory maintenance capabilities related to their capacities to maintain stable attractors during delay periods, but the increased level of inhibition assumed in the dorsolateral PFC may be associated with the capacity of this brain region to support more complex functions. Higher inhibition in the dorsolateral prefrontal cortex might be useful for maintaining several separate short-term memory representations in nearby cortical modules (regions 1–3 mm across as defined by the spread of local high density recurrent collateral connections) and preventing the formation of a global attractor, which could be useful if several items must be held in memory for manipulation (Deco, Rolls and Horwitz 2004).

3.7 Hierarchically organized series of attractor networks

Short-term memory attractor dynamics helps us to understand the underlying mechanisms that implement mappings from one representation to another, for example from sensory to motor, and where short-term memory is required. For example, Deco and Rolls (2003) investigated the 3-level integrate-and-fire hierarchical model of the prefrontal cortex, with sensory neurons, intermediate neurons, and premotor neurons illustrated in Fig. 3.6, in a model of the neuronal activity recorded in the prefrontal cortex during such tasks (Asaad, Rainer and Miller 2000). The hierarchical structure of the pools is to ensure that the mapping is from the sensory pools to the intermediate pools and then to the premotor pools, and is achieved by setting the synaptic weights to be stronger in the forward than in the reverse direction between the thus hierarchically organized 3 levels of attractor networks. In this particular investigation, a top-down rule bias could be applied to the intermediate level neurons, so that the mapping could be selected to be for example according to an object-response or a spatial-response mapping rule. The attractor dynamics of each population of neurons enables the task to be performed when there are delays between the stimuli and the responses.

This concept of a set of attractor networks that are hierarchically organized in this way is formally somewhat similar to the interacting attractors used to solve short-term memory with intervening stimuli described in Section 3.3, and the interacting attractors used to describe attentional processing (Rolls and Deco 2002, Rolls 2008d).

Another application of this concept that shows the temporal organization that can be involved is in delayed response tasks in which there may be separate attractors for the sensory stimuli and for the responses. The sensory attractors may be started by sensory cues, may hold the cue for a time, but there may then be a gradual transition to premotor neuron attractors, so that some pools gradually become quiescent, and others start up, during a delay period (Deco, Ledberg, Almeida and Fuster 2005) (see Fig. 3.18 on page 118). For this, the forward coupling between the sensory and motor attractors must be weak (Rolls 2008d). This whole process will be stochastic, with the transition times, especially of individual neurons in each of the attractors, being probabilistic from trial to trial because the process is being influenced by noise.

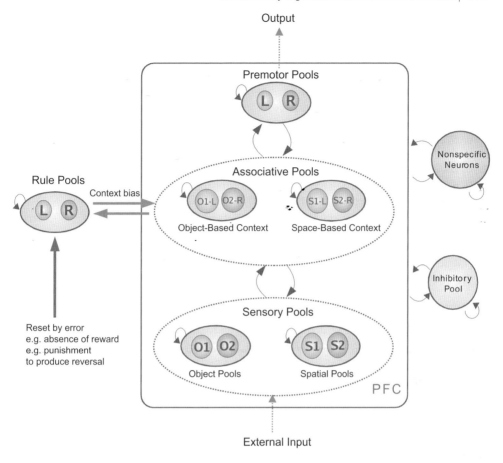

Fig. 3.6 Network architecture of the prefrontal cortex unified model of attention, working memory, and decision-making. There are sensory neuronal populations or pools for object type (O1 or O2) and spatial position (S1 or S2). These connect hierarchically (with stronger forward than backward connections) to the intermediate or 'associative' pools in which neurons may respond to combinations of the inputs received from the sensory pools for some types of mapping such as reversal, as described by Deco and Rolls (2003). For the simulation of the neurophysiological data of Asaad, Rainer and Miller (2000) these intermediate pools respond to O1–L, O2–R, S1–L, or S2–R. These intermediate pools receive an attentional bias from the Rule Pools, which in the case of this particular simulation biases either the object-based context pools or the spatial-based context pools. The Rule Pools are reset by an error to bias the opposite set of associative pools, so that the rule attentional context switches from favouring mapping through the object-based context pools to through the spatial-based context pools, or vice versa. The intermediate pools are connected hierarchically to the premotor pools, which in this case code for a Left or Right response. Each of the pools is an attractor network in which there are stronger associatively modified synaptic weights between the neurons that represent the same state (e.g. object type for a sensory pool, or response for a premotor pool) than between neurons in the other pools or populations. However, all the neurons in the network are associatively connected by at least weak synaptic weights. The attractor properties, the competition implemented by the inhibitory interneurons, and the biasing inputs result in the same network implementing both short-term memory and biased competition, and the stronger feed forward than feedback connections between the sensory, intermediate, and premotor pools results in the hierarchical property by which sensory inputs can be mapped to motor outputs in a way that depends on the biasing contextual or rule input. (After Deco and Rolls 2003.)

3.8 Stochastic dynamics and the stability of short-term memory

We now introduce concepts on how noise produced by neuronal spiking, or noise from other sources, influences short-term memory, with illustrations drawn from Loh, Rolls and Deco (2007a) and Rolls, Loh and Deco (2008a).

The noise caused by the probabilistic firing of neurons as described in Chapter 2 can influence the stability of short-term memory. To investigate this, it is necessary to use a model of short-term memory which is a biophysically realistic integrate-and-fire attractor network with spiking of the neurons, so that the properties of receptors, synaptic currents and the statistical effects related to the noisy probabilistic spiking of the neurons can be analyzed. Loh, Rolls and Deco (2007a) and Rolls, Loh and Deco (2008a) used a minimal architecture, a single attractor or autoassociation network (Hopfield 1982, Amit 1989, Hertz et al. 1991, Rolls and Treves 1998, Rolls and Deco 2002) (see Section 1.10), to investigate how spiking-related stochastic noise influences the stability of short-term memory. They chose a recurrent (attractor) integrate-and-fire network model which includes synaptic channels for AMPA, NMDA and GABA$_A$ receptors (Brunel and Wang 2001). The integrate-and-fire model was necessary to characterize and exploit the effects of the spiking noise produced by the neurons in a finite-sized network. However, to initialize the parameters of the integrate-and-fire model such as the synaptic connection strengths to produce stable attractors, and to ensure that the spontaneous activity is in the correct range, they used a mean-field approximation consistent with the integrate-and-fire network, as described in Section 2.7.2. Both excitatory and inhibitory neurons were represented by a leaky integrate-and-fire model (Tuckwell 1988) described in detail in Section 2.2.3.

The single attractor network contained 400 excitatory and 100 inhibitory neurons, which is consistent with the observed proportions of pyramidal cells and interneurons in the cerebral cortex (Abeles 1991, Braitenberg and Schutz 1991). The connection strengths were adjusted using mean-field analysis (see Brunel and Wang (2001) and Section 2.7.2), so that the excitatory and inhibitory neurons exhibited a spontaneous activity of 3 Hz and 9 Hz respectively (Koch and Fuster 1989, Wilson, O'Scalaidhe and Goldman-Rakic 1994). The recurrent excitation mediated by the AMPA and NMDA receptors is dominated by the long time constant NMDA currents to avoid instabilities during the delay periods (Wang 1999, Wang 2002).

The cortical network model featured a minimal architecture to investigate stability (and also distractibility as described in Chapter 4), and consisted of two selective pools S1 and S2, as shown in Fig. 3.7. We used just two selective pools. Pool S1 is used for the short-term memory item to be remembered, sometimes called the target; and pool S2 is used for the distractor stimulus introduced in Chapter 4. The non-selective pool NS modelled the spiking of cortical neurons and served to generate an approximately Poisson spiking dynamics in the model (Brunel and Wang 2001), which is what is observed in the cortex. The inhibitory pool IH contained 100 inhibitory neurons. There were thus four populations or pools of neurons in the network, and the connection weights were set up using a mean-field analysis to make S1 and S2 have stable attractor properties. The connection weights between the neurons within each selective pool or population were called the intra-pool connection strengths w_+. The increased strength of the intra-pool connections was counterbalanced by the other excitatory connections (w_-) to keep the average input to a neuron constant. The actual synaptic strengths are shown in Tables 3.1 and 3.2 where $w_- = \frac{0.8 - f_{S1} w_+}{0.8 - f_{S1}}$. For these investigations, $w_+ = 2.1$ was selected, because with the default values of the NMDA and GABA conductances this yielded relatively stable dynamics with some effect of the noise being apparent, that is, a relatively stable spontaneous state if no retrieval cue was applied, and a relatively stable state of persistent firing after a retrieval cue had been applied and removed.

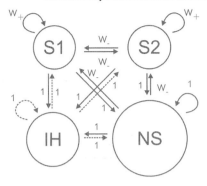

Fig. 3.7 The attractor network model. The excitatory neurons are divided into two selective pools S1 and S2 (with 40 neurons each) with strong intra-pool connection strengths w_+, and one non-selective pool (NS) (with 320 neurons). The other connection strengths are 1 or weak w_-. The network contains 500 neurons, of which 400 are in the excitatory pools and 100 are in the inhibitory pool IH. Each neuron in the network also receives inputs from 800 external neurons, and these neurons increase their firing rates to apply a stimulus or distractor to one of the pools S1 or S2. The synaptic connection matrices are shown in Tables 3.1 and 3.2. (After Loh, Rolls and Deco 2007a; and Rolls, Loh and Deco 2008a.)

Table 3.1 Connection matrix for AMPA and NMDA – [from, to]

	S1	S2	NS	IH
S1	w_+	w_-	1	1
S2	w_-	w_+	1	1
NS	w_-	w_-	1	1
IH	0	0	0	0

Table 3.2 Connection matrix for GABA – [from, to]

	S1	S2	NS	IH
S1	0	0	0	0
S2	0	0	0	0
NS	0	0	0	0
IH	1	1	1	1

Each neuron in the network received Poisson input spikes via AMPA receptors which are envisioned to originate from 800 external neurons at an average spontaneous firing rate of 3 Hz from each external neuron, consistent with the spontaneous activity observed in the cerebral cortex (Wilson et al. 1994, Rolls and Treves 1998, Rolls 2008d) (see Section 2.2.3).

3.8.1 Analysis of the stability of short-term memory

The analyses (Loh, Rolls and Deco 2007a, Rolls, Loh and Deco 2008a) aimed to investigate the stability of the short-term memory implemented by the attractor network. Simulations were performed for many separate trials each run with a different random seed to analyze the statistical variability of the network as the noise varied from trial to trial. We focus here on simulations of two different conditions: the spontaneous, and persistent, with a third condition, the distractor condition, analyzed in Chapter 4.

In spontaneous simulations (see Fig. 3.8), we ran spiking simulations for 3 s without any

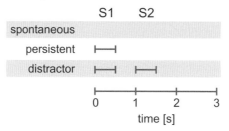

Fig. 3.8 The simulation protocols. Stimuli to either S1 or S2 are applied at different times depending on the type of simulations. The spontaneous simulations include no input. The 'persistent' simulations assess how stably a stimulus is retained in short-term memory by the network. The distractor simulations add a distractor stimulus as described in Chapter 4 to further address the stability of the network activity in the face of a distracting stimulus, and thus both the stability of attention, and the shifting of attention. (After Loh, Rolls and Deco 2007a; and Rolls, Loh and Deco 2008a.)

extra external input. The aim of this condition was to test whether the network was stable in maintaining a low average firing rate in the absence of any inputs, or whether it fell into one of its attractor states without any external input.

In persistent simulations, an external cue of 120 Hz above the background firing rate of 2400 Hz was applied to each neuron in pool S1 during the first 500 ms to induce a high activity state and then the system was run without the additional external cue of 120 Hz for another 2.5 s, which was a short-term memory period. The 2400 Hz was distributed across the 800 synapses of each S1 neuron for the external inputs, with the spontaneous Poisson spike trains received by each synapse thus having a mean rate of 3 Hz. The aim of this condition was to investigate whether once in an attractor short-term memory state, the network can maintain its activity stably, or whether it fell out of its attractor, which might correspond to an inability to maintain short-term memory.

3.8.2 Stability and noise in the model of short-term memory

To clarify the concept of stability, we show examples of trials of spontaneous and persistent simulations in which the statistical fluctuations have different impacts on the temporal dynamics. Fig. 3.9 shows the possibilities, as follows.

In the spontaneous state simulations, no cue was applied, and we are interested in whether the network remains stably in the spontaneous firing state, or whether it is unstable and on some trials due to statistical fluctuations entered one of the attractors, thus falsely retrieving a memory. Figure 3.9 (top) shows an example of a trial on which the network correctly stayed in the low spontaneous firing rate regime, and (bottom) another trial (labelled spontaneous unstable) in which statistical spiking-related fluctuations in the network caused it to enter a high activity state, moving into one of the attractors even without a stimulus.

In the persistent state simulations (in which the short-term memory was implemented by the continuing neuronal firing), a strong excitatory input was given to the S1 neuronal population between 0 and 500 ms. Two such trials are shown in Fig. 3.9. In Fig. 3.9 (top), the S1 neurons (correctly) keep firing at approximately 30 Hz after the retrieval cue is removed at 500 ms. However, due to statistical fluctuations in the network related to the spiking activity, on the trial labelled persistent unstable the high firing rate in the attractor for S1 was not stable, and the firing decreased back towards the spontaneous level, in the example shown starting after 1.5 s (Fig. 3.9 bottom). This trial illustrates a failure to maintain a stable short-term memory state.

When an average was taken over many trials, for the persistent run simulations, in which

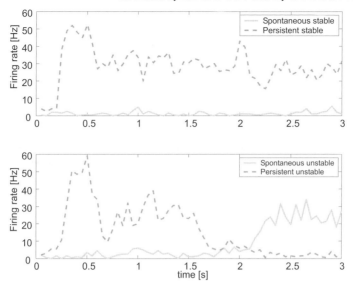

Fig. 3.9 Example trials of the Integrate-and-Fire attractor network simulations of short-term memory. The average firing rate of all the neurons in the S1 pool is shown. Top. Normal operation. On a trial in which a recall stimulus was applied to S1 at 0–500 ms, firing continued normally until the end of the trial in the 'persistent' simulation condition. On a trial on which no recall stimulus was applied to S1, spontaneous firing continued until the end of the trial in the 'spontaneous' simulation condition. Bottom: Unstable operation. On this persistent condition trial, the firing decreased during the trial as the network fell out of the attractor because of the statistical fluctuations caused by the spiking dynamics. On the spontaneous condition trial, the firing increased during the trial because of the statistical fluctuations. In these simulations the network parameter was w_+=2.1. (After Rolls, Loh, and Deco 2008a.)

the cue triggered the attractor into the high firing rate attractor state, the network was still in the high firing rate attractor state in the baseline condition on 88% of the runs. The noise had thus caused the network to fail to maintain the short-term memory on 12% of the runs.

The spontaneous state was unstable on approximately 10% of the trials, that is, on 10% of the trials the spiking noise in the network caused the network run in the condition without any initial retrieval cue to end up in a high firing rate attractor state. This is of course an error that is related to the spiking noise in the network.

We emphasise that the transitions to the incorrect activity states illustrated in Fig. 3.9 are caused by statistical fluctuations (noise) in the spiking activity of the integrate-and-fire neurons. Indeed, we used a mean-field approach with an equivalent network with the same parameters but without noise to establish parameter values where the spontaneous state and the high firing rate short-term memory ('persistent') state would be stable without the spiking noise, and then in integrate-and-fire simulations with the same parameter values examined the effects of the spiking noise (Loh, Rolls and Deco 2007a, Rolls, Loh and Deco 2008a). The mean-field approach used to calculate the stationary attractor states of the network for the delay period (Brunel and Wang 2001) is described in Section 2.6. These attractor states are independent of any simulation protocol of the spiking simulations and represent the behaviour of the network by mean firing rates to which the system would converge in the absence of statistical fluctuations caused by the spiking of the neurons and by external changes. Therefore the mean-field technique is suitable for tasks in which temporal dynamics and fluctuations are negligible. It also allows a first assessment of the attractor landscape and the depths of the basin of attraction which then need to be investigated in detail with stochastical spiking simulations. Part of the utility of the mean-field approach is that it allows the parameter region for the synaptic strengths to be investigated to determine which synaptic strengths will on average produce stable activity in the network, for example of persistent activity in a delay period after the removal of a stimulus. For the spontaneous state, the initial conditions for numerical simulations of the mean-field method were set to 3 Hz for all excitatory pools and 9 Hz for the inhibitory pool. These values correspond to the approximate values of the spontaneous attractors when the network is not driven by stimulus-specific inputs. For the persistent state, the network parameters resulted in a selective pool having a high firing rate value of approximately 30 Hz when in its attractor state (Loh, Rolls and Deco 2007a, Rolls, Loh and Deco 2008a).

We note that there are two sources of noise in the simulated integrate-and-fire spiking networks that cause the statistical fluctuations: the randomly arriving external Poisson spike trains, and the statistical fluctuations caused by the spiking of the neurons in the finite sized network. Some of the evidence that the statistical fluctuations caused by the neuronal spiking do provide an important source of noise in attractor networks in the brain is that factors that affect the noise in the network such as the number of neurons in the network have clear effects on the operation of integrate-and-fire attractor networks, as considered further in Chapter 5. Indeed, the magnitude of these fluctuations increases as the number of neurons in the network becomes smaller (Mattia and Del Giudice 2004).

We further comment that the firing rate distribution may be an important factor that influences the stability of attractor networks. Whereas most studies have been with binary (rectangular) firing rate distributions in which a small proportion of neurons representing any one memory is in a high firing rate state and the other neurons are in a low firing rate state (typically at less than the spontaneous rate because of inhibition due to the high firing rate neurons), it is found that in the brain when a large number of different stimuli must be represented by a network, the firing rate distribution is often approximately exponential, as shown in Section 1.11.3. The influence of the firing rate distribution on the stochastic effects introduced into the operation of cortical networks is now being investigated, and it appears that exponential firing rate distributions introduce more noise for a given mean firing rate and sparseness than a binary distribution (initial results of T.Webb and E.T.Rolls).

3.8.3 Alterations of the stability of short-term memory produced by altering the depths of the basins of attraction

We show here that factors such as alterations in the currents flowing through NMDA and GABA receptor activated ion channels can increase or decrease the stability of short-term memory networks. We note here that the way in which the values for the parameters of these networks are selected during evolution and adjusted during life are very important. Too much stability of a high firing rate attractor state might make it difficult to replace it with another item when something else must be stored in the short-term memory. In the context of attention, it might make it very difficult to switch attention from one item to another. Too little stability of a high firing rate attractor state might make a short-term memory unreliable, and in the context of attention might make the organism very distractible with poor attention. On the other hand, too much stability of the spontaneous state might make it difficult for an input applied to the network to cause the network to fall into a high firing rate attractor state, resulting in a failure of short-term memory, or in the context of attention of a stimulus to capture attention. Too little stability of the spontaneous state might make the system jump into a short-term memory state representing at random any one of the possible attractor states stored in the short-term memory even when there is no input to be remembered. In the context of attention, this could correspond to randomly jumping to an attentional state selected

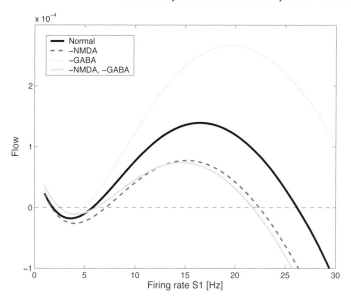

Fig. 3.10 Flow of a single attractor network for different modifications of synaptic conductances. The flow of an attractor network with one selective pool was assessed using a meanfield analysis. The flow represents the force that drives the system towards one of the stable attractors. The stable/unstable attractor states are at crossings with the flow=0 value on the flow axis with a negative/positive derivative respectively. A modulation of the synapses labelled as (−NMDA) and (−GABA) corresponds to a reduction of 4.5% and 9% respectively of the efficacies. A pool cohesion (connectivity) of w_+=1.6 and a selective pool size of 80 neurons were used for the simulations. (After Loh, Rolls and Deco 2007a.)

at random by the noise. Thus the selection of the parameters for the operation of attractor networks in the brain is crucial, and may be a compromise between conflicting advantages provided by different selections. In Sections 8.14 and 8.15 we show how differences in the parameters that control the stability of attractor networks in the face of noise may be related to some brain dysfunctions, including schizophrenia and obsessive-compulsive disorder.

We now illustrate how alterations on the depths of the basins of attraction produced for example by altering the conductances of the synaptically activated ion channels can alter the stability of attractor networks, including potentially those involved in the brain in short-term memory and in episodic memory.

3.8.3.1 The depth of the basins of attraction: flow analysis

First we introduce an analytical approach to the concepts of how changes in transmitters could affect the depth of the basins of attraction in attractor networks, using the mean-field approach.

The depth of the basins of attraction can be assessed by calculating the flow towards the attractors using the mean-field reduction (see Loh, Rolls and Deco (2007a) and Mascaro and Amit (1999)). The flow is the force that draws the dynamical system to one of the attractors, Figure 3.10 shows the flow between the spontaneous and persistent state in a network featuring one selective pool. The curve for (Normal) shows the flow for the reference baseline condition. The crossing of the curve and the 0-flow axis represent the fixed points, which are either stable (negative derivative) or unstable (positive derivative). We used this line as a reference to assess the relative depth of the energy landscape. The system flows either into a spontaneous firing state (of approximately 2 Hz), or a high firing attractor state.

A 4.5% reduction of the conductances of the NMDA receptor activated ion channels

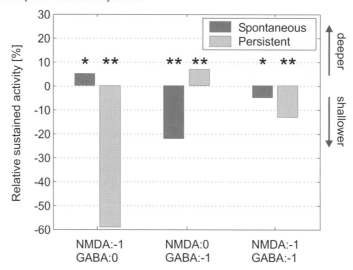

Fig. 3.11 Stability of the spontaneous and persistent state relative to the unmodulated reference state for modulations of the synaptic efficacies in integrate-and-fire simulations. We assessed how often in 100 trials the average activity during the last second (2–3 s) stayed above 10 Hz. The value shows how often it stayed more in the respective state than in the reference state. A negative percentage means that the system was less stable than in the reference state. A modulation of the synapses shown as (−NMDA) and (−GABA) corresponds to a reduction of 4.5% and 9% respectively in their efficacies. We assessed with the Binomial distribution the statistical significance of the effects observed, with P<0.01 relative to the reference state marked by **, and P<0.02 by *. (After Loh, Rolls and Deco 2007a.)

(−NMDA) shows a stronger flow than the unchanged condition at low firing rates towards the spontaneous attractor (at about 2 Hz). The absolute values of the function are higher compared to the normal condition until the first unstable fixed point (at around 6–7 Hz). The basin of attraction towards the persistent attractor at high firing rates yields the reverse picture. Here the (−NMDA) curve is clearly below the unchanged condition and the flow towards the attractor is smaller. Overall, the basin of attraction is deeper for the spontaneous state and shallower for the persistent state compared to the unchanged condition. Also note that the firing rate of the persistent fixed point is reduced in the (−NMDA) condition (crossing with the flow=0 axis).

A 9% reduction of the conductances of the GABA receptor activated ion channels (−GABA) yields the opposite pattern to that in the reduced NMDA condition. Here the basin of attraction of the persistent state is deeper.

However, in the condition in which both the NMDA and the GABA conductances are reduced (−NMDA, −GABA), the persistent state basin of attraction is shallower, and the spontaneous state basin is a little shallower. In particular, in the (−NMDA, −GABA) condition, the system would be less stable in the persistent state, tending to move to another attractor easily, and less stable in the spontaneous state, so tending to move too readily into an attractor from spontaneous activity.

3.8.3.2 Reduced stability of short-term memory by reducing NMDA and GABA receptor activated ion channel conductances

To assess the changes in the stability of the network produced by altering the conductances of the synaptic activated ion channels, integrate-and-fire simulations are needed, as they include the effects of stochastic noise, and show probabilistic behaviour. (The mean-field flow analyses described in Section 3.8.3.1 had no noise, and were deterministic.) For the investigations used as illustrations here, we selected $w_+ = 2.1$, which with the default values of the NMDA and GABA conductances yielded relatively stable dynamics, that is, a stable spontaneous state if no retrieval cue was applied, and a stable state of persistent firing after a retrieval cue had been applied and removed. To investigate the effects of changes (modulations) in the NMDA and GABA conductances, we chose for demonstration purposes a reduction of 4.5% and 9%, respectively, as these could cause instabilities, as illustrated in Fig. 4.8. However, the exact values are not crucial to observe the effects described. The magnitudes of these reductions are smaller than those that can be produced experimentally (Durstewitz and Seamans 2002, Seamans and Yang 2004). The simulations show that even quite small reductions in the synaptic currents can alter the global behaviour of the network, e.g. the stability of its attractors.

We assessed how the stability of both the spontaneous and persistent states changes when NMDA and GABA efficacies are modulated (Loh, Rolls and Deco 2007a). Specifically we ran multiple trial integrate-and-fire network simulations and counted how often the system maintained the spontaneous or persistent state, assessed by the firing rate in the last second of the simulation (2–3 s) of each 3 s trial. Figure 3.11 shows the stability of the spontaneous and persistent attractor relative to the unmodulated reference state (Normal). A negative percentage means that the system was less stable than in the unmodulated state.

A reduction of the NMDA conductance (−NMDA) reduces the stability of the persistent state drastically, while slightly increasing the stability of the spontaneous state (see Fig. 3.11). A reduction of GABA shows the opposite pattern: a slight reduction in the stability of the spontaneous state, and an increased stability of the persistent (i.e. high firing rate attractor) state (see Fig. 3.11).

When both NMDA and GABA are reduced one might think that these two counterbalancing effects (excitatory and inhibitory) would either cancel each other out or yield a tradeoff between the stability of the spontaneous and persistent state. However, this is not the case, and this illustrates the value of performing simulations of this type. The stability of both the spontaneous state and the persistent state is reduced (see Fig. 3.11). In this situation, the depths of both the spontaneous and attractor states are shallow, and the system jumps due to the influence of statistical fluctuations between the different (spontaneous and attractor) states.

To investigate more directly the wandering between spontaneous and several different persistent attractor states, we simulated the condition with decreased NMDA and GABA conductances over a long time period in which no cue stimulus input was given. Figure 3.12 shows the firing rates of the two selective pools S1 and S2. The high activity switches between the two attractors due to the influence of fluctuations, which corresponds to spontaneous wandering in a shallow energy landscape, corresponding for example to sudden jumps between unrelated cognitive processes. These results are consistent with the flow analysis and demonstrate that the changes in the attractor landscape influence the behaviour at the stochastic level. We relate this situation to the symptoms of schizophrenia in Section 8.14.

3.8.3.3 Signal-to-noise ratio

It is possible to express the effects of alterations in the stability of the short-term memory in terms of a signal-to-noise ratio. The signal-to-noise ratio denotes the level of a signal relative to the level of background noise. In an attractor network, a high signal-to-noise ratio indicates that the network will maintain the attractor stably, as it will be unlikely to be disrupted by spiking-related statistical fluctuations that are the source of the noise in the network. Figure 3.13 shows the signal-to-noise ratio of a measure related to the fMRI BOLD signal. [This measure described in the legend to Figure 3.13 and later was used because the experimental data with which we wished to compare the simulation results use fMRI measures (Winterer, Ziller, Dorn, Frick, Mulert, Wuebben, Herrmann and Coppola 2000, Winterer,

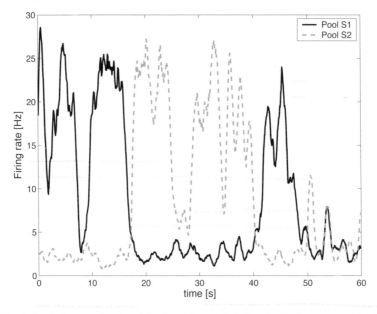

Fig. 3.12 Wandering between attractor states by virtue of statistical fluctuations caused by the randomness of the spiking activity. We simulated a single long trial (60 s) in the spontaneous test condition for the synaptic modification ($-$NMDA, $-$GABA). The two curves show the activity of the two selective pools over time smoothed with a 1 s sliding averaging window. The activity moves noisily between the attractors for the spontaneous state and the two persistent states S1 and S2. (After Loh, Rolls and Deco 2007a.)

Coppola, Goldberg, Egan, Jones, Sanchez and Weinberger 2004, Winterer, Musso, Beckmann, Mattay, Egan, Jones, Callicott, Coppola and Weinberger 2006). The index we used of the activity of the network was the total synaptic current of selective pool 1 averaged over the whole simulation time of 3 s, to take the temporal filtering properties of the BOLD signal into account, given the typical time course which lasts for several seconds of the haemodynamic response function (Deco, Rolls and Horwitz 2004). Further, we subtracted the averages of the spontaneous trial simulations which represent the baseline activity from the persistent trial simulation values. The signal-to-noise ratio was calculated from the mean of this index across trials divided by the standard deviation of the index, both measured using 1000 simulation trials. If the network sometimes had high activity, and sometimes low, then the signal-to-noise measure gave a low value. If the network reliably stayed in the high persistent firing states, then the signal-to-noise ratio measure was high.] As shown in Figure 3.13, we found that in all the cases in which the NMDA, or the GABA, conductance, or both, are reduced, the signal-to-noise ratio, computed by the mean divided by the standard deviation, is also reduced. Expressing the effects in terms of signal-to-noise ratio can be helpful, in that for example it enables comparisons with experimental observations that show a decreased signal-to-noise ratio in schizophrenic patients (Winterer et al. 2000, Winterer et al. 2004, Winterer et al. 2006). Here we directly relate a decrease in the signal-to-noise ratio to changes (in this case decreases) in NMDA and GABA receptor activated synaptic ion channel conductances.

3.8.3.4 Increased stability of short-term memory by increasing NMDA and GABA receptor activated ion channel conductances

It is also possible to investigate how the stability in the face of noise of short-term memory systems is influenced by increases in NMDA, AMPA, and GABA receptor activated ion channels. We illustrate the effects found with examples from Rolls, Loh and Deco (2008a). We

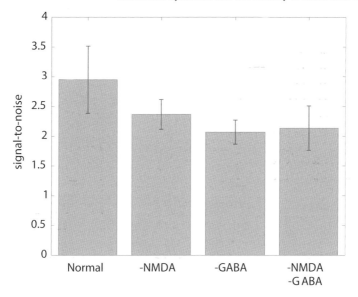

Fig. 3.13 Signal-to-noise ratio of a measure related to the fMRI BOLD signal as a function of the modulations of the synaptic efficacies. We computed the mean and standard deviation of averages of the synaptic currents of the selective pool over the whole simulation period of a persistent condition simulation. The mean of the spontaneous baseline condition was subtracted. We conducted 1000 simulated trials. The signal-to-noise ratio is calculated by division of the mean synaptic current by the standard deviation measured using 1000 trials. The error bars indicate an estimate of the standard deviation measured over 20 epochs containing 50 trials each. A modulation of the synapses shown as (−NMDA) and (−GABA) corresponds to a reduction of 4.5% and 9% respectively in their conductances. (After Loh, Rolls and Deco 2007b.)

chose for demonstration purposes increases of 3% for the NMDA, and 10% for the AMPA and GABA synapses between the neurons in the network shown in Fig. 3.7, as these were found to be sufficient to alter the stability of the attractor network. We assessed how the stability of both the spontaneous and persistent states changes when NMDA and AMPA efficacies are modulated. Specifically we ran multiple trial integrate-and-fire network simulations and counted how often the system maintained the spontaneous or persistent state, assessed by the firing rate in the last second of the simulation (2 s–3 s) of each 3 s trial.

Figure 3.14 shows the stability of the spontaneous and persistent attractor states. We plot the percentage of the simulation runs on which the network during the last second of the simulation was in the high firing rate attractor state. Fig. 3.14 shows that for the persistent run simulations, in which the cue triggered the attractor into the high firing rate attractor state, the network was still in the high firing rate attractor state in the baseline condition on approximately 88% of the runs, and that this had increased to 98% when the NMDA conductances were increased by 3% (+NMDA). Thus increasing the NMDA receptor-activated synaptic currents increased the stability of the network. The effect was highly significant, with the means ± sem shown. Fig. 3.14 also shows that increasing AMPA by 10% (+AMPA) could also increase the stability of the persistent high firing rate attractor state, as did the combination +NMDA +AMPA.

Figure 3.14 shows that in the baseline condition the spontaneous state was unstable on approximately 10% of the trials, that is, on 10% of the trials the spiking noise in the network caused the network run in the condition without any initial retrieval cue to end up in a high firing rate attractor state. This is of course an error that is related to the spiking noise in the

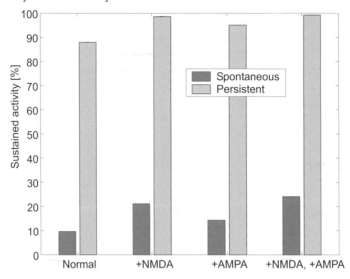

Fig. 3.14 The stability of the spontaneous and persistent attractor states. The percent (%) of the simulation runs on which the network during the last second of the 3 s simulation was in the high firing rate attractor state is shown on the ordinate. For the persistent run simulations, in which the cue triggered the attractor into the high firing rate attractor state, the network was still in the high firing rate attractor state in the baseline condition on approximately 88% of the runs, and this increased to nearly 100% when the NMDA conductances were increased by 3% (+NMDA). The effect was highly significant, with the means ± sem shown. Increasing AMPA by 10% (+AMPA) could also increase the stability of the persistent high firing rate attractor state, as did the combination +NMDA +AMPA. For the spontaneous state simulations, in the baseline condition the spontaneous state was unstable on approximately 10% of the trials, that is, on 10% of the trials the spiking noise in the network caused the network run in the condition without any initial retrieval cue to end up in a high firing rate attractor state. In the +NMDA condition, the spontaneous state had jumped to the high firing rate attractor state on 25% of the runs, that is the low firing rate spontaneous state was present at the end of a simulation on only approximately 75% of the runs. The +AMPA can also make the spontaneous state more likely to jump to a persistent high firing rate attractor state, as can the combination +NMDA +AMPA. (After Rolls, Loh and Deco 2008a.)

network. In the +NMDA condition, the spontaneous state had jumped to the high firing rate attractor state on approximately 22% of the runs, that is the low firing rate spontaneous state was present at the end of a simulation on only approximately 78% of the runs. Thus increasing NMDA receptor activated currents can contribute to the network jumping from what should be a quiescent state of spontaneous activity into a high firing rate attractor state. [We relate this to the symptoms of obsessive-compulsive disorders, in that the system can jump into a state with a dominant memory (which might be an idea or concern or action) even when there is no initiating input, as described in Section 8.15.] Figure 3.14 also shows that +AMPA can make the spontaneous state more likely to jump to a persistent high firing rate attractor state, as can the combination +NMDA +AMPA.

We also investigated to what extent alterations of the GABA-receptor mediated inhibition in the network could restore the system towards more normal activity even when NMDA conductances were high. Figure 3.15 shows that increasing the GABA currents by 10% when the NMDA currents are increased by 3% (+NMDA +GABA) can move the persistent state away from overstability back to the normal baseline state. That is, instead of the system ending up in the high firing rate attractor state in the persistent state simulations on 98% of the runs, the system ended up in the high firing rate attractor state on approximately 88% of the runs, the baseline level. Increasing GABA has a large effect on the stability of the spontaneous state,

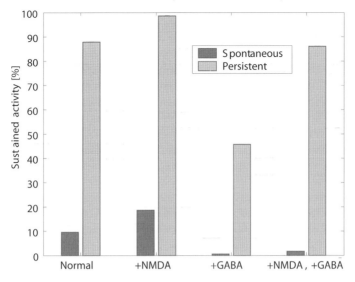

Fig. 3.15 The effect of increasing GABA-receptor mediated synaptic conductances by 10% (+GABA) on the stability of the network. Conventions as in Fig. 3.14. Increasing the GABA currents by 10% when the NMDA currents are increased by 3% (+NMDA +GABA) moved the persistent state away from the overstability produced by +NMDA alone, and returned the persistent state to the normal baseline level. That is, instead of the system ending up in the high firing rate attractor state in the persistent state simulations on 98% of the runs in the +NMDA condition, the system ended up in the high firing rate attractor state on approximately 88% of the runs in the +NMDA +GABA condition. The combination +NMDA +GABA produced a spontaneous state that was less likely than the normal state to jump to a high firing rate attractor (see text). (After Rolls, Loh and Deco 2008a.)

making it less likely to jump to a high firing rate attractor state. The combination +NMDA +GABA produced a spontaneous state in which the +GABA overcorrected for the effect of +NMDA. That is, in the +NMDA +GABA condition the network was very likely to stay in the spontaneous firing rate condition in which it was started, or, equivalently, when tested in the spontaneous condition, the network was less likely than normally to jump to a high firing rate attractor. Increasing GABA thus corrected for the effect of increasing NMDA receptor activated synaptic currents on the persistent type of run when there was an initiating stimulus; and overcorrected for the effect of increasing NMDA on the spontaneous state simulations when there was no initiating retrieval stimulus, and the network should remain in the low firing rate state until the end of the simulation run.

The implications are that agents that increase GABA conductances might reduce and normalize the tendency to remain locked into an idea or concern or action that might be caused by an overactive NMDA system; and would make it much less likely that the quiescent resting state would be left by jumping because of the noisy spiking towards a state representing a dominant memory state. The effects of increasing GABA receptor activated currents alone (+GABA) was to make the persistent simulations less stable (less likely to end in a high firing rate state), and the spontaneous simulations to be more stable (more likely to end up in the spontaneous state). The implications for understanding and treating the symptoms of obsessive-compulsive disorder are considered in Section 8.15.

3.9 Memory for the order of items in short-term memory

How are multiple items held in short-term memory? How is a sequence of items recalled in the correct order? These questions are closely related in at least forms of short-term memory, and this provides an indicator of possible mechanisms. For example in human auditory-verbal short-term memory, when for example we remember a set of numbers, we naturally retrieve the last few items in the order in which they were delivered, and this is termed the recency effect (Baddeley 1986). Another fascinating property of short-term memory is that the number of items we can store and recall is in the order of the 'magic number' 7 ± 2 (Miller 1956).

Investigations on the hippocampus are leading to an approach that potentially may answer these questions.

There is considerable evidence that the hippocampus is involved in spatial memory (Rolls and Kesner 2006, Rolls 2008d). Part of the evidence comes from the effects of CA3 lesions, which impair tasks in which associations must be formed rapidly between an object and the place where it is located, an example of an event or episodic memory (Rolls 2008e, Rolls 2008d, Rolls and Kesner 2006). Some of the evidence also comes from neuronal recording, in which some hippocampal neurons respond to combinations of a position in space and an object, or a position in space and a reward, but not in control tasks such as object-reward association, in both macaques (Rolls and Xiang 2005, Rolls, Xiang and Franco 2005, Rolls and Xiang 2006) and rats (Komorowski, Manns and Eichenbaum 2009). These findings provide support for a computational theory of the hippocampus in which associations between objects and places are formed by an autoassociation process implemented by the hippocampal CA3 recurrent collateral connections (Rolls 1990a, Treves and Rolls 1994, Rolls 1996, Rolls and Kesner 2006). The theory is that different hippocampal neurons code for places and objects in an object-place task (Rolls, Xiang and Franco 2005), and that, by associative synaptic modification in the recurrent collateral synapses, associations can be formed between neurons with such activity to lead to some neurons that respond to particular combinations of objects and places. The attractor property of the CA3 network allows retrieval of the object from the place or vice versa (Rolls, Stringer and Trappenberg 2002).

Some advances include the concept that there is a single attractor network in CA3 allowing associations between any object and any place (Rolls 1990a, Treves and Rolls 1994); quantitative estimates of the storage capacity of the system which can be higher than the number of recurrent collateral synaptic connections onto each neuron (12,000 in the rat) if sparse representations are used (Treves and Rolls 1994, Rolls 1996); a theory of how information can be recalled from the hippocampus to the neocortex with quantitative analyses of the recall capacity (Treves and Rolls 1994); a theory of how the spatial representations in the dentate gyrus and hippocampus are formed from entorhinal cortex grid cells (Rolls, Stringer and Elliot 2006b); and a theory of how spatial view representations are formed in the primate hippocampus (Rolls, Tromans and Stringer 2008d). The theory and approach is of importance in the context of understanding episodic memory and its disorders in humans, for episodic memory appears to require exactly this brain system; typically involves a spatial component; requires storage of arbitrarily co-occurring object, places, people etc; involves recall of a whole memory from a part of the memory; and may often require the retrieval of information from the hippocampus back to the neocortex (Rolls 2008e, Rolls 2008d).

However, there has been for some time evidence that the hippocampus is involved not only in associations between objects and places, but also between objects and time, for example in tasks in which the temporal order of items must be remembered (Rolls and Kesner 2006). Now it is not surprising that associations for the temporal order in which places were visited is impaired by hippocampal dysfunction, given that the hippocampus is involved in memories that include a spatial component (Rolls and Kesner 2006). But very compelling are impairments produced by hippocampal dysfunction in memory tasks in which there is no spatial component, but in which the temporal order of objects (e.g. visual stimuli, odors) must be remembered and retrieved (Rolls and Kesner 2006, Hoge and Kesner 2007).

One computational approach to how the hippocampus could be involved in temporal order memory has revolved around encoding information using the theta cycle, gamma cycles embedded within it, and theta phase precession (Lisman and Redish 2009), but there are difficulties with this approach (Maurer and McNaughton 2007).

Here a different approach is taken, based on recent neurophysiological evidence showing that neurons in the rat hippocampus have firing rates that reflect which temporal part of the task is current. In particular, a sequence of different neurons is activated at successive times during a time delay period (see Fig. 3.16) (MacDonald and Eichenbaum 2009). The tasks used included an object-odor paired associate non-spatial task with a 10 s delay period between the visual stimulus and the odor. The new evidence also shows that a large proportion of hippocampal neurons fire in relation to individual events in a sequence being remembered (e.g. a visual object or odor), and some to combinations of the event and the time in the delay period (MacDonald and Eichenbaum 2009).

These interesting neurophysiological findings indicate that rate encoding is being used to encode time, that is, the firing rates of different neurons are high at different times within a trial, delay period, etc, as illustrated in Fig. 3.16. This provides the foundation for a new computational theory of temporal order memory within the hippocampus (and also the prefrontal cortex) which I outline next, and which utilizes the slow transitions from one attractor to another which are a feature that arises at least in some networks in the brain due to the noise-influenced transitions from one state to another.

First, because some neurons fire at different times in a trial of a temporal order memory task or delay task, the time in a trial at which an object (e.g. a visual stimulus or odor) was presented could become encoded in the hippocampus by an association implemented in the CA3 recurrent collaterals between the neurons that represent the object (already known to be present in the hippocampus for tasks for which the hippocampus is required (Rolls, Xiang and Franco 2005) and the 'time encoding' neurons in the hippocampus (MacDonald and Eichenbaum 2009). This would allow associations for the time at which the object was present to be formed.

Second, these associations would provide the basis for the recall of the object from the time in a trial, or vice versa. The retrieval of object or temporal information from each other would occur in a way that is analogous to that shown for recalling the object from the place, or the place from the object (Rolls, Stringer and Trappenberg 2002), but substituting the details of the properties of the 'time encoding' neurons (MacDonald and Eichenbaum 2009) for what was previously the spatial (place) component. In addition, if the time encoding neurons simply cycled through their normal sequence during recall, this would enable the sequence of objects or events associated with each subset of time encoding neurons to be recalled correctly in the order in which they were presented.

Third, we need a theory of what the origin is of the temporal effect whereby different hippocampal (or potentially prefrontal cortex) neurons fire in different parts of a trial or delay period. The properties of the 'time encoding neurons' (see Fig. 3.16) are key here, and we need to understand how they are generated. Are they generated within the hippocampus, or elsewhere, and in any case, what is the mechanism by which different neurons have high firing rates at different times in a trial? The fundamentally new approach to hippocampal function we are taking here is that rate encoding is being used, that is, the firing rates of different neurons are high at different times within a trial, as illustrated in Fig. 3.16. This is a radically different approach to order encoding than that based on phenomena such a theta and gamma oscillations (Lisman and Redish 2009).

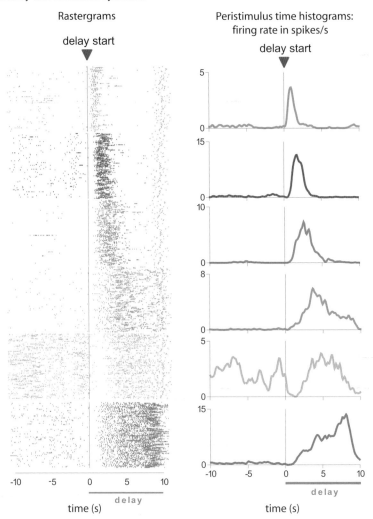

Fig. 3.16 Time encoding neurons. The activity of 6 simultaneously recorded hippocampal neurons each of which fires with a different time course during the 10 s delay in a visual object-delay-odor paired associate task. Each peristimulus time histogram (right) and set of rastergrams (left) is for a different neuron. The onset of the delay is shown. The visual stimulus was shown before the delay period; and the odor stimulus was delivered after the delay period. (After MacDonald and Eichenbaum 2009 with permission.)

We can consider three hypotheses about how the firing of the 'time encoding' hippocampal neurons is produced. All utilize slow transitions between attractor states that can be a property of noisy attractor networks. The first hypothesis is that an attractor network with realistic dynamics (modelled at the integrate-and-fire level with a dynamical implementation of the neuronal membrane and synaptic current dynamics, and with synaptic or neuronal adaptation) can implement a sequence memory (Deco and Rolls 2005c). The hypothesis is that there are several different attractors, and that there are weak connections between the different attractors. In the model, adaptation produces effects whereby whatever sequence (order of stimuli) is presented on an individual trial, that order can be replayed in the same sequence because as one attractor state dies as a result of the adaptation, the next attractor to emerge from the spontaneous firing because of the spiking-related noise is the one that has been ac-

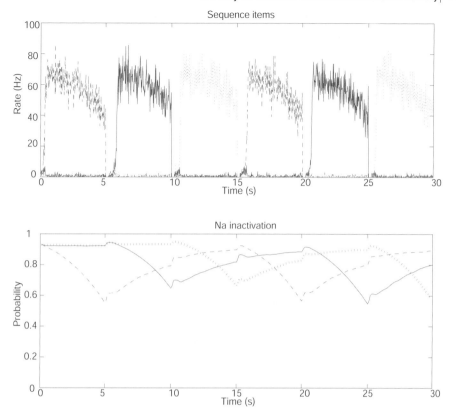

Fig. 3.17 Sequence memory using a sodium inactivation-based spike-frequency-adaptation mechanism to produce adaptation. The firing rate activity and spiking probability due to sodium inactivation as a function of the time for a temporal sequence of three items. The three items were presented in the order item 3 – item 2 – item 1 just at the start of the simulation period shown, and retrieval is demonstrated starting at time=15 s. Activity in the three specific pools corresponding to each item is shown with different linestyles. The lower plot shows the probability of spiking which alters due to a sodium inactivation spike frequency adaptation mechanism. An external signal indicating that the network should move to the next item in the sequence is applied at time = 15, 20 and 25 s. This signal quenches the attractor rapidly, and the next attractor takes some time to develop because of the competitive processes taking place between the neurons in the two populations. Two other adaptation mechanisms, a Ca^{++}-activated K^+ hyper-polarizing current mechanism, and a synaptic adaptation mechanism, also produced similar effects, though at shorter timescales. (After Deco and Rolls 2005c.)

tive least recently, as it is the one that is least adapted (see Fig. 3.17) (Deco and Rolls 2005c). The whole system operates at a rather slow timescale for the transitions between the attractors partly because of the time for the noise to drive the system from one attractor state to another, and the slow time course of the adaptation (Deco and Rolls 2005c). This implements a type of order memory.

The second hypothesis is analogous, and is also implemented in a recurrently connected system such as the hippocampal CA3 system or local recurrent circuits in the neocortex. This second theory is that again there are several attractors, but that each attractor is connected by slightly stronger forward than reverse synaptic weights to the next. In previous work, we have shown that with an integrate-and-fire implementation with spiking noise this allows slow transitions from one attractor state to the next (Deco and Rolls 2003, Deco, Ledberg, Almeida and Fuster 2005). (In the Deco, Ledberg, Almeida and Fuster (2005) system, there

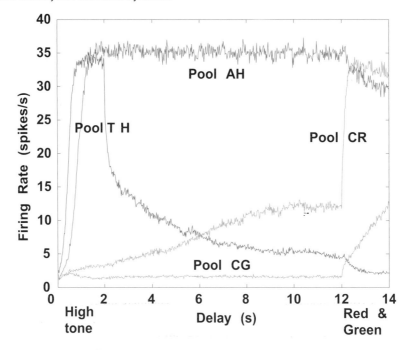

Fig. 3.18 Paired-associate / temporal order memory using stronger forward than reverse connections between attractors. In this spiking network model, input attractor network neuronal pools tone high (TH) and tone low (TL) have stronger forward than reverse synaptic connections to output attractor network neuronal pools colour red (CR) and colour green (CG) respectively. (In addition, there are associative attractor populations of neurons AH and AL that receive stronger forward inputs from TH and TL respectively, and have stronger forward than reverse weights to CR and CG respectively. These associative pools help to keep a short-term memory state initiated by one of the stimuli active throughout a trial.) The mean firing rate of some of these pools is shown in the figure on a trial on which the TH stimulus was shown at time 0–2 s, there was then a delay from 2–12 s, and then red and green stimuli were shown from time 12–14 s, with red being the correct paired associate. On average across trials there is a gradual transition from the stimulus input pool of neurons (in this case TH) to the associated output pool of neurons (in this case CR), which is frequently found in the dorsolateral prefrontal cortex during paired associate tasks. On individual trials the dynamics are noisy, with different time courses averaged across neurons for the transition from the input to the output neuronal population, and in addition somewhat different time courses for different neurons within a pool within a trial. (After Deco, Ledberg, Almeida and Fuster, 2005.)

were in addition associative attractor populations of neurons that received inputs from one of the input pools, and sent outputs to one of the output pools. The concept is the same.) The operation of this type of network is illustrated in Fig. 3.18, and the activity of the different populations of neurons matches those recorded in the prefrontal cortex (Fuster, Bodner and Kroger 2000).

During learning of the synaptic weights in the network, adaptation might lead to each 'time encoding' population of neurons responding for only a limited period, helping to produce multiple sequentially activated populations of time encoding neurons as illustrated in Fig. 3.16. In this scenario, an associative pool of neurons is unlikely to be helpful, and stronger forward that reverse weights between different attractors each consisting of a different population of 'time encoding' neurons would be the essence. It will be of interest to investigate whether this system, because of the noise, is limited to transitions between up to perhaps 7 ± 2 different sequential firing rate states with different neuronal subpopulations for each state, and thus provides an account for the limit of the magical number 7 ± 2 on short-term memory and related types of processing (Miller 1956), and for the recency part of short-term memory in which the items are naturally recalled in the order in which they were presented. This is the most likely model at present of short-term memory and its natural propensity to store and to recall items in the order in which they were received.

A variation on this implementation would be to have short-term attractor memories with different time constants (for example of adaptation), but all started at the same time. This could result in some attractors starting early in the sequence and finishing early, and with other attractors starting up a little later, but lasting for much longer in time. The neurons shown in Fig. 3.16 are not inconsistent with this possibility. This type of time-encoding representation could also be used to associate with items, to implement an item-order memory.

The third hypothesis, very new, is that similar temporal, attractor-based, functionality could be provided in the entorhinal cortex. There is already great interest in the spatial grid cells in the entorhinal cortex (Hafting, Fyhn, Molden, Moser and Moser 2005), and very interestingly, a leading theory about how they are formed is that different adaptation time constants of the neurons at different depths in the entorhinal cortex provide for the generation of larger grid spacing as one progresses more ventrally in the entorhinal cortex (Kropff and Treves 2008). The longer time constants effectively mean that the circular (ring) continuous attractor (Rolls 2008d) takes a longer time to complete the ring more ventrally in the hippocampus, and given a mean movement velocity of the rat in space, this produces larger spatial distance between the peaks of the grid (Kropff and Treves 2008). But this means that, considered in a completely new way, the entorhinal cortex grid cell system can be thought of as a time-encoding system. According to this new approach, time would be encoded in a trajectory through such a grid cell system, and if the system were allowed to run freely (without a spatial motion cue being the driver, and perhaps in a different part of the entorhinal cortex), effectively the grid cells would perform a trajectory through a time state-space. I then propose that this could be decoded to produce the 'time encoding neurons' in the hippocampus (or elsewhere) by exactly the same competitive network computational process that we have proposed accounts for the transform of entorhinal cortex spatial grid cells to dentate / hippocampal place cells (Rolls, Stringer and Elliot 2006b).

This is thus a fundamentally new approach to how noisy networks in the brain could implement the encoding of time using one of several possible mechanisms in which temporal order is an integral component. The approach applies not only to the hippocampus, but also potentially to the prefrontal cortex. It is established that neurons in the prefrontal cortex and connected areas have activity that increases or decreases in a temporally graded and stimulus-specific way during the delay period in short-term memory, sequential decision-making with a delay, and paired associate tasks (Fuster, Bodner and Kroger 2000, Fuster 2008, Brody, Hernandez, Zainos and Romo 2003a, Romo, Hernandez and Zainos 2004, Deco, Ledberg, Almeida and Fuster 2005, Deco, Rolls and Romo 2010), and it is likely too that some prefrontal cortex neurons use time encoding of the type illustrated in Fig. 3.16.

Part of the adaptive value of having a short-term memory in for example the prefrontal cortex that implements the order of the items is that short-term memory is an essential component of planning with multiple possible steps each corresponding to an item in a short-term memory, for plans require that the components be performed in the correct order. Similarly, if a series of events that has occurred results in an effect, the ability to remember the order in which those events occur can be very important in understanding the cause of the effect. This has great adaptive value in very diverse areas of human cognition, from interpreting social situations with several interacting individuals, to understanding mechanisms with a number of components.

We can note that with the above mechanism for order memory, it will be much easier to recall the items in the correct order than in any other order. Indeed, short-term memory becomes working memory when the items in the short-term memory need to be manipulated, for example recalled in reverse order, or with the item that was 2-,3-,n-back in a list. Also, in the type of order short-term memory described, recall can be started anywhere in the list by a recall cue. There are too interesting weaker predictions that the recall will be best if it occurs at the speed with which the items arrived, as during recall the speed of the trajectory through the order state space of short-term memories will be likely to be that inherently encoded in the noisy dynamics of the linked attractors with their characteristic time constants. We can also note that a similar mechanism, involving for example stronger forward than reverse weights between long-term memory representations in attractors, could build a long-term memory with the order of the items implemented.

3.10   Stochastic dynamics and long-term memory

Although we have been concerned in this Chapter with short-term memory, it is the case that attractor networks of the type described in Section 1.10 are part of the way in which long-term memories are stored in some areas of the cortex (see Rolls (2008d)). Part of the foundation for this is the recurrent collateral activity found between cortical pyramidal cells in localized regions of the cerebral cortex described in Chapter 1. Indeed, it is a natural property of this aspect of cortical architecture that networks used to store long-term memories may be able to keep a retrieved state active for a short time provided that there are no intervening stimuli, as described earlier in this Chapter.

The important point to be made here is that the recall of long-term memories is itself a process that is influenced by noise in the brain, and this will be very important when the retrieval cue is weak or the memory itself has become weak, or perhaps, as in forgetting, has not been recently accessed (see Rolls (2008d)). A vivid example that may reflect this type of operation is the recall after perhaps even minutes or hours of a name or fact for which recall was attempted earlier. This operation of long-term memory and thought retrieval in the face of noise makes exactly what is recalled somewhat probabilistic, and this may be important in helping thought processes to vary from time to time, with different trajectories being taken on different occasions through memory or thought state space (see Section 8.2). This may help the system to operate and behave in what is in this sense a non-deterministic way, and this may be an important aspect of creative thinking, as considered further in Section 8.9. Too much instability in the trajectories taken through memory or thought state space may underlie some thought disorders, as described in Section 8.14.

The models of decision-making described in Chapters 5 and 6 and of signal detection described in Chapter 7 are, effectively, models of the dynamical processes involved in the recall of a long-term memory stored in an attractor network. Each memory would correspond for example to one of the possible decisions. (As shown in Section 1.10 many different memories could be stored in an attractor network.) The dynamical processes that are influenced by noise described in Chapters 5–7, including the probabilistic nature of exactly which attractor is recalled on any particular occasion, and the longer time taken for the recall using a weak recall cue or from a weak memory, apply directly to the operation of long-term memory systems. Noise will make the memory recall probabilistic, and if the depth of the basin of attraction is decreased in a memory network, the result will be that the recall cue will need to be closer to the whole memory than would otherwise be the case. This is related to difficulties in episodic memory in aging in Section 8.13.

Overall, we have seen in this Chapter how the operation of memory systems in the brain is probabilistic, and in this sense non-deterministic, and this is because it is influenced by the spiking-related (and other) noise in the brain.

4   Attention and stochastic dynamics

4.1   Introduction

The classical view of attention refers to two functionally distinct stages of visual processing. One stage, termed the pre-attentive stage, implies an unlimited-capacity system capable of processing the information contained in the entire visual field in parallel. The other stage is termed the attentive or focal stage, and is characterized by the serial processing of visual information corresponding to local spatial regions in a 'spotlight of attention' (Treisman 1982, Crick 1984, Broadbent 1958, Kahneman 1973, Neisser 1967).

More recent observations from a number of cognitive neuroscience experiments led to an alternative account of attention termed the '**biased competition hypothesis**', which aims to explain the computational algorithms governing visual attention and their implementation in the brain's neural circuits and neural systems (Duncan and Humphreys 1989, Desimone and Duncan 1995, Duncan 1996, Rolls and Deco 2002). According to this hypothesis, attentional selection operates in parallel by top-down biasing of an underlying competitive interaction between multiple stimuli in the visual field. The massively feedforward and feedback connections that exist in the anatomy and physiology of the cortex (described in Section 1.9) implement the attentional mutual biasing interactions that result in a neurodynamical account of visual perception consistent with the seminal constructivist theories of Helmholtz (1867) and Gregory (1970).

In this Chapter, we show how biased competition can operate as a dynamical process in a precise computational framework that enables multiple interacting brain systems with bottom-up and top-down effects to be analyzed, and show how noise can affect the stability of these attentional processes. More extensive accounts of the empirical data on attention that can be accounted for by the biased competition approach are provided by Rolls and Deco (2002) in *Computational Neuroscience of Vision* and Rolls (2008d) in *Memory, Attention, and Decision-Making*. Here we focus on the neurodynamical model of attention, and how it is influenced by noise in the brain.

We should note that one type of attentional process operates when salient features in a visual scene attract attention (Itti and Koch 2001). This visual processing is described as feedforward and bottom-up, in that it operates forward in the visual pathways from the visual input (see Fig. 4.1 in which the solid lines show the forward pathways). A second type of selective attentional process, with which we are concerned in this Chapter, involves actively maintaining in short-term memory a location or object as the target of attention, and using this by top-down processes to influence earlier cortical processing. Some of the top-down or backprojection pathways in the visual system are shown in Fig. 4.1 by the dashed lines.

4.2   Evidence for the biased competition hypothesis in single neuron studies

A number of neurophysiological experiments (Moran and Desimone 1985, Spitzer, Desimone and Moran 1988, Sato 1989, Motter 1993, Miller, Gochin and Gross 1993a, Chelazzi, Miller,

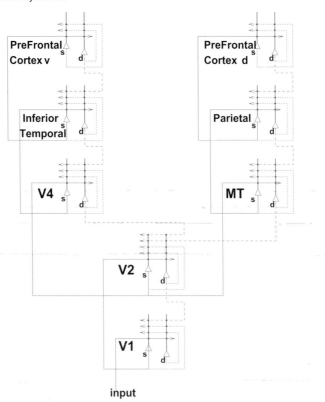

Fig. 4.1 The overall architecture of a model of object and spatial processing and attention, including the prefrontal cortical areas that provide the short-term memory required to hold the object or spatial target of attention active. Forward connections are indicated by solid lines; backprojections, which could implement top-down processing, by dashed lines; and recurrent connections within an area by dotted lines. The triangles represent pyramidal cell bodies, with the thick vertical line above them the dendritic trees. The cortical layers in which the cells are concentrated are indicated by s (superficial, layers 2 and 3) and d (deep, layers 5 and 6). The prefrontal cortical areas most strongly reciprocally connected to the inferior temporal cortex 'what' processing stream are labelled v to indicate that they are in the more ventral part of the lateral prefrontal cortex, area 46, close to the inferior convexity in macaques. The prefrontal cortical areas most strongly reciprocally connected to the parietal visual cortical 'where' processing stream are labelled d to indicate that they are in the more dorsal part of the lateral prefrontal cortex, area 46, in and close to the banks of the principal sulcus in macaques (see Rolls and Deco 2002, and Rolls 2008d).

Duncan and Desimone 1993, Motter 1994, Reynolds and Desimone 1999, Chelazzi 1998) have been performed suggesting biased competition neural mechanisms that are consistent with the theory of Duncan and Humphreys (1989) (i.e. with the role for a top-down memory target template in visual search).

The **biased competition hypothesis** of attention proposes that multiple stimuli in the visual field activate populations of neurons that engage in competitive interactions. Attending to a stimulus at a particular location or with a particular feature biases this competition in favour of neurons that respond to the feature or location of the attended stimulus. This attentional effect is produced by generating signals within areas outside the visual cortex that are then fed back to extrastriate areas, where they bias the competition such that when multiple stimuli appear in the visual field, the cells representing the attended stimulus 'win', thereby suppressing cells representing distracting stimuli (Duncan and Humphreys 1989, Desimone

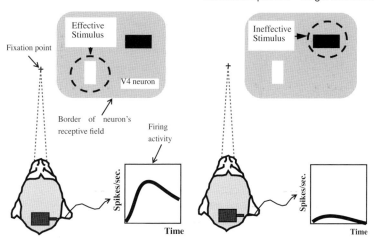

Fig. 4.2 Spatial attention and shrinking receptive fields in single-cell recordings of V4 neurons from the brain of a behaving monkey. The areas that are circled indicate the attended locations. When the monkey attended to effective sensory stimuli, the V4 neuron produced a good response, whereas a poor response was observed when the monkey attended to the ineffective sensory stimulus. The point fixated was the same in both conditions. (Adapted from Moran and Desimone 1985.)

and Duncan 1995, Duncan 1996).

Single-cell recording studies in monkeys from extrastriate areas support the biased competition theory. Moran and Desimone (1985) showed that the firing activity of visually tuned neurons in the cortex was modulated if monkeys were instructed to attend to the location of the target stimulus. In these studies, Moran and Desimone first identified the V4 neuron's classic receptive field and then determined which stimulus was effective for exciting a response from the neuron (e.g. a vertical white bar) and which stimulus was ineffective (e.g. a horizontal black bar). In other words, the effective stimulus made the neuron fire whereas the ineffective stimulus did not (see Fig. 4.2). They presented these two stimuli simultaneously within the neuron's receptive field, which for V4 neurons may extend over several degrees of visual angle. The task required attending to a cued spatial location. When spatial attention was directed to the effective stimulus, the pair elicited a strong response. On the other hand, when spatial attention was directed to the ineffective stimulus, the identical pair elicited a weak response, even though the effective stimulus was still in its original location (see Fig. 4.2).

Based on results of this type, the spatial attentional modulation could be described as a shrinkage of the classical receptive field around the attended location. Similar studies of Luck, Chelazzi, Hillyard and Desimone (1997) and Reynolds, Chelazzi and Desimone (1999) replicated this result in area V4, and showed similar attentional modulation effects in area V2 as well. Even in area V1 a weak attentional modulation has been described (McAdams and Maunsell 1999).

Maunsell (1995) and many others (Duhamel, Colby and Goldberg 1992, Colby, Duhamel and Goldberg 1993, Gnadt and Andersen 1988) have shown that the modification of neural activity by attentional and behavioural states of the animal is not only true for the extrastriate areas in the ventral stream, but is also true for the extrastriate areas of the dorsal stream as well. In particular, Maunsell (1995) demonstrated that the biased competitive interaction due to spatial and object attention exists not only for objects within the same receptive field, but also for objects in spatially distant receptive fields. This suggests that mechanisms exist to provide biased competition of a more global nature.

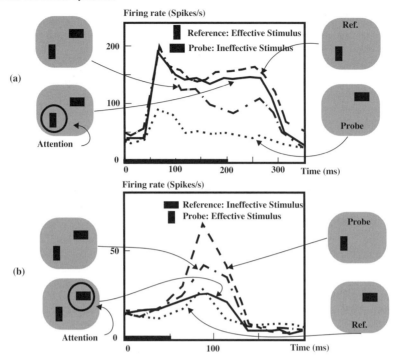

Fig. 4.3 Competitive interactions and attentional modulation of responses of single neurons in area V2. (a) Inhibitory suppression by the probe and attentional compensation. (b) Excitatory reinforcement by the probe and attentional compensation. The black horizontal bar at the bottom indicates stimulus duration. (Adapted from Reynolds, Chelazzi and Desimone 1999.)

In order to test at the neuronal level the biased competition hypothesis more directly, Reynolds, Chelazzi and Desimone (1999) performed single-cell recordings of V2 and V4 neurons in a behavioural paradigm that explicitly separated sensory processing mechanisms from attentional effects. They first examined the presence of competitive interactions in the absence of attentional effects by having the monkey attend to a location far outside the receptive field of the neuron that they were recording. They compared the firing activity response of the neuron when a single reference stimulus was within the receptive field, with the response when a probe stimulus was added to the field. When the probe was added to the field, the activity of the neuron was shifted toward the activity level that would have been evoked if the probe had appeared alone. This effect is shown in Fig. 4.3. When the reference was an effective stimulus (high response) and the probe was an ineffective stimulus (low response), the firing activity was suppressed after adding the probe (Fig. 4.3a). On the other hand, the response of the cell increased when an effective probe stimulus was added to an ineffective reference stimulus (Fig. 4.3b).

These results are explained by assuming that V2 and V1 neurons coding the different stimuli engage in competitive interactions, mediated for example through intermediary inhibitory neurons, as illustrated in Fig. 4.4.

The neurons in the extrastriate area receiving input from the competing neurons respond according to their input activity after the competition; that is, the response of a V4 neuron to two stimuli in its field is not the sum of its responses to both, but rather is a weighted average of the responses to each stimulus alone.

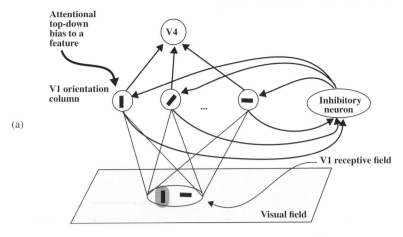

(a)

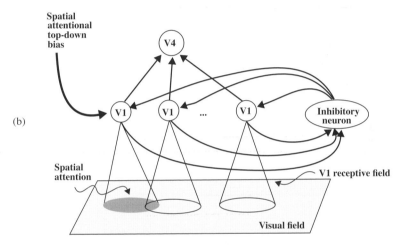

(b)

Fig. 4.4 Biased competition hypothesis at the neuronal level. (a) Attention to a particular top-down pre-specified feature. Intermediate inhibitory interneurons provide an inhibitory signal to V1 neurons in an orientation column (with receptive fields at the same spatial location) that are sensitive to an orientation different from the to-be-attended orientation. (b) Attention to a particular top-down prespecified spatial location. Inhibitory interneurons provide an inhibitory signal to V1 neurons with receptive fields corresponding to regions that are outside the to-be-attended location. In this schematic representation, we assume for simplicity direct connections from V1 to V4 neurons without intermediate V2 neurons. (After Rolls and Deco 2002.)

Attentional modulatory effects have been independently tested by repeating the same experiment, but now having the monkey attend to the reference stimulus within the receptive field of the recorded neuron. The effect of attention on the response of the V2 or V4 neuron was to almost compensate for the suppressive or excitatory effect of the probe (Reynolds, Chelazzi and Desimone 1999). That is, if the probe caused a suppression of the neuronal response to the reference when attention was outside the receptive field, then attending to the reference restored the neuron's activity to the level corresponding to the response of the neuron to the reference stimulus alone (Fig. 4.3a). Symmetrically, if the probe stimulus had increased the neuron's level of activity, attending to the reference stimulus compensated the response by shifting the activity to the level that had been recorded when the reference was presented alone (Fig. 4.3b). As is shown in Fig. 4.4, this attentional modulation can be un-

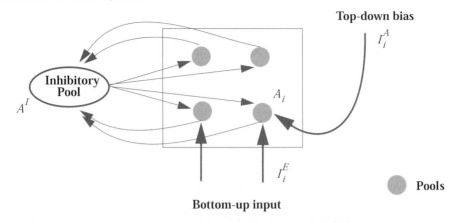

Fig. 4.5 Basic computational module for biased competition: a competitive network with external top-down bias. Excitatory pools (i.e. populations) of neurons with activity A_i for the ith pool are connected with a common inhibitory pool with activity A^I in order to implement a competition mechanism. I_i^E is the external bottom-up sensory input to the cells in pool i, and I_i^A attentional top-down bias, an external input coming from higher modules. The external top-down bias can shift the competition in favour of a specific pool or group of pools. (After Rolls and Deco 2002.)

derstood by assuming that attention biases the competition between V1 neurons in favour of a specific location or feature.

Models of attention, and their relation to neuronal, psychophysical, and fMRI data, are considered in detail by Rolls and Deco (2002) and Rolls (2008d). Here we focus on how attention and its non-linear properties are implemented at the level of spiking neurons, and on how noise in the brain influences attention.

4.3 A basic computational module for biased competition

We have seen in Fig. 4.4 how given forward (bottom-up) inputs to a network, competition between the principal (excitatory) neurons can be implemented by inhibitory neurons which receive from the principal neurons in the network and send back inhibitory connections to the principal neurons in the network. This competition can be biased by a top-down input to favour some of the populations of neurons, and this describes the biased competition hypothesis.

We can make the model of the biased competition hypothesis of attention more precise so that it can be investigated computationally by formulating it as shown in Fig. 4.5. Each population or pool of neurons is tuned to a different stimulus property, e.g. spatial location, or feature, or object. The more pools of the module that are active, the more active the common inhibitory pool will be and, consequently, the more feedback inhibition will affect the pools in the module, such that only the most excited group of pools will survive the competition. The external top-down bias can shift the competition in favour of a specific group of pools. Some non-linearity in the system, implemented for example by the fact that neurons have a threshold non-linearity below which they do not fire, can result in a small top-down bias having quite a large attentional effect, especially for relatively weak bottom-up inputs, as shown in a realistic spiking neuron simulation by Deco and Rolls (2005b) described in Section 4.4.

The architecture of Fig. 4.5 is that of a competitive network (described in more detail by Rolls and Deco (2002) and Rolls (2008d)) but with top-down backprojections. However, the network could equally well include associatively modifiable synapses in recurrent collateral connections between the excitatory (principal) neurons in the network, making the network into an autoassociation or attractor network capable of short-term memory, as described in Section 1.10. The competition between the neurons in the attractor network is again implemented by the inhibitory feedback neurons.

The implementation of biased competition in a computational model can be at the mean-field level, in which the average firing rate of each population of neurons is specified. Equations to implement the dynamics of such a model are given by Rolls and Deco (2002). The advantages of this level of formulation are relative simplicity, and speed of computation.

The implementation can also be at the level of the spiking neuron by using an *integrate-and-fire simulation* (see Sections 2.2 and 4.4). These more realistic dynamics allow the use of realistic biophysical constants (such as conductances, delays, etc.) in a thorough study of the realistic time scales and firing rates involved in the time course of the neural activity underlying cognitive processes, for comparison with experimental data. It is important for a *biologically plausible* model that the different time scales involved are properly described, because the dynamical system is sensitive to the underlying different spiking and synaptic time courses, and the non-linearities involved in these processes. For this reason, it is convenient to include a thorough description of the different time courses of the synaptic activity, by including fast and slow excitatory receptors (AMPA and NMDA) and GABA-inhibitory receptors.

A second reason why this temporally realistic and detailed level of description of synaptic activity is required is the goal to perform realistic fMRI-simulations. These involve the realistic calculation of BOLD-signals (see Section 6.5) that are intrinsically linked with the synaptic dynamics, as shown by Logothetis, Pauls, Augath, Trinath and Oeltermann (2001).

A third reason is that one can consider the influence of neurotransmitters and pharmacological manipulations, e.g. the influence of dopamine on the NMDA and GABA receptor dynamics (Zheng, Zhang, Bunney and Shi 1999, Law-Tho, Hirsch and Crepel 1994), to study the effect on the global dynamics and on the related cortical functions (e.g. on working memory, Deco, Rolls and Horwitz (2004), and Deco and Rolls (2003)).

A fourth reason for analysis at the level of spiking neurons is that the computational units of the brain are the neurons, in the sense that they transform a large set of inputs received from different neurons into an output spike train, that this is the single output signal of the neuron which is connected to other neurons, and that this is therefore the level at which the information is being transferred between the neurons, and thus at which the brain's representations and computations can be understood (Rolls and Treves 1998, Rolls and Deco 2002, Rolls 2008d).

Fifth, the integrate-and-fire level also allows statistical effects related to the probability that spikes from different neurons will occur at different times to be investigated (which can be very important in decision-making as shown in Chapter 5), and allows single neuron recording studies to be simulated. Integrate-and-fire simulations are more difficult to set up, and are slow to run. Equations for such simulations are given in Sections 2.2 and 2.2.3, and as applied to a biased competition model of attention by Deco and Rolls (2005b).

The computational investigations can be extended to systems with many interacting modules, as described by Rolls and Deco (2002), Deco and Rolls (2003), and Deco and Rolls (2005b). These could be arranged in a hierarchy, with repeated layers of bottom-up and top-down connections as illustrated in Fig. 4.1.

p. 122

4.4 The neuronal and biophysical mechanisms of attention: linking computational and single-neuron data

The theoretical framework described by Rolls and Deco (2002) and Deco and Lee (2004) provides an account of a potential functional role for the backprojection connections in the visual cortex in visual attention. We now consider how closely this approach can account for the responses of single neurons in the system, and how well a detailed implementation of the model at the neuronal level can lead to the overall functional properties described earlier and by Rolls and Deco (2002). We also show that the same theoretical framework and model can also be directly related to psychophysical data on serial vs parallel processing, and to neuropsychological data (Rolls and Deco 2002). The framework and model operate in the same way to account for findings at all these levels of investigation, and in this sense provide a unifying framework (Rolls and Deco 2002, Deco and Rolls 2005a, Rolls 2008d).

Deco and Lee (2004) and Corchs and Deco (2002) (see Rolls and Deco (2002)) showed a long-latency enhancement effect on V1 units in the model under top-down attentional modulation which is similar to the long-latency contextual modulation effects observed in early visual cortex (Lamme 1995, Zipser, Lamme and Schiller 1996, Lee, Mumford, Romero and Lamme 1998, Roelfsema, Lamme and Spekreijse 1998, Lee, Yang, Romero and Mumford 2002). Interestingly, in our simulation, we found that the observed spatial or object attentional enhancement is stronger for weaker stimuli. This predicted result has been confirmed neurophysiologically by Reynolds, Pastemak and Desimone (2000). The mechanism for this may be that the top-down attentional influence can dominate the firing of the neurons relatively more when there are weaker feedforward forcing and shunting effects on the neurons. An extension of this model (Deco and Rolls 2004) can account for the reduced receptive fields of inferior temporal cortex neurons in natural scenes (Rolls, Aggelopoulos and Zheng 2003a), and makes predictions about how the receptive fields are affected by interactions of features within them, and by object-based attention.

The model has also been extended to the level of spiking neurons which allows biophysical properties of the ion channels affected by synapses, and of the membrane dynamics, to be incorporated, and shows how the non-linear interactions between bottom-up effects (produced for example by altering stimulus contrast) and top-down attentional effects can account for neurophysiological results in areas MT and V4 (Deco and Rolls 2005b) (see Fig. 4.6). The model and simulations show that attention has its major modulatory effect at intermediate levels of bottom-up input, and that in particular the effect of attention acting on a competing stimulus occurs most clearly when the contrast of the competing stimulus is relatively low.

The model assumed no kind of multiplicative attentional effects on the gain of neuronal responses. Instead, in the model, both top-down attention and bottom-up input information (contrast) are implemented in the same way, via additive synaptic effects in the postsynaptic neurons. There is of course a non-linearity in the effective activation function of the integrate-and-fire neurons, and this is what we (Deco and Rolls 2005b) identify as the source of the apparently multiplicative effects (Martinez-Trujillo and Treue 2002) of top-down attentional biases on bottom-up inputs. The relevant part of the effective activation function of the neurons (the relation between the firing and the injected excitatory currents) is the threshold non-linearity, and the first steeply rising part of the activation function, where close to threshold the firing increases markedly with small increases in synaptic inputs (cf. Amit and Brunel (1997), Brunel and Wang (2001) and Fig. 2.4). Attention could therefore alternatively be interpreted as a phenomenon that results from purely additive synaptic effects, non-linear effects in the neurons, and cooperation–competition dynamics in the network, which together

Measurement

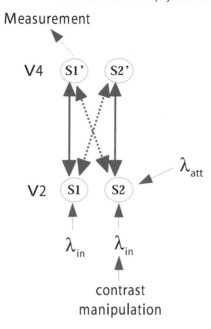

Fig. 4.6 Interaction between salience (contrast, a bottom-up influence) and attention (a top-down influence) in an integrate-and-fire model of biased competition. The architecture shows that used to model the neurophysiological experiment of Reynolds and Desimone (2003). In the model we measured neuronal responses from neurons in pool $S1'$ in V4 to a preferred and a non-preferred stimulus simultaneously presented within the receptive field. We manipulated the contrast of the stimulus that was non-preferred for the neurons $S1'$ (in the simulation by altering λ_{in} to S2). We analysed the effects of this manipulation for two conditions, namely without spatial attention or with spatial attention on the non-preferred stimulus S2 (implemented in these simulations by applying an extra bias λ_{att} to S2 from an external top-down source). We observed that the attentional suppressive effect implemented through λ_{att} on the responses of neurons $S1'$ of the competing non-preferred stimulus S2 occurs most clearly when the contrast of the non-preferred stimulus S2 is relatively low, as in the original neurophysiological experiments. (After Deco and Rolls 2005b.)

yield a variety of modulatory effects, including effects that appear (Martinez-Trujillo and Treue 2002) to be multiplicative. In addition, we were able to show that the non-linearity of the NMDA receptors may facilitate non-linear attentional effects, but is not necessary for them. This was shown by disabling the voltage-dependent non-linearity of the NMDA receptors in the simulations (Deco and Rolls 2005b).

More detail is now provided about the integrate-and-fire model of attention described by Deco and Rolls (2005b), as it shows how attentional processes can be understood at the neuronal and biophysical level. The design of the experiments and the different measures as implemented in our simulations are shown graphically in Fig. 4.6. Figure 4.6 shows the experimental design of Reynolds and Desimone (2003) and the architecture we used to simulate the results. They measured the neuronal response in V4, manipulating the contrast of the non-preferred stimulus and comparing the response to both stimuli when attention was allocated to the poor stimulus. They observed that the attentional suppressive effect of the competing non-preferred stimulus is most evident when the contrast of the stimulus is in the low to intermediate range. In our simulations we measured neuronal responses from neurons in pool $S1'$ in V4 to a preferred and a non-preferred stimulus simultaneously presented within the receptive field. We manipulated the contrast of the stimulus that was non-preferred for the

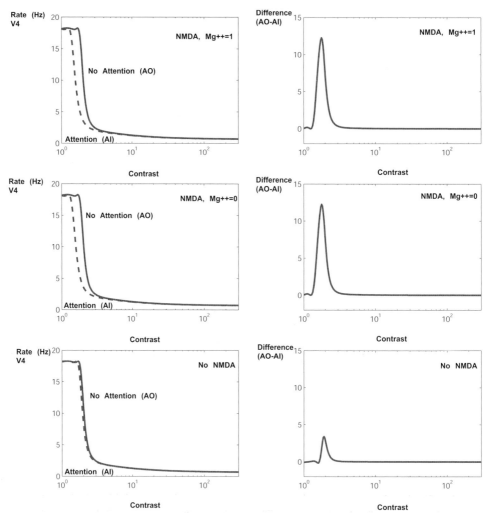

Fig. 4.7 (Top): Results of our simulations for the effect of interaction between contrast and attention after the design of Reynolds and Desimone (2003). The left figure shows the response of V4 neurons S1′ to the preferred stimulus S1 (see Fig. 4.6) as a function of different log contrast levels of the non-preferred stimulus S2 (abscissa) in the no-attention condition (AO: attending outside the S2 receptive field), and in the attention condition (AI: attending inside the S2 receptive field) (see legend to Fig. 4.6). The right figure shows the difference between both conditions. As in the experimental observations the suppressive effect of the competing non-preferred stimulus S2 on the response S1′ is influenced most by whether attention is being paid to S2 (which it is in the AI condition) when the contrast of S2 is low or intermediate, but at higher levels of salience (contrast of S2) the attentional modulation effect disappears, and suppression by S2 of S1′ occurs without as well as with attention to S2. Middle: as top, but with the NMDA receptor non-linearity removed by setting $[Mg^{2+}]=0$. Bottom: as top, but with the NMDA receptor time constants set to the same values as those of the AMPA receptors (see text). (After Deco and Rolls 2005b.)

neurons S1′ (in the simulation by altering λ_{in} to S2). We analysed the effects of this manipulation for two conditions, namely without spatial attention, or with spatial attention on the non-preferred stimulus S2 implemented by adding an extra bias λ_{att} to S2.

Figure 4.7 (top) shows the results of our simulations for the design of Reynolds and Desimone (2003). We observed that the attentional suppressive effect on the responses of neurons S1′ of the competing non-preferred stimulus implemented through λ_{att} is high when the contrast of the non-preferred stimulus S2 plotted on the abscissa of Fig. 4.7 is at low to intermediate levels, as in the original neurophysiological experiments. The top figure shows the response of a V4 neuron to different log contrast levels of S2 (abscissa) in the no-attention condition (AO: attending outside the receptive field of S2) and in the attention condition (AI: attending inside the receptive field of S2). The top right part of Fig. 4.7 shows the difference between both conditions. As in the experimental observations the suppressive effect of the competing non-preferred stimulus S2 is modulated by attention to S2 at low to intermediate levels of the contrast of S2, but at higher levels of salience (contrast of S2), the top-down attentional effect disappears, and S2 always suppresses the responses of neurons S1′ to S1.

In order to study the relevance of NMDA synapses for the inter-area cortical dynamics of attention, we repeated the analysis shown in Fig. 4.7 (top), but with the voltage-dependent non-linearity removed from the NMDA receptors (in the feedforward, the feedback, and the recurrent collateral connections) by setting $[\mathrm{Mg}^{2+}] = 0$ (which corresponds to removing the non-linear dependence of the NMDA synapses on the postsynaptic potential, see Deco and Rolls (2005b)). (We compensated for the effective change of synaptic strength by rerunning the mean field analysis to obtain the optimal parameters for the simulation, which was produced with $J_{\mathrm{f}} = 1.6$ and $J_{\mathrm{b}} = 0.42$, and then with these values, we reran the simulation.) The results are shown in Fig. 4.7 (middle), where it is clear that the same attentional effects can be found in exactly the same qualitative and even quantitative form as when the non-linear property of NMDA receptors is operating. The implication is that the non-linearity of the effective activation function (firing rate as a function of input current to a neuron) of the neurons (both the threshold and its steeply rising initial part) implicit in the integrate-and-fire model with AMPA and other receptors is sufficient to enable non-linear attentional interaction effects with bottom-up inputs to be produced. The non-linearity of the NMDA receptor may facilitate this process, but this non-linearity is not necessary.

Deco and Rolls (2005b) further studied the relevance of the NMDA receptors by repeating the analysis in Fig. 4.7 (top), but with the time constants of the NMDA receptors set to be the same values as those of the AMPA receptors (and as in Fig. 4.7 (middle) with the NMDA voltage-dependent effects disabled by setting $[\mathrm{Mg}^{2+}] = 0$). (We again compensated for the effective change of synaptic strength by rerunning the mean field analysis to obtain the optimal parameters for the simulation.) The results are shown in Fig. 4.7 (bottom), where it is shown that top-down attentional effects are now very greatly reduced. (That is, there is very little difference between the no-attention condition (AO: attending outside the receptive field of S2), and the attention condition (AI: attending inside the receptive field of S2).) This effect is not just because the NMDA receptor system with its long time constant may play a generic role in the operation of the integrate-and-fire system, by facilitating stability and helping to prevent oscillations, for a similar failure of attention to operate normally was also found with the mean-field approach, in which the stability of the system is not an issue. Thus the long time constant of NMDA receptors does appear to be an important factor in enabling top-down attentional processes to modulate correctly the bottom-up effects to account for the effects of attention on neuronal activity. This is an interesting result which deserves further analysis. Deco and Rolls (2005b) did show that the mean-field equation that effectively defines the non-linear transfer function of the neurons will be affected by the long time constant of the NMDA receptors.

These investigations thus show how attentional processes can be understood by the interactions between top-down and bottom-up influences on non-linear processes operating within neurons that become explicit with integrate-and-fire analyses, and how these processes in turn impact the performance of large separate populations of interacting neurons.

4.5   Stochastic dynamics and attention

So far in this Chapter and elsewhere (Rolls and Deco 2002, Rolls 2008d) we have set out a
foundation for understanding top-down attention, but the implementation has been largely at
the deterministic level of a mean-field analysis with systems of coupled differential equations
that has no stochastic noise. In this Section we show how noise introduced for example by
neuronal spiking is an important factor in understanding the operation of attentional systems
in the brain.

   We have seen that a short-term memory is needed to hold the object of attention active or
'in mind'. The short-term memory with a particular set of neurons held active in a high firing
rate attractor state is the source of the top-down bias that implements the biased competition in
other cortical networks. A key to understanding the stability of attention, its distractibility, and
how attention can be switched is therefore the stability of the short-term memory that holds
the object of attention active. We therefore introduce investigation of the stability of attention
by further investigation of the integrate-and-fire short-term memory model described and
analyzed in Section 3.8.

   The model of short-term memory and aspects of its operation are shown in Figs. 3.7–3.9.
Three different conditions were simulated as shown in Fig. 3.8 on page 104, and investigations with all three are relevant to understanding the stability of attention. The three conditions
are summarized next, and the third condition was not considered previously.

   In spontaneous simulations, we ran spiking simulations for 3 s without any extra external
input. The aim of this condition is to test whether the network is stable in maintaining a low
average firing rate in the absence of any inputs, or whether it falls into one of its attractor states
without any external input. Attention to spurious stimuli could arise if the spontaneous state
is too unstable. A failure for attention to be attracted by an incoming stimulus could occur
if the spontaneous state is too stable. To clarify further the concept of stability, we show examples of trials of spontaneous and persistent simulations in which the statistical fluctuations
have different impacts on the temporal dynamics. Figure 4.8 shows some examples (with a
difference from the examples shown in Fig. 3.9 being that the currents through the NMDA
synapses have been decreased by 4.5% and in the GABA synapses by 9% as described in
more detail in Section 8.14). Averaging across a large number of trials (in the baseline condition, with the standard synaptic conductances), the spontaneous state was maintained on
approximately 90% of the trials, with stochastic noise causing the network to jump incorrectly to a high firing rate attractor state on approximately 10% of the trials. This shows how,
probabilistically, attention could be generated to spurious stimuli that are not present. This
corresponds to situations found in real life.

   In persistent simulations, an external cue of 120 Hz above the background firing rate of
2400 Hz is applied to each neuron in pool S1 during the first 500 ms to induce a high activity
state and then the system is run for another 2.5 s. The 2400 Hz is distributed across the 800
synapses of each S1 neuron for the external inputs, with the spontaneous Poisson spike trains
received by each synapse thus having a mean rate of 3 Hz. The aim of this condition is to
investigate whether once in an attractor short-term memory state, the network can maintain
its activity stably, or whether it falls out of its attractor, which might correspond to an inability
to maintain attention. It was found that on approximately 90% of the trials the short-term
memory state, and thus attention, was correctly maintained. However, on approximately 10%
of the trials the high firing rate state was lost due to the spiking-related noise. This would
correspond to a lapse of attention in real life.

   With aging, the maintenance of attention could be impaired because the attractor basins
become less deep, perhaps because of reduced synaptic modification, or because receptor
activated ion currents become reduced, in ways analogous to those described in Section 8.14.

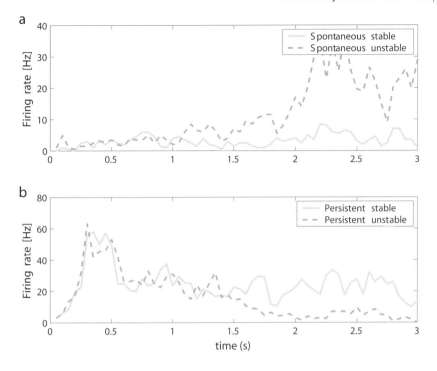

Fig. 4.8 Example trials of the integrate-and-fire attractor network simulations. (a). Trials in which no recall stimulus was applied to S1, and spontaneous firing continued until the end of the trial in the 'spontaneous' simulation condition. However, on one of the trials illustrated, the firing increased as the network climbed into one of the attractors because of the statistical fluctuations caused by the spiking dynamics. (b). Trials in which a recall stimulus was applied to S1 at 0–500 ms, and continuing firing normally occurred until the end of the trial in the 'persistent' simulation condition. However, on one of the trials illustrated, the firing decreased as the network fell out of the attractor because of the statistical fluctuations caused by the spiking dynamics. In these simulations the network parameter is $w_+ = 2.1$ and the modulation of the synapses is a 4.5% reduction of NMDA and 9% reduction of GABA receptor activated channel conductances. The average firing rate of all the neurons in the S1 pool is shown. (After Loh, Rolls and Deco 2007a.)

Indeed, the fact that schizophrenia typically does not appear until after the late teens may be related to these concepts of a reduced depth in the basins of attraction being a contributory factor to some of the symptoms of schizophrenia (see Section 8.14), and to some of the cognitive symptoms of aging such as less good maintenance of attention and short-term memory. If the depth of the basins of attraction was decreased in aging (whether by reduced synaptic strengths, or reduced ion channel conductances), the firing rates of neurons when in the short-term memory attractor state could be lower, and this could also impair attention because the source of the top-down bias would be less effective (see further Section 8.13).

The distractor simulations start off like the persistent simulations with a 500 ms input to pool S1 to start the S1 short-term memory attractor state, but between 1 s and 1.5 s we apply a distracting input to pool S2 with varying strengths. The aim of this condition is to measure how distractible the network is. The degree of distractibility is measured parametrically by the strength of the input to S2 required to remove the high activity state of the S1 population.

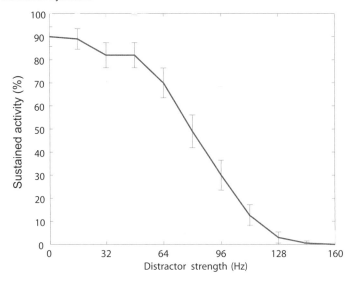

Fig. 4.9 Stability and distractibility as a function of the distractor strength. We show how often in 200 trials the average activity during the last second (2–3 s) stayed above 10 Hz in the S1 attractor, instead of being distracted by the S2 distractor. The strength of the distractor stimulus applied to S2 is an increase in firing rate above the 2.4 kHz background activity which is distributed among 800 synapses per neuron. The higher the sustained activity is in S1 the higher is the stability and the lower is the distractibility by S2. The standard deviations were approximated with the binomial distribution. (After Rolls, Loh and Deco 2008a.)

To investigate the distractibility of the network, the system was set up with two high firing rate attractors, S1 and S2. Then as shown in Fig. 3.8, the network was started in an S1 attractor state with S1 applied at t=0–0.5 s, and then a distractor S2 was applied at time t=1.0–1.5 s to investigate how strong S2 had to be to distract the network out of its S1 attractor. Rolls, Loh and Deco (2008a) assessed how often in 200 trials the average activity during the last second (2 s–3 s) stayed above 10 Hz in the S1 attractor (percentage sustained activity shown on the ordinate of Fig. 4.9). The strength of the distractor stimulus applied to S2 was an increase in firing rate above the 2.4 kHz background activity distributed among the 800 synapses per neuron. A distractor strength of 0 Hz corresponds to the persistent condition, and in this state, pool S1 continued to maintain its attractor firing without any distractor throughout the delay period on almost 90% of the trials.

Fig. 4.9 shows that as the distractor current was increased, the network was more likely to move out of the S1 state. With no distractor, the S1 state was maintained on 88% of the trials, and this proportion gradually decreased as S2 was increased. *The point to be emphasized is that the whole operation of the system is probabilistic, with attention sometimes being maintained, and sometimes not, with the proportion of trials on which attention is maintained to S1 gradually decreasing as the strength of the distracting stimulus S2 is increased.* This is thus a model of distractibility (with respect to S1), but also a model of the capture of attention (by S2) when attention is already being paid to another stimulus (S1).

We can note that although noise is an important factor in altering the stability of attention-related attractor networks, synaptic adaptation over a period of seconds or minutes could weaken the strength of an ongoing attractor, and make the attractor more susceptible to disruption by noise, as described in Section 8.4. This gradual weakening of attention over time might have been selected for its evolutionary adaptive value in promoting increasing sensitivity over a number of seconds to potentially new events in the environment that it would be useful to detect.

We also note that the constraints on the evolutionary and life-time setting of the values of the parameters that affect the stability of short-term memory networks as described in Section 3.8.2 apply just as much to top-down attention. The parameters must be set so that in the face of noise the system remains stable in a low firing rate state but is then sufficiently sensitive so that it will fall into an attractor when a stimulus is applied; and when in a high firing rate state must maintain that state reliably, yet be sufficiently sensitive and not locked into the low energy high firing rate state so deeply that it cannot be shifted if another stimulus is applied to which attention needs to be paid.

4.6   Disengagement of attention, and neglect

An important computational concept is that once an attractor state has been started, it becomes quite stable because of the recurrent collateral positive feedback effect, and a strong external stimulus is needed to disengage or distract it from that stable state. In fact, if a network is in the spontaneous state (not already in a high firing rate attractor), then the input firing rates and induced synaptic currents necessary to push the network into an attractor state are smaller than those shown in Fig. 4.9 needed to push the attractor network out of its high firing rate state, and thus to disengage attention.

This process helps to understand symptoms such as spatial neglect and extinction found in some patients with neurological damage. In a typical spatial attention experiment using a precueing method, the subject is asked to respond, by pressing a key, to the appearance of a target at a peripheral location (Posner, Walker, Friedrich and Rafal 1984). The target is preceded by a precue that indicates to which visual hemifield the subject is to covertly attend. The cue is then followed by a target in either the correctly (valid cue) or the incorrectly (invalid cue) cued location. Performance is worse with an incorrect cue, as attention cannot be disengaged from the location to which the precue has already drawn attention. If a patient with a parietal cortex lesion is tested on this task, performance is especially impaired if the distracting cue must enter the visual system through a brain hemisphere with a parietal lesion (Posner et al. 1984, Posner, Walker, Friedrich and Rafal 1987, Baynes, Holtzman and Volpe 1986, Morrow and Ratcliff 1988, Rafal and Robertson 1995). This 'failure to disengage attention' can be accounted for by the fact that once attention has been precued and there is a high firing rate stable attractor, it will be very difficult to distract it if the distracting stimulus is impaired in strength because of brain damage to the hemisphere through which the distracting spatial stimulus would normally be processed (Deco and Zihl 2004, Rolls and Deco 2002, Rolls 2008d).

The present approach emphasizes that all these effects are probabilistic, with trial by trial variation, because the processes are influenced by the spiking-related and other noise in the attractor networks.

4.7   Decreased stability of attention produced by alterations in synaptically activated ion channels

Ways in which the distractibility of attention could be increased by decreases in the NMDA receptor activated ion channel conductances and changes in other conductances making attention less stable and more distractible are described next, and applications to cognitive symptoms are described in Sections 8.14 and 8.13.

To illustrate how distractibility is affected by changes in synaptically activated ion channel conductances, we show the results of simulations specifically to assess this property using

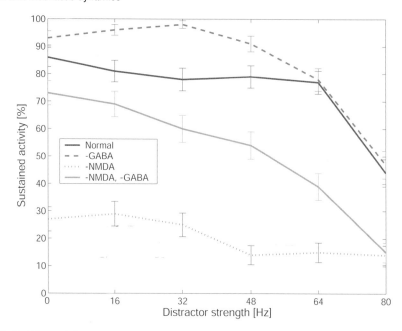

Fig. 4.10 Stability and distractibility as a function of the distractor strength and modulations of the synaptic efficacies. We assessed how often in 100 trials the average activity during the last second (2–3 s) stayed above 10 Hz in the S1 attractor. The modulation of the synapses NMDA:-1 and GABA:-1 corresponds to a reduction of 4.5% and 9%, respectively. The strength of the distractor stimulus applied to S2 is an increase in firing rate above the 2.4 kHz background activity which is distributed among 800 synapses per neuron. The lower the sustained activity in S1 the higher is the distractibility. The standard deviations were approximated with the binomial distribution. (After Loh, Rolls and Deco 2007a.)

persistent and distractor simulations (see Fig. 3.8 on page 104). A distractor strength of 0 Hz corresponds to the persistent condition described in the preceding Section 3.8.2. Figure 4.10 shows the stability and distractibility for reductions of NMDA and GABA currents, which correspond to decreases of synaptic strength or efficacy. The reference state is labelled 'Normal'. In this state, pool S1 continued to maintain its attractor firing without any distractor (distractor strength = 0 Hz) throughout the delay period on almost 90% of the trials. In both conditions that reduce the NMDA current (labelled −NMDA), the network was less and less able to maintain the S1 attractor firing as the distractor stimulus strength was increased through the range 0–80 Hz. The stability of the persistent state was reduced, and also the distractibility was increased, as shown by the fact that increasing distractor currents applied to S2 could move the attractor away from S1.

Reducing only the GABA currents (−GABA) reduces the distractibility for low distractor strengths and coincides with the reference (Normal) condition at high values of the distractor strengths.

An implication is that a reduction of the NMDA currents could cause the cognitive symptoms of schizophrenia, by making short-term memory networks less stable and more distractible, thereby reducing the ability to maintain attention, as described in Section 8.14.

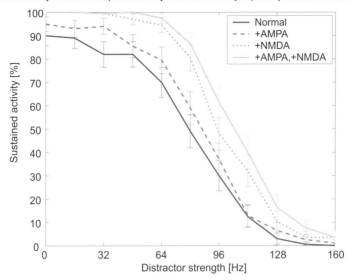

Fig. 4.11 Stability and distractibility as a function of the distractor strength and the synaptic efficacies. We assessed how often in 200 trials the average activity during the last second (2-3 s) stayed above 10 Hz in the S1 attractor, instead of being distracted by the S2 distractor. The modulation of the synapses +NMDA and +AMPA corresponds to an increase of 5% and 10%, respectively. The strength of the distractor stimulus applied to S2 is an increase in firing rate above the 2.4 kHz background activity which is distributed among 800 synapses per neuron. The higher the sustained activity is in S1 the higher is the stability and the lower is the distractibility by S2. The standard deviations were approximated with the binomial distribution. (After Rolls, Loh and Deco 2008a.)

4.8 Increased stability of attention produced by alterations in synaptically activated ion channels

Ways in which the distractibility of attention could be decreased by increases in the NMDA receptor activated ion channel and other conductances and could make attention more stable and less distractible are described next, and applications to cognitive symptoms are described in Section 8.15.

We now illustrate how an increase of NMDA currents might make an attractor system involved in attentional control less distractible, and overstable in remaining in an attractor (Rolls, Loh and Deco 2008a). This is illustrated as shown in Fig. 3.8 by setting up a system with two high firing rate attractors, S1 and S2, then starting the network in an S1 attractor state with S1 applied at t=0–0.5 s, and then applying a distractor S2 at time t=1.0–1.5 s to investigate how strong S2 had to be to distract the network out of its S1 attractor. We assessed how often in 200 trials the average activity during the last second (2 s–3 s) stayed above 10 Hz in the S1 attractor (percentage sustained activity shown on the ordinate of Fig. 4.11). The strength of the distractor stimulus applied to S2 was an increase in firing rate above the 2.4 kHz background activity which is distributed among 800 synapses per neuron.

Fig. 4.11 shows that increasing the NMDA receptor activated currents by 5% (+NMDA) means that larger distractor currents must be applied to S2 to move the system away from S1 to S2. That is, the +NMDA condition makes the system more stable in its high firing rate attractor, and less able to be moved to another state by another stimulus (in this case S2). This can be related to the symptoms of obsessive-compulsive disorder, in that once in an attractor state (which might reflect an idea or concern or action), it is very difficult to get the system to move to another state (see Section 8.15). Increasing AMPA receptor activated synaptic

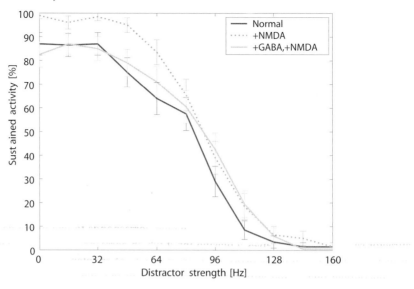

Fig. 4.12 Stability and distractibility: the effect of increasing GABA by 10% (+GABA) on the overstability to S1 / decrease in distractibility by S2 produced by increasing NMDA synaptic conductances by 3% (+NMDA). We assessed how often in 200 trials the average activity during the last second (2 s–3 s) stayed above 10 Hz in the S1 attractor, instead of being distracted to the S2 attractor state by the S2 distractor. The modulation of the synapses +NMDA and +GABA corresponds to an increase of 3% and 10%, respectively. The strength of the distractor stimulus applied to S2 is an increase in firing rate above the 2.4 kHz background activity which is distributed among 800 synapses per neuron. The higher the sustained activity is in S1 the higher is the stability to S1 and the lower is the distractibility by S2. The standard deviations were approximated with the binomial distribution. (After Rolls, Loh and Deco 2008a.)

currents (by 10%, +AMPA) produces similar, but smaller, effects.

Fig. 4.12 shows that increasing GABA activated synaptic conductances (by 10%, +GABA) can partly normalize the overstability and decrease of distractibility that is produced by elevating NMDA receptor activated synaptic conductances (by 3%, +NMDA). The investigation and conventions used the same protocol as for the simulations shown in Fig. 4.11. We assessed how often in 200 trials the average activity during the last second (2 s–3 s) stayed above 10 Hz in the S1 attractor, instead of being distracted by the S2 distractor. The strength of the distractor stimulus applied to S2 is an increase in firing rate above the 2.4 kHz background activity which is distributed among 800 synapses per neuron. The higher the sustained activity is in S1 the higher is the stability to S1 and the lower is the distractibility by S2. What was found (as shown in Fig. 4.12) is that the increase of GABA was able to move the curve in part back towards the normal state away from the overstable and less distractible state produced by increasing NMDA receptor activated synaptic currents (Rolls, Loh and Deco 2008a). The effect was particularly evident at low distractor currents.

In this Chapter, we have explored the neurodynamical systems involved in attention, have shown how noise can affect their stability, and have paved the way for investigations about how noise interacting with alterations in the attractor landscape may produce neurocomputational accounts, with implications for treatment, of the symptoms of aging (Section 8.13), schizophrenia (Section 8.14), and obsessive-compulsive disorder (Section 8.15).

5 Probabilistic decision-making

5.1 Introduction

In this Chapter, we show how an attractor network can model probabilistic decision-making. Attractor or autoassociation memory networks that can implement short-term memory are described in Section 1.10. For decision-making, the attractor network is trained to have two (or more) attractor states, each one of which corresponds to one of the decisions. Each attractor set of neurons receives a biasing input which corresponds to the evidence in favour of that decision. When the network starts from a state of spontaneous firing, the biasing inputs encourage one of the attractors to gradually win the competition, but this process is influenced by the Poisson-like firing (spiking) of the neurons, so that which attractor wins is probabilistic. If the evidence in favour of the two decisions is equal, the network chooses each decision probabilistically on 50% of the trials. The model not only shows how probabilistic decision-making could be implemented in the brain, but also how the evidence can be accumulated over long periods of time because of the integrating action of the attractor short-term memory network; how this accounts for reaction times as a function of the magnitude of the difference between the evidence for the two decisions (difficult decisions take longer); and how Weber's law appears to be implemented in the brain. Details of the model are provided by Deco and Rolls (2006), and in Section 2.2.3 for the integrate-and-fire implementation, and Sections 2.7.2 and 2.7.3 for the mean-field implementation.

p 71

It is very interesting that the model of decision-making is essentially the same as an attractor model of long-term memory or short-term memory in which there are competing retrieval cues. This makes the approach very unifying, and elegant, and consistent with the presence of well-developed recurrent collateral excitatory connections in the neocortex which with the same architecture and functionality can be put to different uses. This provides for economy and efficiency in evolution and in the genetic prescription of a type of cortical architecture that can be used for many functions (see further Section 8.18 on page 253).

The link between perception and action can be conceptualized by a chain of neural operations, which leads a stimulus to guide behaviour to make a decision to choose a particular action or motor response. For example, when subjects discriminate two stimuli separated by a time interval, the chain of neural operations encompasses mechanisms from the encoding of sensory stimuli, the attentional filtering of relevant features, their maintenance in working memory, to the comparison leading to a motor response (Romo and Salinas 2001, Romo and Salinas 2003). The comparison step is a crucial operation in the decision-making process. A number of neurophysiological experiments on decision-making are providing information on the neural mechanisms underlying perceptual comparison, by analyzing the responses of neurons that correlate with the animal's behaviour (Werner and Mountcastle 1965, Talbot, Darian-Smith, Kornhuber and Mountcastle 1968, Salzman, Britten and Newsome 1990, Kim and Shadlen 1999, Gold and Shadlen 2000, Schall 2001, Hernandez, Zainos and Romo 2002, Romo, Hernandez, Zainos, Lemus and Brody 2002, Romo, Hernandez, Zainos and Salinas 2003, Glimcher 2003, Glimcher 2004, Romo et al. 2004, Smith and Ratcliff 2004, Sugrue, Corrado and Newsome 2005, Gold and Shadlen 2007). An important finding is that cortical areas involved in generating motor responses also show activity reflecting a gradual accumulation of evidence for choosing one or another decision, such that the process of making a decision and action generation often can not be differentiated (see for example, Gold and Shadlen (2000) and Romo, Hernandez and Zainos (2004); but see also Section 8.12, where it is argued that different types of categorical choice take place in different cortical areas).

Complementary theoretical neuroscience models are approaching the problem by designing biologically realistic neural circuits that can perform the comparison of two signals (Wang 2002, Brody, Romo and Kepecs 2003b, Machens, Romo and Brody 2005, Wang 2008). These models involve two populations of excitatory neurons, engaged in competitive interactions mediated by inhibition, and external sensory inputs that bias this competition in favor of one of the populations, producing a binary choice that develops gradually. Consistent with the neurophysiological findings, this neurodynamical picture integrates both the accumulation of perceptual evidence for the comparison, and choice, in one unifying network. Even more, the models are able to account for the experimentally measured psychometric and neurometric curves, and reaction times (Marti et al. 2008).

The comparison of two stimuli for which is the more intense becomes more difficult as they become more similar. The 'difference-threshold' or 'just-noticeable difference' (jnd) is the amount of change needed for us to recognize that a change has occurred. Weber's law (enunciated by Ernst Heinrich Weber 1795–1878) states that the ratio of the difference-threshold to the background intensity is a constant. Most theoretical models of decision-making (Wang 2002, Brody et al. 2003b, Machens et al. 2005) have not investigated Weber's law, and therefore a thorough understanding of the neural substrate underlying the comparison process has been missing until recently (Deco and Rolls 2006).

In this Chapter, the neurodynamical mechanisms engaged in the process of comparison in a decision-making paradigm is investigated, and these processes are related to the psychophysics, as described for example by Weber's law. That is, the probabilistic behaviour of the neural responses responsible for detecting a just noticeable stimulus difference is part of what is described.

5.2   Decision-making in an attractor network

Let us consider the attractor network architecture again, but this time as shown in Fig. 5.1a with two competing inputs λ_1 and λ_2, each encouraging the network to move from a state of spontaneous activity into the attractor corresponding to λ_1 or to λ_2. These are separate attractor states that have been set up by associative synaptic modification, one attractor for the neurons that are coactive when λ_1 is applied, and a second attractor for the neurons that are coactive when λ_2 is applied. When λ_1 and λ_2 are both applied simultaneously, each attractor competes through the inhibitory interneurons (not shown), until one wins the competition, and the network falls into one of the high firing rate attractors that represents the decision. The noise in the network caused by the random spiking of the neurons means that on some trials, for given inputs, the neurons in the decision 1 attractor are more likely to win, and on other trials the neurons in the decision 2 attractor are more likely to win. This makes the decision-making probabilistic, for, as shown in Fig. 5.1b, the noise influences when the system will jump out of the spontaneous firing stable (low energy) state S, and whether it jumps into the high firing state for decision 1 or decision 2 (D).

The operation and properties of this model of decision-making (Wang 2002, Deco and Rolls 2006) are described first in this Chapter (5), and then in Section 5.8 we build on what

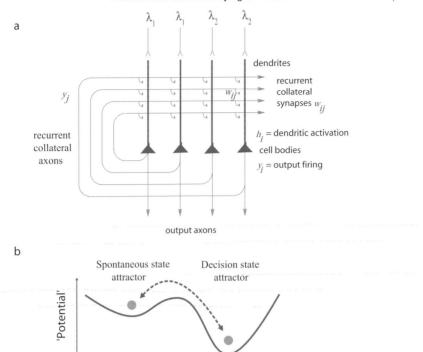

Fig. 5.1 (a) Attractor or autoassociation network architecture for decision-making. The evidence for decision 1 is applied via the λ_1 inputs, and for decision 2 via the λ_2 inputs. The synaptic weights w_{ij} have been associatively modified during training in the presence of λ_1 and at a different time of λ_2. When λ_1 and λ_2 are applied, each attractor competes through the inhibitory interneurons (not shown), until one wins the competition, and the network falls into one of the high firing rate attractors that represents the decision. The noise in the network caused by the random spiking of the neurons means that on some trials, for given inputs, the neurons in the decision 1 attractor are more likely to win, and on other trials the neurons in the decision 2 attractor are more likely to win. This makes the decision-making probabilistic, for, as shown in (b), the noise influences when the system will jump out of the spontaneous firing stable (low energy) state S, and whether it jumps into the high firing state for decision 1 or decision 2 (D).

has been described, and introduce the different stable states of this type of model, and the effects of noise in such a system.

5.3   The neuronal data underlying a vibrotactile discrimination task

A good paradigm for studying the mechanisms of decision-making is the vibrotactile sequential discrimination task, because evidence on the neuronal basis is available. In the two-alternative, forced-choice task used, subjects must decide which of two mechanical vibrations applied sequentially to their fingertips has the higher frequency of vibration (see Fig. 7.1 on page 214). Neuronal recording and behavioural analyses (Romo and Salinas 2003) have provided sufficient detail about the neuronal bases of these decisions for a quantitative model to be developed. In particular, single neuron recordings in the ventral premotor cortex (VPC) reveal neurons whose firing rate was dependent only on the difference between the two applied frequencies, the sign of that difference being the determining factor for correct task performance (Romo, Hernandez and Zainos 2004, de Lafuente and Romo 2005). We consider other potential applications of this approach to decision-making in Chapter 8.

Deco and Rolls (2006) modelled the activity of these VPC neurons by means of a theoretical framework first proposed by Wang (2002), but investigated the role of finite-size fluctuations in the probabilistic behaviour of the decision-making neurodynamics, and especially the neural encoding of Weber's law. (The finite size fluctuations are the statistical effects caused by the chance firing of different numbers of spiking neurons in a short time period, which are significant in networks of less that infinite size, and which can influence the way in which a network operates or settles, as described further below.) The neurodynamical formulation is based on the principle of biased competition/cooperation that has been able to simulate and explain, in a unifying framework, attention, working memory, and reward processing in a variety of tasks and at different cognitive neuroscience experimental measurement levels (Rolls and Deco 2002, Deco and Lee 2002, Corchs and Deco 2002, Deco, Pollatos and Zihl 2002, Corchs and Deco 2004, Deco and Rolls 2002, Deco and Rolls 2003, Deco and Rolls 2004, Deco, Rolls and Horwitz 2004, Szabo, Almeida, Deco and Stetter 2004, Deco and Rolls 2005b, Deco and Rolls 2005a, Rolls and Deco 2006, Rolls, Loh, Deco and Winterer 2008b, Rolls, Grabenhorst and Deco 2010b).

The neuronal substrate of the ability to discriminate two sequential vibrotactile stimuli has been investigated by Romo and colleagues (Romo and Salinas 2001, Hernandez, Zainos and Romo 2002, Romo, Hernandez, Zainos, Lemus and Brody 2002, Romo, Hernandez, Zainos and Salinas 2003, Romo and Salinas 2003, Romo, Hernandez and Zainos 2004, de Lafuente and Romo 2005). They used a task where trained macaques (*Macaca mulatta*) must decide and report which of two mechanical vibrations applied sequentially to their fingertips has the higher frequency of vibration by pressing one of two pushbuttons. This decision-making paradigm requires therefore the following processes: (1) the perception of the first stimulus, a 500 ms long vibration at frequency f1; (2) the storing of a trace of the f1 stimulus in short-term memory during a delay of typically 3 s; (3) the perception of the second stimulus, a 500 ms long vibration at frequency f2; and (4) the comparison of the second stimulus f2 to the trace of f1, and choosing a motor act based on this comparison (f2-f1). The vibrotactile stimuli f1 and f2 utilized were in the range of frequencies called *flutter*, i.e. within approximately 5–50 Hz.

Deco and Rolls (2006) were particularly interested in modelling the responses of ventral premotor cortex (VPC) neurons (Romo, Hernandez and Zainos 2004). The activity of VPC neurons reflects the current and the remembered sensory stimuli, their comparison, and the motor response, i.e the entire cascade of decision-making processing linking the sensory evaluation to the motor response. Many VPC neurons encode f1 during both the stimulus presentation and the delay period. During the comparison period, the averaged firing rate of VPC neurons after a latency of a few hundred milliseconds reflects the result of the comparison, i.e. the sign of (f2-f1), and correlates with the behavioural response of the monkey. In particular, we are interested in VPC neurons which show the strongest response only during the comparison period and reflect the sign of the comparison f2-f1, i.e. these neurons are only activated during the presentation of f2, with some responding to the condition f1<f2, and others to the condition f1>f2. These neurons, which are shown in figure 2(G, H, I) of Romo, Hernandez and Zainos (2004) (see example in Fig. 5.2), reflect the decision-making step of the comparison, and therefore we will describe here their probabilistic dynamical behaviour as reported by the experimental work; and through theoretical analyses we will relate their behaviour to how decisions are taken in the brain, and to Weber's law.

Earlier brain areas provide inputs useful to the VPC. In the primary somatosensory area S1 the average firing rates of neurons in S1 convey information about the vibrotactile freq-

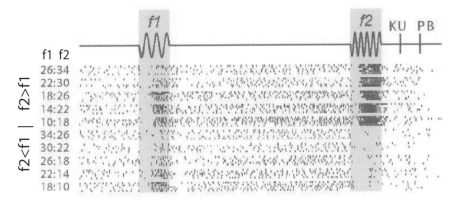

Fig. 5.2 A neuron in the ventral premotor cortex that responded to a vibrotactile stimulus during the f1 stimulation and during the delay period. However, the strongest response was for condition f2 > f1 during the f2 period. In the raster plot, each row of ticks is a trial, and each tick is an action potential. Trials were delivered in random order with 5 trials per stimulus pair shown. The labels on the left show the f1, f2 stimulus pairs for each set of 5 trials illustrated. At KU (key up) the monkey releases the key and presses either a lateral or a medial push button (PB) to indicate whether the comparison frequency (f2) was higher or lower than the base frequency (f1). (After Romo, Hernandez and Zainos 2004.)

uency f1 or f2 during the stimulation period. (The neuronal responses stop reflecting information about the stimuli immediately after the end of the stimulus.) The firing rates increase monotonically with stimulus frequency (Romo and Salinas 2003). Neurons in the secondary somatosensory area S2 respond to f1 and show significant delay activity for a few hundred milliseconds after the end of the f1 stimulus (Romo et al. 2002). Some neurons have positive and others negative monotonic relationships between their firing rate and the vibrotactile stimulus frequency. During the initial part of f2 (ca. 200 ms) the firing rate reflects either f1 or f2; later, during the last 300 ms, the firing rate reflects the comparison (f2-f1), and therefore the result of the decision. Prefrontal cortex (PFC) neurons (Brody et al. 2003a) also have a positive or negative monotonic firing rate relationship with f1. Furthermore, PFC neurons convey information about f1 into the delay period, with some neurons carrying it only during the early part of the delay period (*early neurons*), others only during the late part of the delay period (*late neurons*), and others persistently throughout the entire delay period (*persistent neurons*). During the presentation of the second stimulus f2, PFC neurons also respond like S2 neurons. Some PFC neurons respond as a function of f2 during the initial part of the comparison period, whereas other neurons show a firing rate dependency only on f1 before and at the onset of the second stimulus. In the latter part of the comparison, the firing rate reflects the comparison f2-f1. Medial premotor cortex (MPC) neurons respond similarly to PFC neurons, i.e. MPC neurons respond during f1 itself, with either positive or negative monotonic firing rate relationships, during the late part of the delay period in an f1-dependent manner in the same way as the *late* PFC neurons, and during the comparison period reflecting the comparison f2-f1 (Hernandez et al. 2002).

In summary, in the sequential vibrotactile discrimination task, S1 is predominantly sensory and the primary motor cortex (M1) is predominantly motor. A number of other cortical areas have activity that reflects the encoding, short-term memory, and comparison functions involved, perhaps as a result of information exchange between these cortical areas: the differences between S2, MPC and VPC are reflected mainly in their different latencies. In a detection task, the activity of S1 neurons codes for the stimulus but not for the behavioural choice made, whereas neuronal activity in MPC correlates with behavioural choice and detection (de Lafuente and Romo 2005). Within this context, VPC (and MPC) neurons seem to reflect the core of the processing that links sensory information with action, and therefore they may represent the decision-making process itself, rather than the representation of the stimulus. Consequently VPC neurons are excellent candidates for encoding also the probabilistic behavioural response as expressed in Weber's law.

Key questions are how ventral premotor cortex (VPC) neurons (or neurons with similar activity in connected areas such as MPC) implement the decision-making process. What are the principles by which the probabilistic decisions are taken? How is the processing implemented by the neurons?

5.4   Theoretical framework: a probabilistic attractor network

The theoretical framework within which the new model was developed was utilized by Wang (2002), which is based on a neurodynamical model first introduced by Brunel and Wang (2001), and which has been recently extended and successfully applied to explain several experimental paradigms (Rolls and Deco 2002, Deco and Rolls 2002, Deco and Rolls 2003, Deco and Rolls 2004, Deco, Rolls and Horwitz 2004, Szabo, Almeida, Deco and Stetter 2004, Deco and Rolls 2005b, Rolls, Grabenhorst and Deco 2010b, Rolls, Grabenhorst and Deco 2010c). In this framework, we model probabilistic decision-making by a single attractor network of interacting neurons organized into a discrete set of populations, as depicted in Fig. 5.3. Populations or pools of neurons are defined as groups of excitatory or inhibitory neurons sharing the same inputs and connectivities. The network contains N_E (excitatory) pyramidal cells and N_I inhibitory interneurons. In the simulations, we used $N_E = 800$ and $N_I = 200$, consistent with the neurophysiologically observed proportion of 80% pyramidal cells versus 20% interneurons (Abeles 1991, Rolls and Deco 2002). The neurons are fully connected (with synaptic strengths as specified later).

The specific populations have specific functions in the task. In our minimal model, we assumed that the specific populations encode the categorical result of the comparison between the two sequentially applied vibrotactile stimulation, f1 and f2, i.e. the result that f1>f2 or the result that f1<f2. Each specific population of excitatory cells contains rN_E neurons (in our simulations $r = 0.1$). In addition there is one non-specific population, named 'Non-specific', which groups all other excitatory neurons in the modelled brain area not involved in the present task, and one inhibitory population, named 'Inhibitory', grouping the local inhibitory neurons in the modelled brain area. The latter population regulates the overall activity and implements competition in the network by spreading a global inhibition signal.

Because we were mainly interested in the non-stationary probabilistic behaviour of the network, the proper level of description at the microscopic level is captured by the spiking and synaptic dynamics of one-compartment *Integrate-and-Fire* (IF) neuron models (see Section 2.2). At this level of detail the model allows the use of realistic biophysical time constants, latencies and conductances to model the synaptic current, which in turn allows a thorough study of the realistic time scales and firing rates involved in the time evolution of the neural activity. Consequently, the simulated neuronal dynamics, that putatively underlie cognitive processes, can be quantitatively compared with experimental data. For this reason, it is very useful to include a thorough description of the different time constants of the synaptic activity. The IF neurons are modelled as having three types of receptor mediating the synaptic currents flowing into them: AMPA, NMDA (both activated by glutamate), and GABA receptors. The excitatory recurrent postsynaptic currents (EPSCs) are considered to be mediated by AMPA (fast) and NMDA (slow) receptors; external EPSCs imposed onto the network

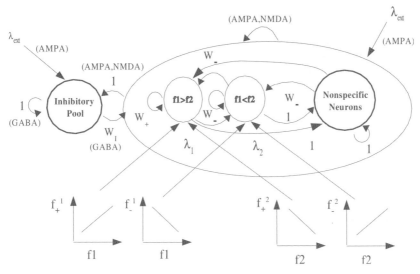

Fig. 5.3 The architecture of the neurodynamical model for a probabilistic decision-making network. The single attractor network has two populations or pools of neurons (f1>f2) and (f1<f2) which represent the decision states. One of these pools becomes active when a decision is made. If pool (f1>f2) is active, this corresponds to the decision that stimulus f1 is greater than stimulus f2. There is also a population of non-specific excitatory neurons, and a population of inhibitory neurons. Pool (f1>f2) is biased by λ_1 which reflects the strength of stimulus f1, and pool (f2>f1) is biased by λ_2 which reflects the strength of stimulus f2. (In the simulations performed f1 is the frequency of vibrotactile stimulus 1, f2 is the frequency of vibrotactile stimulus 2, and the stimuli must be compared to decide which is the higher frequency.) The integrate and fire network is subject to finite size noise, and therefore probabilistically settles into either an attractor with the population (f1>f2) active, or with the population (f1<f2) active, depending on the biasing inputs λ_1 and λ_2. The network is thus a biased competition model of decision-making. The weights connecting the different populations of neurons are shown as w_+, w_-, w_I, and 1, and the values found in the mean field analysis are given in the text. All neurons receive a small random Poisson set of input spikes λ_{ext} from other neurons in the system. The nonspecific excitatory neurons are connected to pool (f1>f2) as well as to pool (f1<f2). (After Deco and Rolls 2006.)

from outside are modelled as being driven only by AMPA receptors. Inhibitory postsynaptic currents (IPSCs) to both excitatory and inhibitory neurons are mediated by GABA receptors. The details of the mathematical formulation are summarized in previous publications (Brunel and Wang 2001, Deco and Rolls 2005b, Deco and Rolls 2006), and are provided in Section 2.2.3.

We modified the conductance values for the synapses between pairs of neurons by synaptic connection weights, which can deviate from their default value 1. The structure and function of the network was achieved by differentially setting the weights within and between populations of neurons. The labelling of the weights is defined in Fig. 5.3. We assumed that the connections are already formed, by for example earlier self-organization mechanisms, as if they were established by Hebbian learning, i.e. the coupling will be strong if the pair of neurons have correlated activity (i.e. covarying firing rates), and weak if they are activated in an uncorrelated way. We assumed that the two decisions 'f1>f2' and 'f1<f2', corresponding to the two categories, are already encoded, in the sense that the monkey is already trained that pushing one or the other button, but not both, might produce a reward. As a consequence of this, neurons within a specific excitatory population are mutually coupled with a strong weight w_+, and each such population thus forms an attractor. Furthermore, the populations encoding these two decisions are likely to have anti-correlated activity in this behavioural

context, resulting in weaker than average connections between the two different populations. Consequently, we choose a weaker value $w_- = 1 - r(w_+ - 1)/(1 - r)$, so that the overall recurrent excitatory synaptic drive in the spontaneous state remains constant as w_+ is varied (Brunel and Wang 2001). Neurons in the inhibitory population are mutually connected with an intermediate weight $w = 1$. They are also connected with all excitatory neurons in the same layer with the same intermediate weight, which for excitatory-to-inhibitory connections is $w = 1$, and for inhibitory-to-excitatory connections is denoted by a weight w_I. Neurons in a specific excitatory population are connected to neurons in the non-selective population in the same layer with a feedforward synaptic weight $w = 1$ and a feedback synaptic connection of weight w_-.

Each individual population is driven by two different kinds of input. First, all neurons in the model network receive spontaneous background activity from outside the module through $N_{ext}=800$ external excitatory connections. Each connection carries a Poisson spike train at a spontaneous rate of 3 Hz, which is a typical spontaneous firing rate value observed in the cerebral cortex. This results in a background external input with a rate summed across all 800 external synapses onto each neuron of 2.4 kHz for each neuron. Second, the neurons in the two specific populations additionally receive external inputs encoding stimulus-specific information. They are assumed to originate from the somatosensory area S2 and from the PFC, encoding the frequency of both stimuli f1 (stored) and f2 (present) to be compared during the comparison period, i.e. when the second stimulus is applied. (Stimuli f1 and f2 influence λ_1 and λ_2 as shown in Fig. 5.3. The way in which the different S2 and PFC neurons described by Romo et al. (2004) are combined linearly to produce λ_1 and λ_2 is described by Deco and Rolls (2006).) These inputs which convey the evidence for each of the decisions are added to the background external inputs being applied via the 800 synapses to each neuron.

In summary, f1 is the frequency of vibrotactile stimulus 1, f2 is the frequency of vibrotactile stimulus 2, and the stimuli must be compared to decide which is the higher frequency. The single attractor network has two populations or pools of neurons (f1>f2) and (f1<f2) which represent the decision states (see Fig. 5.3). One of these pools becomes active when a decision is made. If pool (f1>f2) is active, this corresponds to the decision that stimulus f1 is greater than stimulus f2. Pool (f1>f2) is biased by λ_1 which reflects the frequency of stimulus f1, and pool (f2>f1) is biased by λ_2 which reflects the frequency of stimulus f2. The integrate and fire network is subject to finite size noise, and therefore probabilistically settles into an attractor either with the population (f1>f2) active, or with the population (f1<f2) active, depending on the biasing inputs λ_1 and λ_2. The network is thus a biased competition model of decision-making. All neurons receive a small random Poisson set of input spikes λ_{ext} via their 800 synapses that receive external inputs.

5.5 Stationary multistability analysis: mean-field

A first requirement for using the network described earlier as a probabilistic decision-making neurodynamical framework is to tune its connectivity such that the network operates in a regime of multistability. This means that at least for the stationary conditions, i.e. for periods after the dynamical transients, different possible attractors are stable. The attractors of interest for our task correspond to high activity (high spiking rates) or low activity (low spiking rates) of the neurons in the specific populations (f1>f2) and (f1<f2). The activation of the specific population (f1>f2) and the simultaneous lack of activation of the specific population (f1<f2) corresponds to encoding associated with a motor response of the monkey reporting the categorical decision f1>f2. The opposite decision corresponds to the opposite attractor states in the two specific neuronal populations. Low activity in both specific populations (the 'spontaneous state') corresponds to encoding that no decision has been made, i.e. the monkey does not answer or generates a random motor response. The same happens if both specific populations are activated (the 'pair state'). Because the monkey responds in a probabilistic way depending on the different stimuli, the operating working point of the network should be such that both possible categorical decisions, i.e. both possible single states, and sometimes (depending on the stimuli) the pair and spontaneous states, are possible stable states.

The network's operating regimes just described can all occur if the synaptic connection weights are appropriate. To determine the correct weights a mean field analysis was used (Deco and Rolls 2006) as described in Section 2.7.2 on page 85. Although a network of integrate-and-fire neurons with randomness in the spikes being received is necessary to understand the dynamics of the network, and how these are related to probabilistic decision-making, this means that the spiking activities fluctuate from time-point to time-point and from trial to trial. Consequently, integrate-and-fire simulations are computationally expensive and their results probabilistic, which makes them rather unsuitable for systematic parameter explorations. To solve this problem, we simplified the dynamics via the *mean-field* approach at least for the stationary conditions, i.e. for periods after the dynamical transients, and then analyzed the bifurcation diagrams of the dynamics. The essence of the mean-field approximation is to simplify the integrate-and-fire equations by replacing after the diffusion approximation (Tuckwell 1988, Amit and Brunel 1997, Brunel and Wang 2001), the sums of the synaptic components by the average D.C. component and a fluctuation term. The stationary dynamics of each population can be described by the *population transfer function*, which provides the average population rate as a function of the average input current. The set of stationary, self-reproducing rates ν_i for the different populations i in the network can be found by solving a set of coupled self-consistency equations using the formulation derived by Brunel and Wang (2001) (see Section 2.7.2). The equations governing the activities in the mean-field approximation can hence be studied by standard methods of dynamical systems. The formulation departs from the equations describing the dynamics of one neuron to reach a stochastic analysis of the mean-first passage time of the membrane potentials, which results in a description of the population spiking rates as functions of the model parameters, in the limit of very large N. Obtaining a mean-field analysis of the stationary states that is consistent with the network when operating dynamically as an integrate-and-fire network is an important part of the approach used by Deco and Rolls (see Sections 2.6 and 2.7).

To investigate how the stable states depend on the connection parameters w_+ and w_I, Deco and Rolls (2006) solved the mean-field equations for particular values of these parameters starting at different initial conditions. For example, to investigate the stability of the state described by population (f1>f2) being in an active state and all other populations inactive, we initialize the system with that population at 10 Hz, all other excitatory populations (including the non-specific ones) at 3 Hz, and the inhibitory population at 9 Hz. If and only if, after solving the equations numerically[1], the population (f1>f2) is still active (meaning that they have a firing rate ≥ 10 Hz) but no other excitatory population is active, we conclude that the state is stable. This procedure is then repeated for all other combinations of w_+ and w_I to find the region where the active population (f1>f2) is stable. The stable regions of the other states are found in the same way.

Figure 5.4 presents the bifurcation diagrams resulting from the mean-field analysis, for a particular case where the behavioural decision-making is hardest and is in fact purely random (i.e. at chance) as f1 and f2 are equal. Figure 5.4 shows how the stable states of average firing vary as a function of the strength of w_+ and w_I for the case: f1=f2=17.5 Hz corresponding to a

---

[1]For all simulation periods studied, the mean-field equations were integrated using the Euler method with step size 0.1 and 4000 iterations, which always allowed for convergence.

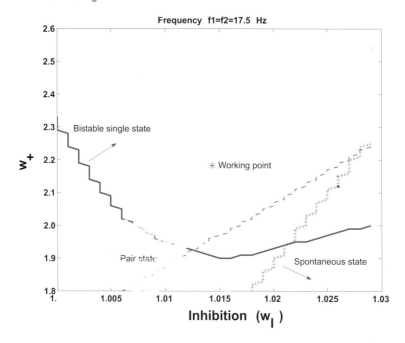

Fig. 5.4 Mean-field analysis to determine suitable values of the synaptic weights for the decision-making network. The bifurcation diagram is for the particular case where the behavioural decision-making is at chance due to f1 and f2 being equal. The diagram shows how the stable states of the average firing rate vary as a function of the synaptic strengths w_+ and w_I for the case: f1=f2=17.5 Hz corresponding to a low frequency of vibrotactile stimulation. The different regions where single states, a pair state, and a spontaneous firing rate state are stable are shown. In the following simulations we focus on the region of multistability (i.e. where either one or the other pool of neurons wins the competition, but where the spontaneous firing state is also a stable state), so that a probabilistic decision is possible, and therefore a convenient working point is one corresponding to a connectivity given by w_+=2.2 and w_I=1.015. (After Deco and Rolls 2006.)

low frequency of vibrotactile stimulation. In these cases, the specific populations (f1>f2) and (f1<f2) received an extra stimulation of λ_1 and λ_2, respectively, encoding the two vibrotactile stimuli to be compared. (The way in which these λ values were calculated simply reflects the neurons recorded in the VPC and connected areas, as described by Deco and Rolls (2006).) The different regions where single states, a pair state, and a spontaneous state are stable are shown. In the simulations, Deco and Rolls (2006) focused on a region of multistability, in which both the possible decision states, and the spontaneous firing state, were stable (see further Section 5.8), so that a probabilistic decision is possible, and therefore a convenient working point (see Fig. 5.4) is one corresponding to a connectivity given by $w_+ = 2.2$ and $w_I = 1.015$.

Overall, Fig. 5.4 shows very large regions of stability, so that the network behaviour described here is very robust.

5.6 Non-stationary probabilistic analysis: spiking dynamics

5.6.1 Integrate-and-fire simulations of decision-making

A full characterization of the dynamics, and especially of its probabilistic behaviour, including the non-stationary regime of the system, can only be obtained through computer simulations of the spiking network model. Moreover, these simulations enable comparisons between the model in which spikes occur and neurophysiological data. The simulations of the spiking dynamics of the network were integrated numerically using the second order Runge–Kutta method with a step size 0.05 ms. Each simulation was started by a period of 500 ms where no stimulus was presented, to allow the network to stabilize. The non-stationary evolution of spiking activity was averaged over 200 trials initialized with different random seeds. In all cases, Deco and Rolls (2006) aimed to model the behaviour of the VPC neurons as shown in Fig. 5.2 from Romo, Hernandez and Zainos (2004) which reflect the decision-making performed during the comparison period. Therefore, Deco and Rolls (2006) studied the non-stationary probabilistic behaviour of the spiking network defined in Fig. 5.3, during this comparison period (during the presentation of f2), by stimulating the network simultaneously with f1 and f2. This was done by increasing the rate of the Poisson train to the neurons of both specific populations (f1>f2) and (f1<f2) by an extra value of λ_1 and λ_2, respectively, as these encode the two vibrotactile stimuli to be compared.

5.6.2 Decision-making on a single trial

Figure 5.5 shows for a single trial a typical time course of the network of VPC neurons during the decision period when the two stimuli are being compared for the case of f1=35 Hz and f2=25 Hz. The top part of Fig. 5.5 plots the time course of the mean firing rate of the populations (f1>f2), (f1<f2), and the inhibitory population. The bin widths used for the simulations were 20 ms. The transition shown corresponds to a correct trial, i.e. a transition to the correct final attractor encoding the result of the discrimination f1>f2. We observe that after 200 ms the populations (f1>f2) and (f1<f2) start to separate in such a way that the population (f1>f2) wins the competition and the network performs a transition to a single-state final attractor corresponding to a correct discrimination (i.e. high activity in the population (f1>f2) and low activity in the population (f1<f2)). The bottom part of Fig. 5.5 plots the corresponding rastergrams of 10 randomly selected neurons for each pool in the network. Each vertical line corresponds to the generation of a spike. The spiking activity shows how the firing makes the transition to the correct final single-state attractor. Further examples of the neuronal decision-making process on individual trials are shown in Figs. 5.6 and 6.2.

Romo et al. (2004) analysed the responses of VPC neurons performing the comparison as a function of both f1 and f2 (and showed the neurophysiological findings in their Figure 2G,H,I, see Fig. 5.2). Figure 5.6 shows the simulation results that correspond to those cases. Figure 5.6A shows rastergrams of a single neuron of the population (f1<f2) during the decision period starting when the f1 and f2 stimuli are applied at time = 0 ms. (The neuron should thus fire at a high rate only on trials when f2 is greater than f1, and data are shown only for correct trials.) The labels on the left indicate the vibrotactile frequencies of the f1, f2 stimulus pairs between which a decision was being made. Each row of ticks is a trial (with 10 trials of the single neuron shown for each case), and each tick is a spike. All neurons were tested with 10 trials per stimulus pair, selecting only trials where the network performed correctly. The top 5 cases correspond to a situation where f1<f2 and therefore the population (f1<f2) is highly activated after 100–200 ms. The lower 5 cases correspond to a situation where f1>f2 and therefore the population (f1<f2) is not activated. (The population (f1>f2), not shown in

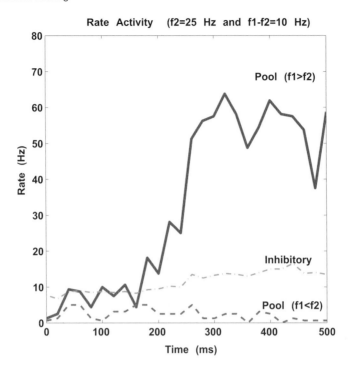

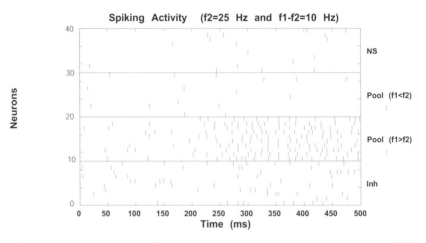

Fig. 5.5 The time course of a decision. The activity of the simulated ventral premotor cortex (VPC) neurons is shown during the decision period for the case of f1=35 Hz and f2=25 Hz. The stimuli were applied continuously starting at time = 0 ms. The top part plots the time course of the average spiking rate of the populations (f1>f2), (f1<f2), and the inhibitory population. The bin widths used for the simulations were 20 ms. The bottom part plots the corresponding rastergrams of 10 randomly selected neurons for each pool in the network. Each vertical line corresponds to the generation of a spike. In the rastergram, NS refers to neurons in the non-specific population, and Inh to neurons in the inhibitory population. The spiking activity shows the transition to the correct final single-state attractor, i.e. a transition to the correct final attractor encoding the result of the discrimination (f1>f2) (see text). (After Deco and Rolls 2006.)

Fig. 5.6, won the competition in the lower 5 cases shown in Fig. 5.6A, and therefore inhibited the (f1<f2) population in the correctly discriminated trials selected for this Figure.) Figure

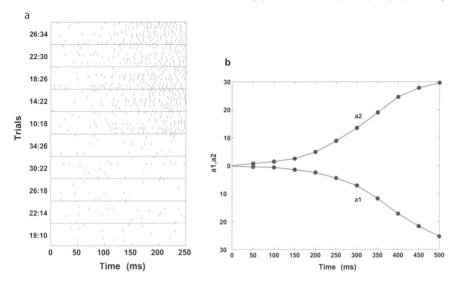

Fig. 5.6 Responses of a single neuron of the population (f1<f2) during the decision period between f1 and f2. The simulations corresponds to the experimental cases measured and studied by Romo et al. (2004) (see figure 2G,H,I of that paper, and Fig. 5.2). The different stimulation cases labelled on the left indicate f1, f2 pairs of the vibrotactile stimulation frequencies. (a) Rastergrams. Each row of ticks is a trial, and each tick is a spike. Ten correct trials are shown for each stimulus pair case. (b) Evolution of fitting coefficients $a1(t)$ and $a2(t)$ of the firing rates (see text) during the decision period. Both coefficients evolve in an antisymmetrical fashion, indicating that the average firing rate $r(t)$ is dependent primarily on the sign of the difference between f1 and f2 (i.e. $a1(t) = -a2(t)$), and not on the magnitude of the difference. (After Deco and Rolls 2006.)

5.6A shows that the firing rate at the end of the simulation period shown, at 250 ms, is high only when f2 is greater than f1, and that the final firing rate on these correct trials is relatively independent of the exact values of f1 and f2 (including the difference between f2 and f1), and responds just depending on the sign of the difference between f2 and f1, reaching a high rate only when f2 is greater than f1.

To analyse more quantitatively the dependence of the average firing rate of the VPC neurons encoding the comparisons as a function of f1 and f2, Romo et al. (2004) fitted and plotted the time evolution of the coefficients that directly measure the dependence on f1 and f2. Let us denote by $r(t)$ the trial averaged firing rate of the population (f1<f2) at time t. We determined the coefficients $a1(t), a2(t)$ and $a3(t)$ that fit the firing rate of the population (f1<f2) according to $r(t) = a1(t)f1 + a2(t)f2 + a3(t)$. We considered for all 10 f1,f2 pair cases shown in Fig. 5.6A, 50 correct trials, using a bin width of 50 ms. Figure 5.6B plots the timecourse of $a1(t)$ and $a2(t)$ during the comparison period. In this Figure, both coefficients evolve in an antisymmetrical fashion, indicating that the average firing rate $r(t)$ depends primarily on the difference between f1 and f2 (i.e. $a1(t) = -a2(t)$) (see further Chapter 6). Even more, in the range of vibrotactile flutter frequencies, the simulation and neurophysiological results show that the final firing rate depends significantly primarily on the sign of the difference and not on the magnitudes of the vibrotactile frequency values.

5.6.3 The probabilistic nature of the decision-making

The decision-making implemented by this attractor model is probabilistic. We now investigate this systematically.

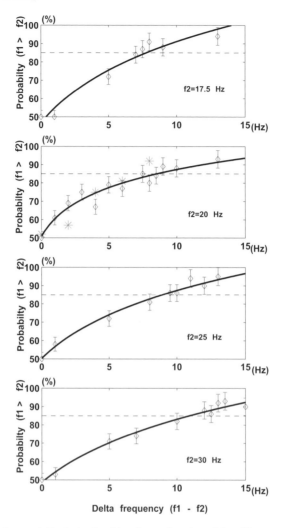

Fig. 5.7 Probability of correct discrimination (± sd) as a function of the difference between the two presented vibrotactile frequencies to be compared. In the simulations, we assume that f1>f2 by a Δ-value (labelled 'Delta frequency (f1-f2)'), i.e. f1=f2+Δ. The points correspond to the trial averaged spiking simulations. The line interpolates the points with a logarithmic function. The horizontal dashed line represents the threshold of correct classification for a performance of 85% correct discrimination. The second panel down includes actual neuronal data (indicated by *) described by Romo and Salinas (2003) for the f2=20 Hz condition. (After Deco and Rolls 2006.)

Figure 5.7 shows the probability of correct discrimination as a function of the difference between the two presented vibrotactile frequencies to be compared. We assume that f1>f2 by a Δ-value, i.e. f1=f2+Δ. (In Fig. 5.7 this value is called 'Delta frequency (f1-f2)'.) Each diamond-point in the Figure corresponds to the result calculated by averaging 200 trials of the full spiking simulations. The lines were calculated by fitting the points with a logarithmic function. A correct classification occurs when during the 500 ms comparison period, the network evolves to a 'single-state' attractor that shows a high level of spiking activity (larger than 10 Hz) for the population (f1>f2), and simultaneously a low level of spiking activity for the population (f1<f2) (at the level of the spontaneous activity). Figure 5.7 shows that the

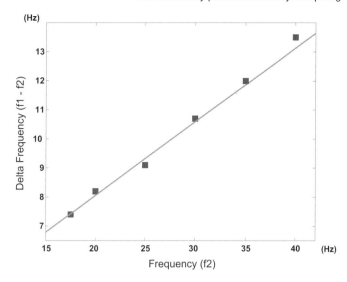

Fig. 5.8 Weber's law for the vibrotactile discrimination task. The critical discrimination Δ-value ('difference-threshold') is shown corresponding to an 85% correct performance level as a function of the base frequency f2. The 'difference-threshold' increases linearly as a function of the base frequency. (After Deco and Rolls 2006.)

decision-making is probabilistic, and that the probability of a correct discrimination increases as Δf, the difference between the two stimuli being compared, increases. When Δf is 0, the network performs at chance, and its choices are 50% correct. The second panel of Fig. 5.7 shows a good fit between the actual neuronal data described by Romo and Salinas (2003) for the f2=20 Hz condition (indicated by *), and the results obtained with the model (Deco and Rolls 2006).

One can observe from the different panels in Fig. 5.7 corresponding to different base vibrotactile frequencies f2, that to reach a threshold of correct classification of for example 85% correct (horizontal dashed line in Fig. 5.7), the difference between f1 and f2 must become larger as the base frequency f2 increases.

5.6.4 Probabilistic decision-making and Weber's law

Figure 5.8 plots the critical discrimination Δ-value corresponding to an 85% correct performance level (the 'difference threshold') as a function of the base frequency f2. The 'difference threshold' increases linearly as a function of the base frequency, that is, Δf/f is a constant. This corresponds to Weber's law for the vibrotactile discrimination task. (Weber's law is often expressed as $\Delta I/I$ is a constant, where I stands for stimulus intensity, and ΔI for the smallest difference of intensity that can just be discriminated, sometimes called the just noticeable difference. In the case simulated, the stimuli were vibrotactile frequencies, hence the use of f to denote the frequencies of the stimuli.)

The analysis shown in Figs. 5.7 and 5.8 suggests that Weber's law, and consequently the ability to discriminate two stimuli, is encoded in the probability of performing a transition to the correct final attractor. To reinforce this hypothesis, Fig. 5.9 shows that Weber's law is not encoded in the firing rates of the VPC decision-making neurons that were modelled. Deco and Rolls (2006) simulated again a situation corresponding to the cases where f1 is larger than f2 by a Δ-value, and therefore the network will perform correctly when the dynamics perform a transition to the final single attractor corresponding to high activity in the population (f1>f2)

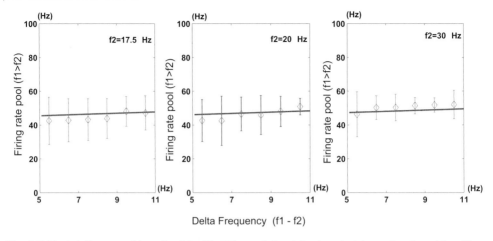

Fig. 5.9 Final stationary spiking rate of the (f1<f2) population (after transients) as a function of the difference between the two vibrotactile frequencies to be compared, for the cases where this population correctly won the competition and the network has performed a transition to the proper attractor. The plots show the results obtained with the spiking simulations. (The diamond-points correspond to the average values over 200 trials, and the error bars to the standard deviation.) The lines correspond to the mean-field calculations. The firing rate of the population encoding the result of the comparison is relatively independent of f2 and of the difference between f1 and f2, for the range of parameter values shown. (After Deco and Rolls 2006.)

and low activity in the population (f1<f2). Figure 5.9 plots for three different frequencies f2, the firing rate of the (f1<f2) population (for the cases where this population correctly won the competition and the network has performed a transition to the proper attractor) as a function of the difference between the two vibrotactile frequencies to be compared. The plots show the results obtained with the spiking simulations. The diamond-points correspond to the average values over 200 trials, and the error bars to the standard deviation. The lines show the results of the mean-field calculations. A good agreement between the spiking simulations and the corresponding mean-field results is observed. The most interesting observation is the fact that the firing rate of the population (f1>f2) in the correct attractor, for different base frequencies f2 and for different differences between f1 and f2 (Δf), is practically constant. Thus the firing rate of the population encoding the result of the comparison does not encode Weber's law. What happens is that the attractor dynamics makes a binary choice. The decision is reflected in a high firing rate of one population, and a low firing rate of the other population. These high and low firing rates thus reflect the binary choice, and depend rather little on Δf. Thus even for an equal value of f1 and f2, the attractor network makes a binary choice, with the probability of which choice is made reflecting Δf, but the firing rate that is reached as a result of the choice reflecting Δf very little. (In fact a small increase in firing rate is evident in Fig. 5.9 as Δf increases, and we show in Chapter 6 that this is related to decision confidence.)

The model also gives further insights into the mechanisms by which Weber's law is implemented. We hypothesized that because Δf/f is practically constant in the model, the difference of frequencies Δf required to push the single attractor network towards an attractor basin might increase with f because as f increases, shunting (divisive) inhibition produced by inhibitory feedback inputs (from the inhibitory interneurons) might act divisively on the pyramidal cells in the attractor network to shunt the excitatory inputs f1 and f2. In more detail, as the base frequency f increases, more excitation will be provided to the network by the inputs λ_1 and λ_2, this will tend to increase the firing rates of the pyramidal cells which will in turn provide a larger excitatory input to the inhibitory neurons. This will tend to make the

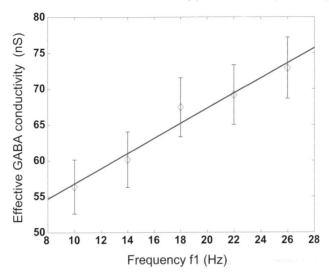

Fig. 5.10 The conductance in nS (mean $\pm$ sd) produced by the GABA inputs to the pyramidal cells as a function of the base frequency f1. The effective conductance produced through the GABA synapses (i.e. $I_{\text{GABA}}/(V - V_{\text{I}})$) was averaged over the time window in which the stimuli were presented in one of the excitatory neuron pools, when the base frequency was f1, and f2-f1 was set to 8 Hz. (After Deco and Rolls 2006.)

inhibitory neurons fire faster, and their GABA synapses onto the pyramidal cells will be more active. Because these GABA synapses open chloride channels and act with a driving potential $V_I = -70$ mV which is relatively close to the membrane potential (which will be in the range $V_L = -70$ mV to $V_{\text{thr}} = -50$ mV), a large part of the GABA synaptic input to the pyramidal cells will tend to shunt, that is to act divisively upon, the excitatory inputs to the pyramidal cells from the vibrotactile biasing inputs λ_1 and λ_2. To compensate for this current shunting effect, f1 and f2 are likely to need to increase in proportion to the base frequency f in order to maintain the efficacy of their biasing effect. To assess this hypothesis, we measured the change in conductance produced by the GABA inputs as a function of the base frequency. Figure 5.10 shows that the conductance increases linearly with the base frequency (as does the firing rate of the GABA neurons, not illustrated). The shunting effect does appear therefore to be dividing the excitatory inputs to the pyramidal cells in the linear way as a function of f that we hypothesized.

Deco and Rolls (2006) therefore proposed that Weber's law is implemented by shunting effects acting on pyramidal cells that are produced by inhibitory neuron inputs which increase linearly as the base frequency increases, so that the difference of frequencies Δf required to push the network reliably into one of its decision attractors must increase in proportion to the base frequency. We checked the excitatory inputs to the pyramidal cells (for which $V_E = 0$ mV), and found that their conductances were much smaller (in the order of 5 nS for the AMPA and 15 nS for the NMDA receptors) than those produced by the GABA receptors, so that it is the GABA-induced conductance changes that dominate, and that produce the shunting inhibition.

Further properties of the attractor network model of decision-making that enable it to implement Weber's law include stability of the spontaneous firing rate condition, even when the decision cues are applied, so that the system depends on the noise to escape this starting state, as described in Section 5.8.

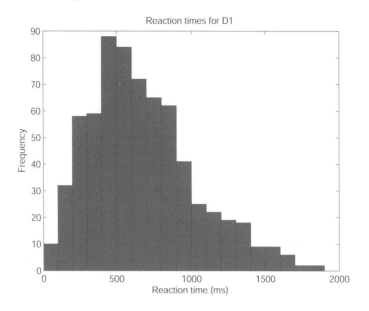

Fig. 5.11 Reaction time distribution for the decision-making attractor network described in Chapter 6. The difference between the two stimuli was relatively small (ΔI=16 Hz, though sufficient to produce 81% correct choices). The criterion for the reaction time was the time after application of both stimuli at which the firing rate of the neurons in the correct attractor became 25 Hz or more greater than that of the neurons in the incorrect attractor, and remained in that state for the remainder of the trial.

5.6.5 Reaction times

Because of the noise-driven stochastic nature of the decision-making, the reaction times even for one set of parameters vary from trial to trial. An example of the distribution of reaction times of the attractor network are shown in Fig. 5.11. This distribution is for a case when the difference between the two stimuli is relatively small ($\Delta I = 16$ Hz for the simulations described in Chapter 6, though sufficient to produce 81% correct choices). The considerable variability of the reaction times across trials, and the long tail of the probability distribution, provide a very useful basis for understanding the variability from trial to trial that is evident in human choice reaction times (Welford 1980). Indeed, human studies commonly report skewed reaction time distributions with a long right tail (Luce 1986, Ratcliff and Rouder 1998, Ratcliff, Zandt and McKoon 1999, Usher and McClelland 2001, Marti et al. 2008).

The reaction times of this model of decision-making are faster when the discrimination is easy, in this case when the difference between f2 and f1 is large. This is analysed in Fig. 5.12. We calculated for a fixed f2=25 Hz and different f1>f2 (from 1 Hz to 13 Hz), the probability of correct or incorrect discrimination, and the corresponding reaction time. The reaction time was the time that the winning population ((f1>f2) for the correct cases, and (f1<f2) for the incorrect cases) took to cross a threshold of a firing rate of 20 Hz. (We considered the averaged reaction time over 200 trials.) Figure 5.12A plots the relation between the reaction time and the probability of correct classification. The larger the probability of correct classification, the faster is the decision-making, which is reasonable and consistent with the decision-making literature. Figure 5.12B plots the reaction time of an incorrect decision, as a function of the probability of performing a misclassification. The behaviour is now the converse, in that a low probability of incorrect discrimination implies also shorter reaction times. This means that the reaction time of a correct or incorrect classification is relatively similar, and therefore

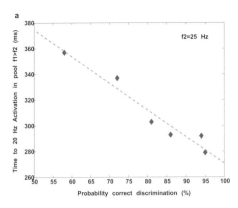
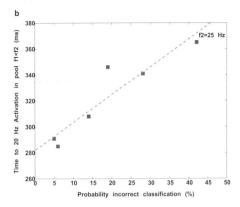

Fig. 5.12 Reaction time as a function of the probability of being correct or incorrect. Results are shown for a fixed f2=25 Hz and different f1>f2 (from 1 Hz to 13 Hz). The reaction time was the time that the winning population ((f1>f2) for the correct cases, and (f1<f2) for the incorrect cases) took to cross a threshold of 20 Hz. (The reaction time was averaged across 200 trials.) (a) Reaction time as a function of the probability of correct classification. The larger the probability the faster is the decision-making. (b) Reaction time of an incorrect decision, as a function of the probability of performing a misclassification. (After Deco and Rolls 2006.)

the dependence on the probability of a correct or incorrect classification is inverted.

Further analyses of reaction times as a function of the easiness of the decision are shown in Figs. 6.2, 6.3b, and 6.11.

5.6.6 Finite-size noise effects

The results described earlier indicate that the probabilistic settling of the system is related to the finite size noise effects of the spiking dynamics of the individual neurons with their Poisson-like spike trains in a network of limited size. The concept here is that the smaller the network, the greater will be the statistical fluctuations (i.e. the noise) caused by the random spiking times of the individual neurons in the network. In an infinitely large system, the statistical fluctuations would be smoothed out, and reach zero. To investigate this further, and to show what sizes of network are in practice influenced by these finite-size related statistical fluctuations, an important issue when considering the operation of real neuronal networks in the brain, Deco and Rolls (2006) simulated networks with different numbers of neurons, N. The noise due to the finite size effects is expected to increase as the network becomes smaller, and indeed to be proportional to $1/\sqrt{N}$. We show in Fig. 5.13 the effects of altering N on the operation of the network, where $N = N_E + N_I$, and $N_E : N_I$, was held at 4:1 as in the simulations shown earlier. The simulations were for f1=30 Hz and f2=22 Hz. Figure 5.13 shows overall that when N is larger than approximately 1,000, the network shows the expected settling to the (f1 > f2) attractor state on a proportion of occasions that is in the range 85–93%, increasing only a little as the number of neurons reaches 4,000 (top panel). The settling remains probabilistic, as shown by the standard deviations in the probability that the (f1 > f2) attractor state will be reached (top panel). When N is less than approximately 1,000, the finite size noise effects become very marked, as shown by the fact that the network reaches the correct attractor state (f1>f2) much less frequently, and in that the time for a decision to be reached can be premature and fast, as the large fluctuations in the stochastic noise can cause the system to reach the criterion [in this case of a firing rate of 20 Hz in the pool (f1>f2)] too quickly.

The overall conclusion of the results shown in Fig. 5.13 is that the size of the network,

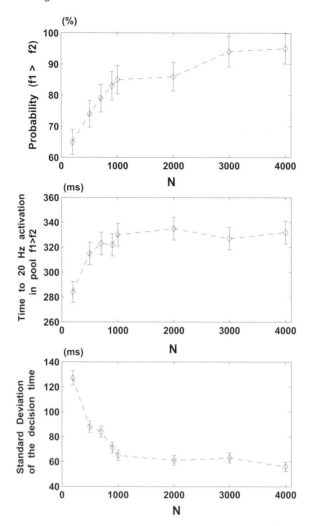

Fig. 5.13 The effects of altering N, the number of neurons in the network, on the operation of the decision-making network. The simulations were for f1=30 Hz and f2=22 Hz. The top panel shows the probability that the network will settle into the correct (f1>f2) attractor state. The mean ± the standard deviation is shown. The middle panel shows the time for a decision to be reached, that is for the system to reach a criterion of a firing rate of 20 Hz in the pool (f1>f2). The mean ± the standard deviation of the sampled mean is shown. The bottom panel shows the standard deviation of the reaction time. (After Deco and Rolls 2006.)

N, does influence the probabilistic settling of the network to the decision state. None of these probabilistic attractor and decision-related settling effects would of course be found in a mean-field or purely rate simulation, without spiking activity. The size of N in the brain is likely to be greater than 1,000 (and probably in the neocortex in the range 4,000–12,000) (see Table 1.1). With diluted connectivity, the relevant parameter is the number of connections per neuron (Amit and Brunel 1997).

It will be of interest to investigate further this scaling as a function of the number of neurons in a population with a firing rate distribution that is close to what is found in the brain, namely close to exponential (Franco, Rolls, Aggelopoulos and Jerez 2007) (see Section

1.11.3.2), as the firing rate distribution, and the sparseness of the representation, will influence the noise in the network. The results shown in Fig. 5.13 are for a sparseness of 0.1, and a binary firing rate probability distribution (i.e. one in which all the neurons in the winning attractor have the same average high firing rate, and the other neurons in the network have a low firing rate). Initial results of E.T.Rolls and T.Webb suggest that there is more noise with a biologically plausible approximately exponential firing rate distribution of the type illustrated in Fig. 1.14. The underlying cause may be the large noise contributed by the few neurons with rather high firing rates, given that for a Poisson distribution the variance increases with (and is equal to) the mean rate.

5.7 Properties of this model of decision-making

Key properties of this biased attractor model of decision-making are now described.

5.7.1 Comparison with other models of decision-making

In the attractor network model of decision-making described in this book and elsewhere (Wang 2002, Deco and Rolls 2006, Wang 2008, Deco, Rolls and Romo 2009b, Rolls, Grabenhorst and Deco 2010b, Rolls, Grabenhorst and Deco 2010c), the decisions are taken probabilistically because of the finite size noise due to spiking activity in the integrate-and-fire dynamical network, with the probability that a particular decision is made depending on the biasing inputs provided by the sensory stimuli f1 and f2.

The model described here is different in a number of ways from accumulator or counter models which may include a noise term and which undergo a random walk in real time, which is a diffusion process (Ratcliff, Zandt and McKoon 1999, Carpenter and Williams 1995) (see further Wang (2002), Wang (2008), and Usher and McClelland (2001)). First, in accumulator models, a mechanism for computing the difference between the stimuli is not described, whereas in the current model this is achieved, and scaled by f, by the feedback inhibition included in the attractor network.

Second, in the current attractor network model the decision corresponds to high firing rates in one of the attractors, and there is no arbitrary threshold that must be reached.

Third, the noise in the current model is not arbitrary, but is accounted for by finite size noise effects of the spiking dynamics of the individual neurons with their Poisson-like spike trains in a system of limited size.

Fourth, because the attractor network has recurrent connections, the way in which it settles into a final attractor state (and thus the decision process) can naturally take place over quite a long time, as information gradually and stochastically builds up due to the positive feedback in the recurrent network, the weights in the network, and the biasing inputs, as shown in Figs. 5.5 and 5.6.

Fifth, the recurrent attractor network model produces longer response times in error trials than in correct trials (Wong and Wang 2006, Rolls et al. 2010c) (see Chapter 6), consistent with experimental findings (Roitman and Shadlen 2002). Longer reaction times in error trials can be realized in the diffusion model only with the additional assumption that the starting point varies stochastically from trial to trial (Ratcliff and Rouder 1998).

Sixth,the diffusion model never reaches a steady state and predicts that performance can potentially improve indefinitely with a longer duration of stimulus processing, e.g. by raising the decision bound. In the recurrent attractor network model, ramping activity eventually stops as an attractor state is reached (Fig. 6.2). Consequently performance plateaus at sufficiently long stimulus-processing times (Wang 2002, Wang 2008).

Seventh, the attractor network model has been shown to be able to subtract negative signals as well as add positive evidence about choice alternatives, but the influence of newly arriving inputs diminishes over time, as the network converges towards one of the attractor states representing the alternative choices (Wang 2002, Wang 2008). This is consistent with experimental evidence (Wong, Huk, Shadlen and Wang 2007). This violation of time-shift invariance cannot be accounted for by the inclusion of a leak in the linear accumulator model. In fact, in contrast to the recurrent attractor network model, the linear leaky competing accumulator model, which takes into account a leakage of integration and assumes competitive inhibition between accumulators selective for choice alternatives (Usher and McClelland 2001), actually predicts that later, not earlier, signals influence more the ultimate decision, because an earlier pulse is gradually 'forgotten' due to the leak and does not affect significantly the decision that occurs much later (Wong et al. 2007).

The approach described here offers therefore a new, alternative, approach to this type of linear diffusion model, in the sense that the new attractor network model is nonlinear (due to the positive feedback in the attractor network), and is derived from and consistent with the underlying neurophysiological experimental data. The model thus differs from traditional linear diffusion models of decision-making used to account for example for human reaction time data (Luce 1986, Ratcliff and Rouder 1998, Ratcliff et al. 1999, Usher and McClelland 2001). The non-linear diffusion process that is a property of the attractor network model is analyzed further in Section A.6.1.

The model of decision-making described here is also different to a model suggested by Sugrue, Corrado and Newsome (2005) in which it is suggested that the probabilistic relative value of each action directly dictates the instantaneous probability of choosing each action on the current trial. The present model shows how probabilistic decisions could be taken depending on the two biasing inputs (λ_1 and λ_2 in Fig. 5.3, which could be equal) to a biased competition attractor network subject to statistical fluctuations related to finite size noise in the dynamics of the integrate-and-fire network.

5.7.2 Integration of evidence by the attractor network, escaping time, and reaction times

An interesting aspect of the model is that the recurrent connectivity, and the relatively long time constant of the NMDA receptors (Wang 2002), may together enable the attractor network to accumulate evidence over a long time period of several hundred milliseconds. Important aspects of the functionality of attractor networks are that they can accumulate and maintain information.

A more detailed analysis suggests that there are two scenarios that are needed to understand the time course of decision-making (Marti, Deco, Mattia, Gigante and Del Giudice 2008).

First, in the scenario investigated by Wang (2002), the spontaneous state is unstable when the decision cues are applied. The network, initially in the spontaneous state, is driven to a competition regime by an increase of the external input (that is, upon stimulus presentation) that destabilizes the initial state. The decision process can then be seen as the relaxation from an unstable stationary state towards either of the two stable decision states (see Fig. 5.14 (right)). When the system is completely symmetric, i.e. when there is no bias in the external inputs that favours one choice over the other, this destabilization occurs because the system undergoes a pitchfork bifurcation for sufficiently high inputs. The time spent by the system to evolve from the initial state to either of the two decision states is determined by the actual stochastic trajectory of the system in the phase space. In particular, the transition time increases significantly when the system wanders in the vicinity of the saddle that appears when the spontaneous state becomes unstable. Reaction times in the order of hundreds of ms may be produced in this way, and are strongly influenced by the long time constants of the NMDA receptors (Wang 2002). The transition can be further slowed down by setting the external input slightly above the bifurcation value. This tuning can be exploited to obtain realistic decision times.

Second, there is a scenario in which the stimuli do not destabilize the spontaneous state, but rather increase the probability for a noise-driven transition from a stable spontaneous state to one of the decision states (see Fig. 5.14 (centre) and Section 5.8). Due to the presence of finite-size noise in the system there is a nonzero probability that this transition occurs and hence a finite mean transition rate between the spontaneous and the decision states. It has been shown that in this scenario mean decision times tend to the Van't Hoff-Arrhenius exponential dependence on the amplitude of noise in the limit of infinitely large networks. As a consequence, in this limit, mean decision times increase exponentially with the size of the network (Marti, Deco, Mattia, Gigante and Del Giudice 2008). Further, the decision events become Poissonian in the limit of vanishing noise, leading to an exponential distribution of decision times. For small noise a decrease in the mean input to the network leads to an increase of the positive skewness of decision-time distributions.

These results suggest that noise-driven decision models as in this second scenario provide an alternative dynamical mechanism for the variability and wide range of decision times observed, which span from a few hundred milliseconds to more than one second (Marti et al. 2008). In this scenario (see Section 5.8), there is an escaping time from the spontaneous firing state. In this time the information can be thought of as accumulating in the sense that the stochastic noise may slowly drive the firing rates in a diffusion-like way such than an energy barrier is jumped over and escaped (see Fig. 2.5 on page 73 and Fig. 2.6). In this situation, a landscape can be specified by a combination of the synaptic weights and external decision-related input evidence that biases the firing rates of the decision attractors, as described in Section 2.3. The noise introduced into the network by for example the random neuronal spiking can be conceptualized as influencing the way that the system flows across this fixed landscape shaped by the synaptic weights, and by the external inputs if they remain on during operation of the network.

The model for the second scenario, with a stable spontaneous state even when the decision cues are being applied, makes specific predictions about reaction times. One, in relation to $\Delta f / f$, is shown in Fig. 5.12. Further predictions about reaction times are shown in Figs. 6.3 and 6.11, and a fuller analysis showing that there will be a gamma-like distribution with an exponential tail of long reaction times in the reaction time distribution with this second scenario is provided by Marti et al. (2008).

5.7.3 Distributed decision-making

Although the model described here is effectively a single attractor network, we note that the network need not be localized to one brain region. Long-range connections between cortical areas enable networks in different brain areas to interact in the way needed to implement a single attractor network. The requirement is that the synapses between the neurons in any one pool be set up by Hebb-like associative synaptic modification, and this is likely to be a property of connectivity between areas (using forward and backprojections, see Section 1.9), as well as within areas (Rolls and Treves 1998, Rolls and Deco 2002). In this sense, the decision could be thought of as distributed across different brain areas. Consistent with this, Romo and colleagues have found neurons related to vibrotactile decisions not only in the ventral premotor cortex (VPC), but in a number of connected brain areas including the medial premotor cortex, as described in Section 5.3.

In order to achieve the desired probabilistic settling behaviour, the network we describe must not have very high inhibition, and, related to this, may sometimes not settle into one of its attractor states. In a forced choice task in which a decision must be reached on every trial, a possible solution is to have a second decision-making network, with parameters adjusted so that it will settle into one of its states (chosen at chance) even if a preceding network in the decision-making chain has not settled. This could be an additional reason for having a series of networks in different brain regions involved in the decision-making process.

In any case, we believe that there are decision-making networks, that is, networks that can reach a categorical state, in many cortical areas, each specializing in taking a decision about the information represented in that region (see Section 8.12). In this situation, a behavioural decision may reflect the operation of a number of partly separate, and partly sequential, decision-making processes.

5.7.4 Weber's law

Deco and Rolls (2006) showed that with this attractor-based model of decision-making, the relevant parameters for the decision to be made to a criterion of a given per cent correct about whether f1 is different from f2 by the network are found not to be the absolute value of f1 or f2, but the difference between them scaled by their absolute value. If the difference between the two stimuli at which they can be discriminated $\Delta f = f1\text{-}f2$, then it is found that Δf increases linearly as a function of the base frequency f2, which is Weber's law. The results show that Weber's law does not depend on the final firing rates of neurons in the attractor, but instead reflects the nature of the probabilistic settling into a decision-related attractor, which depends on the statistical fluctuations in the network, the synaptic connectivity, and the difference between the bias input frequencies f1 and f2 scaled by the baseline input f2.

Weber's law is usually formulated as $\Delta f / (f_0 + f) = a$ constant, where f_0 allows the bottom part of the curve to asymptote at f_0. In vision, f_0 is sometimes referred to as 'dark light'. The result is that there is a part of the curve where Δf is linearly related to f, and the curve of Δf vs f need not go through the origin. This corresponds to the data shown in Fig. 5.8.

An analysis of the non-stationary evolution of the dynamics of the network model, performed by explicit full spiking simulations, shows that Weber's law is implemented in the probability of transition from the initial spontaneous firing state to one of the two possible attractor states. In this decision-making paradigm, the firing rates of neurons in the VPC encode the outcome of the comparison and therefore the decision and motor response, but not how strong the stimuli are, i.e. what Weber called 'sensation' (as described for example in a detection task by de Lafuente and Romo (2005)). The probability of obtaining a specific decision, i.e. of detecting a just noticeable difference, is encoded in the stochastic dynamics of the network. More specifically, the origin of the fluctuations that will drive the transitions towards particular decisions depends on the connectivity between the different populations, on the size of the populations, and on the Poisson-like spike trains of the individual neurons in the system. In other words, the neural code for the outcome of the decision is reflected in the high rate of one of the populations of neurons, but whether the rate of a particular population becomes high is probabilistic. This means that an essential part of how the decision process is encoded is contained in the *synapses*, in the *finite size* of the network, and in the Poisson-like firing of individual neurons in the network.

The statistical fluctuations in the network are due to the finite size noise, which approximates to the square root of the (firing rate / number of neurons in the population) (see Mattia and Del Giudice (2002)), as shown in Fig. 5.13. This is the first time we know when the implementation of a psychophysical law is not the firing rate of the neurons, nor the spike timing, nor is single neuron based, but instead is based on the synaptic connectivity of the network and on statistical fluctuations due to the spiking activity in the network.

The way in which the system settles (i.e. the probability of reaching one attractor state vs the other from the initial spontaneous state, and the time it takes) depends on factors that include the distortion of the attractor landscapes produced by the biasing inputs λ_1 and λ_2 which will influence both the shapes and the depth of the attractor basins, and the finite size noise effects. Of particular importance in relation to Weber's law is likely to be that when λ_1 and λ_2 increase, the increased firing of the neurons in the two attractors results in more activity of the inhibitory feedback neurons, which then produce effectively divisive inhibition on the principal cells of the attractor network. This is reflected in the conductance change produced by the GABA inputs to the pyramidal cells shown in Fig. 5.10. The inhibitory feedback is mainly divisive because the GABA-activated channels operate primarily as a current shunt, and do not produce much hyperpolarization, given that V_I is relatively close to the membrane potential. After the division implemented by the feedback inhibition, the differential bias required to push the network reliably into one of the attractors must then be larger, and effectively the driving force ($\lambda_1 - \lambda_2$ or $\Delta\lambda$) must get larger in proportion to the inhibition. As the inhibition is proportional to λ, this produces the result that $\Delta\lambda/\lambda$ is approximately a constant. We thus propose that Weber's law, $\Delta I/I$ is a constant, is implemented in part by shunting effects acting on pyramidal cells that are produced by inhibitory neuron inputs which increase linearly as the baseline input I increases, so that the difference of intensities ΔI required to push the network reliably into one of its attractors must increase in proportion to the base input I. Another part of the mechanism appears to be the noise-driven stochastic process of jumping out of a stable state of spontaneous firing in what we describe as a multistable operating regime, as described in Section 5.8.

We emphasize that this account (Deco and Rolls 2006) of Weber's law is intended to be a general account, and is not restricted to the particular data set or brain system against which the development of the model described here was validated.

A prediction of the model is that Weber's law for frequency discrimination could be implemented not by the firing rate of a given population of neurons (which reflects just the discrete decision taken), but by the probability that a particular population will be activated, which depends on $\Delta f / f$. This prediction could be tested by a trial-by-trial analysis of the neurophysiological data in which the firing of neurons at different base frequencies f and for different Δf is measured, to investigate whether the type of result shown in Fig. 5.7, and thereby in Fig. 5.8, which are derived from the model, are also found in the neuronal data from the experiments, and this would also usefully confirm that Weber's law holds at the neuronal level in this particular vibrotactile task with a delay. In particular, although Romo, Hernandez and Zainos (2004) in their figure 5 show that choice probability and neuronal activity increases as a function of f2-f1, we predict that neurons should follow the functions shown in Figs. 5.7 and 5.8 for different values of Δf and f, and it would be of interest to test this prediction.

We note that Weber's law holds in most though not all discrimination situations, and to the extent that Weber's law does generally hold, the model described here provides a computational neuroscience-based account of how it arises. This is the first time we know when the implementation of a psychophysical law is not the firing rate of the neurons, nor the spike timing, nor is single neuron based, but instead is based on the synaptic connectivity of the network and on statistical fluctuations due to the spiking activity in the network.

5.7.5 Unifying principles

In summary, we now have a model for how attractor networks could operate by biased competition to implement probabilistic decision-making. This type of model could operate in very many brain areas, as described in Section 8.12. This model of decision-making is part of the larger conceptual issue of how memory networks retrieve information. In this case the short-term memory property of the attractor network helps the network to integrate information over time to reach the decision. Although retrieval of information from attractor networks has been intensively studied using *inter alia* approaches from theoretical physics (see Section 1.10, Amit (1989), and Rolls and Treves (1998)), the way in which the retrieval is probabilistic when an integrate-and-fire implementation with spiking neurons is considered has been less studied. The approaches of Brunel and Wang (2001) and Deco and Rolls (2006) do open up the issue of the probabilistic operation of memory networks with spiking dynamics.

We may raise the conceptually important issue of why the operation of what is effectively memory retrieval is probabilistic. Part of the answer is shown in Fig. 5.13, in which it is seen that even when a fully connected recurrent attractor network has 4,000 neurons, the operation of the network is still probabilistic. In this network of 4,000 neurons, there were 3,600 excitatory neurons and 360 neurons represented each pattern or decision (that is, the sparseness a was 0.1). The firing rates of the neurons corresponded to those found in VPC, with rates above 20 spikes/s considered to be a criterion for the attractor state being reached, and 40–50 spikes/s being typical when fully in the attractor state (see Fig. 5.9). Under these conditions, the probabilistic spiking of the excitatory (pyramidal) cells is what introduces noise into the network. (Deco and Rolls (2006) showed that it is this noise in the recurrent collateral firing, rather than external noise due to variability in the inputs, which makes the major contribution to the probabilistic behaviour of the network.) Thus, once the firing in the recurrent collaterals is spike implemented by integrate-and-fire neurons, the probabilistic behaviour seems inevitable, even up to quite large attractor network sizes.

To investigate the 'decision-related' switching of these systems, it may be important to use a firing rate distribution of the type found in the brain, in which few neurons have high rates, more neurons have intermediate rates, and many neurons have low rates (see Section 1.11.3). It is also important to model correctly the proportion of the current that is being passed through the NMDA receptors (which are voltage-dependent), as these receptors have a long time-constant, which will tend to smooth out short-term statistical fluctuations caused by the stochastic firing of the neurons (cf. Wang (1999)), and this will affect the statistics of the probabilistic switching of the network. This can only be done by modelling integrate-and-fire networks with the firing rates and the firing rate distributions found in a cortical area.

Reasons why the brain is inherently noisy are described in Section 2.4.

Applications of this model of decision-making are described in Chapter 8.

5.8 A multistable system with noise

In the situation illustrated in Figs. 2.5 and 5.1, there is multistability, in that the spontaneous state and a large number of high firing rate persistent states are stable. More generally, and depending on the network parameters including the strengths of the inputs, a number of different scenarios can occur. These are illustrated in Fig. 5.14. Let us consider the activity of a given neuronal population while inputs are being applied.

In Fig. 5.14 (left) we see a situation in which only the spontaneous state S is stable. This might occur if the external inputs λ_1 and λ_2 are weak.

On the right we have a situation in which our neuronal population is either in a high firing rate stable state C2, or in a low firing rate state C1 because another population is firing fast and inhibiting our neuronal population. There is no stable spontaneous state.

Spontaneous
stable

Multistable

Bistable

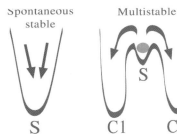

Fig. 5.14 Computational principles underlying the different dynamical regimes of the decision-making attractor network (see text). The x-axis represents the neuronal activity of one of the populations (ν_i) and the landscape represents an energy landscape regulating the evolution of the system. S is a stable state of spontaneous activity, C2 is a high firing rate state of this neuronal population corresponding to the choice implemented by this population, and C1 is a low firing rate state present when the other population wins the competition.

In the middle of Fig. 5.14 we see a situation in which our population may be either in C1, or in C2, or in a spontaneous state of firing S when no population has won the competition. We emphasize that this can be a scenario even when the decision cues λ_1 and λ_2 are being applied during the decision-making period. We refer to this system as a multistable system.

The differences between these scenarios are of interest in relation to how noise influences the decision-making. In the scenario shown in the middle of Fig. 5.14 we see that there are three stable states when the inputs λ_1 and λ_2 are being applied, and that it is the stochastic noise that influences whether the system jumps from the initial spontaneous state to a high firing rate state in which one of the decision-state populations fires fast, producing either C2 if our population wins, or C1 if our population loses. The statistical properties of the noise (including its amplitude and frequency spectrum), and the shape of the different basins in the energy landscape, influence whether a decision will be taken, the time when it will be taken, and which high firing rate decision attractor wins. In contrast, in the scenario shown in Fig. 5.14 (right) the energy landscape when the stimuli are being applied is such that there is no stable spontaneous state, so the system moves to one of the high firing rate decision attractors without requiring noise. In this case, the noise, and the shape of the energy landscape, influence which high firing rate decision state attractor will win.

In an exploration of the neural network described in this Chapter that models vibrotactile flutter frequency discrimination in humans, Deco, Scarano and Soto-Faraco (2007b) found that the neurodynamical mechanisms and computational principles underlying the decision-making processes in this perceptual discrimination task are consistent with a fluctuation - driven scenario in a multistable regime of the type illustrated in Fig. 5.14 (middle) as incorporated in the model of Deco and Rolls (2006). Deco et al. (2007b) used a mean-field analysis with a system of nonlinear coupled differential equations of the Wilson-Cowan type (Renart et al. 2003, La Camera et al. 2004) to describe the evolution of the average firing rate of each population, and added a noise fluctuation term to drive the transitions.

Fig. 5.14 represents schematically the different ways in which the network could operate, depending on the value of λ, the input, as shown by a mean-field analysis. The x-axis represents the neuronal activity of one of the populations (ν_i) and the landscape represents an energy landscape regulating the evolution of the system. The energy landscape reflects the synaptic values and the effects of the incoming sensory information (the λ_1 and λ_2 values in Fig. 5.1). For values of $\lambda < \lambda_{c1}$=20 Hz (Fig. 5.14, left panel), only the spontaneous state is stable, and no decision states appear (for the unbiased case, i.e. when $\lambda_1 = \lambda_2$). For increasing $\Delta\lambda$ (biased case), one decision state (corresponding to the choice where the increased value of $\lambda + \Delta\lambda$ is applied) emerges, attracting the dynamics towards this decision state. For values of λ between λ_{c1}=20, λ_{c2}=40 Hz (Fig. 5.14, middle panel) there is a region of multi-stability, with the spontaneous state and each of the two decision states all stable. In this λ interval, the fluctuations are responsible for driving the system from the initial stable spontaneous state to one of the two decision states corresponding to the two possible response choices. (C2 is a high firing rate of the population shown in Fig. 5.14 middle that corresponds to the decision implemented by that population of neurons.) Thus, in this scenario, fluctuations play a crucial role in the computation of decision-making. It is only in this region that Weber's law is found to apply to the network, as described in this Chapter. For values of $\lambda > \lambda_{c2}$=40 Hz (Fig. 5.14, right panel) a region of bistability is found where the initial spontaneous state is unstable, and only the two decision states are stable. In this regime, the spontaneous state destabilizes, so that the dynamics rapidly evolves towards one of the two decision states, resembling therefore a pure diffusion integrating the relative evidence for one choice over another. The fact that the network can implement Weber's law, and does so only in a range of values for λ in which the network operates as a fluctuation-driven multistable system, provides further evidence to support the hypothesis that decision-making is implemented by a multistable fluctuation-driven attractor system, where there are in the unbiased case stable states for the spontaneous firing state, and for each of the decisions.

With respect to the multistable scenario (in the middle of Fig. 5.14), the attractor network acts as a short-term memory that can accumulate evidence over time, and usually gradually though stochastically the firing rates of the two groups of neurons corresponding to the two choices diverge, one set of neurons stochastically increasing their firing rate, and the other set being inhibited by the first set of neurons. A situation like this is probably occurring in Fig. 5.5, and is analyzed in more detail by Marti et al. (2008). We can illustrate this point by the middle landscape in Fig. 5.14, and note that the accumulation of evidence corresponds to the position in the space indicated by the ball moving noisily in a direction towards for example C2, but not yet jumping over the energy barrier into the C2 attractor.

It is important to stress that the effect of the noise is particularly relevant in the multistable regime (middle of Fig. 5.14), because the fluctuations are the driving force that allow the system to escape the decision barriers around the stable spontaneous state. In the multistable scenario, the choices are associated with stable attractors, and the starting condition is also given by a stable spontaneous state. To make a decision, the system has to escape the stable spontaneous state towards one of the choice-attractors. (This is related to the so called 'Kramers' escape problem' (Kramers 1940)). On the other hand, in the bistable regime (right of Fig. 5.14), the so called 'ballistic' regime), the noise is of course relevant as the basis of the diffusion process, but it is not the main driving force. This is because in the bistable scenario the spontaneous state is not a stable state, and therefore with or without noise, the system will necessarily evolve to one or the other decision-attractor just because of the neurodynamical flow (Deco and Marti 2007a, Deco and Marti 2007b).

Another interesting aspect of the model is that the recurrent connectivity, and the relatively long time constant of the NMDA receptors (Wang 2002), together enable the attractor network to accumulate evidence over a long time period of several hundred milliseconds. This is an important aspect of the functionality of attractor networks. Nevertheless, long reaction times can also be obtained without NMDA receptors, using the alternative scenario of multistability (Marti, Deco, Mattia, Gigante and Del Giudice 2008). In this case, the level of noise is the main variable that drives the escape from the stable spontaneous firing rate state, and low levels of noise can produce long reaction times.

6   Confidence and decision-making

In this Chapter we consider how probabilistic decision-making is influenced by the easiness vs the difficulty of the decision, and how confidence in a decision emerges as a property of the neuronal attractor network decision-making process that we describe in Chapter 5. Indeed, in that Chapter we showed how the difficulty of the decision, set quantitatively by the magnitude of the difference between the stimuli ΔI, influences the decision-making network. Here we extend this approach considerably. A link from discriminability to confidence can be made, for it is well established that subjective decision confidence increases with discriminability, ΔI (Vickers 1979, Vickers and Packer 1982, Jonsson, Olsson and Olsson 2005). Consistently, in rats too the probability that a trial will be aborted reflecting low decision confidence also increases with ΔI on error trials (Kepecs, Uchida, Zariwala and Mainen 2008).

Confidence in a decision is higher after a correct decision than after an incorrect decision has been made (Vickers and Packer 1982), and we also show how this emerges as a property of the neuronal attractor network decision-making process. In more detail, if one makes a correct decision, consistent with the evidence, then one's confidence is higher than when one makes an incorrect decision (as shown by confidence ratings) (Vickers and Packer 1982), and consistent with this the probability that a rat will abort a trial is higher if a decision just made is incorrect (Kepecs et al. 2008). We also show (in Section 6.8) how being correct vs making an error influences, as different functions of ΔI, the neuronal activity involved in making the choice, and also the fMRI BOLD (functional magnetic resonance imaging blood oxygenation level) signals that are associated with choice decision-making. This leads to the conclusion that decision confidence is an *emergent property* of the 'mechanistic' (i.e. biologically plausible at the integrate-and-fire level) model of decision-making described. We then show how fMRI experiments can be used to test the predictions from the model.

We then show how if a decision must be made based on one's confidence about a decision just made, a second decision-making network can read the information encoded in the firing rates of the first decision-making network to make a decision based on confidence (Section 6.9).

Noise in the brain generated by the almost random (Poisson) spike timings on neurons (for a given mean rate) influences all these decision-related processes, in ways that we elucidate (Rolls, Grabenhorst and Deco 2010b, Rolls, Grabenhorst and Deco 2010c, Insabato, Pannunzi, Rolls and Deco 2010).

A fundamental issue in understanding how the brain takes decisions is to identify a neural signature for decision-making processes and to explain how this signature arises as a result of the operation of dynamic cortical neural networks. Evidence from single neuron recordings in monkeys (Kim and Shadlen 1999) shows that neuronal responses in a motion decision-making task occur earlier on easy vs difficult trials in a decision-related brain region, the dorsolateral prefrontal cortex. In the human dorsolateral prefrontal cortex, higher fMRI BOLD signals can be observed on easy trials vs difficult trials in a similar decision-making task (Heekeren, Marrett, Bandettini and Ungerleider 2004). On the basis of these findings it has been suggested (Heekeren et al. 2004) that the dorsolateral prefrontal cortex implements a general mechanism for decision-making in the brain which is thought to involve the comparison of sensory input signals until a decision criterion is reached and a choice is made. Thus a neural signature of decision-making seems to involve higher levels of activity when an easy compared to a difficult decision is taken. However, the fact that some areas of the brain thought to be involved in decision-making show earlier or larger neuronal responses on easy trials (Kim and Shadlen 1999) and reflect decision confidence (Kiani and Shadlen 2009) does not by itself provide a theoretical understanding about what is special about this effect as an indicator of decision-making, or lead to an understanding of how confidence is computed.

To address these issues, we can capitalize on the recent advances in theoretical understanding of how choice decisions are made using attractor networks that make probabilistic decisions because of the noise in the brain to which an important contributor is the random spiking times of the neurons (Wang 2002, Deco and Rolls 2006, Wang 2008, Deco, Rolls and Romo 2009b) (see Chapter 5). We simulated integrate-and-fire models of attractor-based choice decision-making, and predicted from them the fMRI BOLD signals on easy vs difficult trials (Rolls, Grabenhorst and Deco 2010b). We then described two fMRI investigations of decision-making about the reward value and subjective pleasantness of thermal and olfactory stimuli, and showed that areas implicated by other analyses of the same datasets in decision-making (Rolls and Grabenhorst 2008, Grabenhorst, Rolls and Parris 2008, Rolls, Grabenhorst and Parris 2009c) do show the predicted difference between activations on easy vs difficult trials (Rolls, Grabenhorst and Deco 2010b), found also by other investigators (Heekeren et al. 2004, Heekeren, Marrett and Ungerleider 2008).

The model also makes detailed predictions about the neuronal activity that underlies the decision-making. This provides a fundamental and unifying approach to decision-making in the brain that specifies how probabilistic decisions influenced by neuronal noise are taken, and how activity at the synaptic, neuronal, network, neuroimaging, behavioural choice, and subjective confidence levels of investigation are related to the decision-making process so that further predictions can be tested.

6.1 The model of decision-making

The theoretical framework of the model used here (Rolls, Grabenhorst and Deco 2010b, Rolls, Grabenhorst and Deco 2010c) was introduced by Wang (2002) and developed further (Deco and Rolls 2006, Deco, Scarano and Soto-Faraco 2007b, Marti, Deco, Mattia, Gigante and Del Giudice 2008, Deco, Rolls and Romo 2009b). In this framework, we model probabilistic decision-making by a network of interacting neurons organized into a discrete set of populations, as depicted in Fig. 6.1.

The model simulated for the investigations described in this Chapter was similar to that described in Section 5.4, and the description here focusses on the differences. The network contains N_E (excitatory) pyramidal cells and N_I inhibitory interneurons. In the simulations, we used N_E=400 and N_I=100, consistent with the neurophysiologically observed proportion of 80% pyramidal cells versus 20% interneurons (Abeles 1991, Rolls and Deco 2002). The neurons are fully connected (with synaptic strengths as specified below). In the model, the specific populations D1 (for decision 1) and D2 encode the categorical result of the choice between the two stimuli that activate each of these populations. Each specific population of excitatory cells contains rN_E neurons (in our simulations r=0.1). In addition there is one non-specific population, named 'Non-specific', which groups all other excitatory neurons in the modelled brain area not involved in the present tasks, and one inhibitory population, named 'Inhibitory', grouping the local inhibitory neurons in the modelled brain area. The latter population regulates the overall activity and implements competition in the network by spreading a global inhibition signal using GABA synapses.

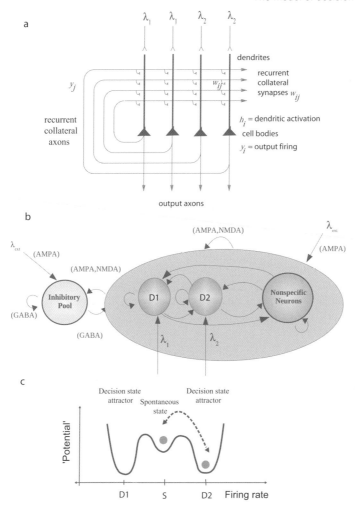

Fig. 6.1 (a) Attractor or autoassociation single network architecture for decision-making. The evidence for decision 1 is applied via the λ_1 inputs, and for decision 2 via the λ_2 inputs. The synaptic weights w_{ij} have been associatively modified during training in the presence of λ_1 and at a different time of λ_2. When λ_1 and λ_2 are applied, each attractor competes through the inhibitory interneurons (not shown), until one wins the competition, and the network falls into one of the high firing rate attractors that represents the decision. The noise in the network caused by the random spiking times of the neurons (for a given mean rate) means that on some trials, for given inputs, the neurons in the decision 1 (D1) attractor are more likely to win, and on other trials the neurons in the decision 2 (D2) attractor are more likely to win. This makes the decision-making probabilistic, for, as shown in (c), the noise influences when the system will jump out of the spontaneous firing stable (low energy) state S, and whether it jumps into the high firing state for decision 1 (D1) or decision 2 (D2). (b) The architecture of the integrate-and-fire network used to model decision-making (see text). (c) A multistable 'effective energy landscape' for decision-making with stable states shown as low 'potential' basins. Even when the inputs are being applied to the network, the spontaneous firing rate state is stable, and noise provokes transitions from the low firing rate spontaneous state S into the high firing rate decision attractor state D1 or D2.

We modify the conductance values for the synapses between pairs of neurons by synaptic connection weights, which can deviate from their default value 1. The structure and function of the network are achieved by differentially setting the weights within and between populations of neurons. The labelling of the weights is defined in Fig. 5.3. We assume that the connections are already formed, by for example earlier self-organization mechanisms, as if they were established by Hebbian learning, i.e. the coupling will be strong if the pair of neurons have correlated activity (i.e. covarying firing rates), and weak if they are activated in an uncorrelated way. As a consequence of this, neurons within a specific excitatory population (D1 and D2) are mutually coupled with a strong synaptic weight w_+, set to 2.1 for the simulations described here. Furthermore, the populations encoding these two decisions are likely to have anti-correlated activity in this behavioural context, resulting in weaker than average connections between the two different populations. Consequently, we choose a weaker value $w_- = 1 - r(w_+ - 1)/(1 - r)$, so that the overall recurrent excitatory synaptic drive in the spontaneous state remains constant as w_+ is varied (Brunel and Wang 2001). Neurons in the inhibitory population are mutually connected with an intermediate weight $w = 1$. They are also connected with all excitatory neurons in the same layer with the same intermediate weight, which for excitatory-to-inhibitory connections is $w = 1$, and for inhibitory-to-excitatory connections is denoted by a weight w_I. Neurons in a specific excitatory population are connected to neurons in the nonselective population in the same layer with a feedforward synaptic weight $w = 1$ and a feedback synaptic connection of weight w_-.

Each individual population is driven by two different kinds of input. First, all neurons in the model network receive spontaneous background activity from outside the module through $N_{ext} = 800$ external excitatory connections. Each synaptic connection carries a Poisson spike train at a spontaneous rate of 3 Hz, which is a typical spontaneous firing rate value observed in the cerebral cortex. This results in a background external input summed over all 800 synapses of 2400 Hz for each neuron. Second, the neurons in the two specific populations additionally receive added firing to the external inputs that encodes the evidence for the decision to be made. When stimulating, the rate of the Poisson train to the neurons of the specific population D1 is increased by an extra value of λ_1, and to population D2 by λ_2, as these encode the two stimuli to be compared. The simulations were run for 2 s of spontaneous activity, and then for a further 2 s while the stimuli were being applied.

During the spontaneous period, the stimuli applied to D1 and D2 (and to all the other neurons in the network) had a value of 3 Hz. (This 3 Hz is the firing rate being applied by Poisson spikes to all 800 external synaptic inputs of each neuron in the network, so the total synaptic bombardment on each neuron is 2400 spikes/s.) During the decision period, the mean input to each external synapse of each neuron of D1 and D2 was increased to 3.04 Hz per synapse (an extra 32 Hz per neuron). For $\Delta I = 0$, 32 extra Hz to the spontaneous was applied to each neuron of both D1 and D2. For $\Delta I = 16$, 32 + 8 Hz was the extra applied to D1 and corresponds to λ_1 in Fig. 6.1, and 32 − 8 Hz was the extra applied to D2, etc. The firing rates, and the absolute value of the sum of the synaptic currents (AMPA, NMDA, and GABA, defined in Section 2.2.3), in all four populations of neurons were saved every 50 ms for later analysis (Rolls, Grabenhorst and Deco 2010b, Rolls, Grabenhorst and Deco 2010c).

The criterion for which population won, that is for which decision was taken, was a mean rate for the last second of the simulation that was 10 Hz greater than that of the other population. (This is in the context that the spontaneous rate was typically close to 3 spikes/s, and that the winning population typically had a mean firing rate of 35–40 Hz, as will be shown.)

The latency of a decision was measured by the time of the first 50 ms bin of three consecutive bins at which the mean rate of one population was more than 25 Hz higher than that of the other population.

The parameters for the synaptic weights and input currents were chosen using the mean-field equivalent of this network (Brunel and Wang 2001, Deco and Rolls 2006) so that in the absence of noise when the input stimuli are being applied there were three stable states, the spontaneous firing rate state (with a mean firing for the pyramidal cells of approximately 3 spikes/s), and two high firing rate attractor states (with a mean firing for the pyramidal cells of approximately 40 spikes/s), with one neuronal population (D1) representing decision 1, and the other population (D2) decision 2. In particular, w_+ was set to 2.1. We describe this as a multistable system, and show its 'effective energy landscape' in Fig. 6.1 and discuss it in Section 2.3.

This multistable regime is one with many interesting decision-making properties, including implementing Weber's law $\Delta I/ (I_0 + I) = $ a constant, where ΔI is the just noticeable difference in the intensity I (and where I_0 allows the bottom part of the curve to asymptote at I_0, i.e. the change of intensity I that can be reliably detected divided by the intensity value has a linear component) (Deco and Rolls 2006, Deco, Rolls and Romo 2009b). With operation in this regime, it is the approximately random spiking times of the neurons (i.e. approximately Poisson firing at a given mean rate) that causes statistical fluctuations that makes the network jump from the spontaneous firing state into one of the high firing rate attractor (decision) states. The randomness of the firing dynamically and probabilistically provokes transitions from the spontaneous firing state to one of the high firing rate attractor basins that represent a decision (see Fig. 6.1c). The stability of an attractor is characterized by the average time in which the system stays in the basin of attraction under the influence of noise. The noise provokes transitions to other attractor states. It results from the interplay between the Poissonian character of the spikes and the finite-size effect due to the limited numbers of neurons in the network.

The parameters were also chosen so that if after a decision had been reached and the network had fallen into the D1 or D2 attractor state the decision-related inputs λ_1 and λ_2 were removed, the attractor would remain stable, and maintain its decision-related high firing rate in the short-term memory implemented by the attractor network. *This continuing firing and short-term memory is an extremely important property and advantage of this type of decision-making network, for it enables the decision state to be maintained for what could be a delay before and while an action is being implemented to effect the decision taken.* No further, separate, special-purpose, memory of the decision is needed, and the memory of a recent decision state emerges as a property of the decision-related mechanism. This is an important way in which the approach described in this book provides a unifying approach to understanding many properties of neocortical function, implemented by its special type of connectivity, short-range excitatory recurrent collateral synapses (see further Section 8.18). The continuing short-term memory-related firing of the choice decision just taken also very usefully retains the information needed for any subsequent decision that might be taken about the first decision, such as whether the change the first decision, and perhaps abort or reverse any action taken as a result of the first decision (see Section 6.9).

6.2 Neuronal responses on difficult vs easy trials, and decision confidence

Fig. 6.2a and e show the mean firing rates of the two neuronal populations D1 and D2 for two trial types, easy trials (ΔI=160 Hz) and difficult trials (ΔI=0) (where ΔI is the difference in spikes/s summed across all synapses to each neuron between the two inputs, λ_1 to population D1, and λ_2 to population D2). The results are shown for correct trials, that is, trials on which the D1 population won the competition and fired with a rate of > 10 spikes/s more than the rate of D2 for the last 1000 ms of the simulation runs. Figure 6.2b shows the mean firing rates of the four populations of neurons on a difficult trial, and Fig. 6.2c shows the rastergrams for the same trial, for which the energy landscape is also shown in Fig. 6.1d. Figure 6.2d shows the firing rates on another difficult trial (ΔI=0) to illustrate the variability shown from trial

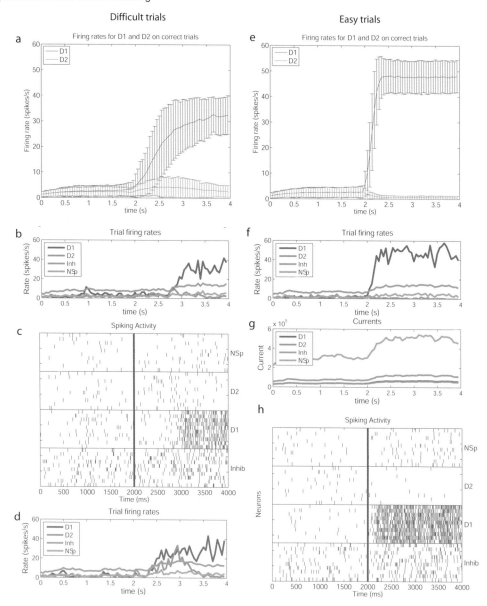

Fig. 6.2 (a) and (e) Firing rates (mean ± sd) for difficult (ΔI=0) and easy (ΔI=160) trials. The period 0–2 s is the spontaneous firing, and the decision cues were turned on at time = 2 s. The mean was calculated over 1000 trials. D1: firing rate of the D1 population of neurons on correct trials on which the D1 population won. D2: firing rate of the D2 population of neurons on the correct trials on which the D1 population won. A correct trial was one in which in which the mean rate of the D1 attractor averaged > 10 spikes/s for the last 1000 ms of the simulation runs. (Given the attractor nature of the network and the parameters used, the network reached one of the attractors on >90% of the 1000 trials, and this criterion clearly separated these trials, as indicated by the mean rates and standard deviations for the last 1 s of the simulation as shown.) (b) The mean firing rates of the four populations of neurons on a difficult trial. Inh is the inhibitory population that uses GABA as a transmitter. NSp is the non-specific population of neurons (see Fig. 6.1). (c) Rastergrams for the trial shown in b. 10 neurons from each of the four pools of neurons are shown. (d) The firing rates on another difficult trial (ΔI=0) showing prolonged competition between the D1 and D2 attractors until the D1 attractor finally wins after approximately 1100 ms. (f) Firing rate plots for the 4 neuronal populations on a single easy trial (ΔI=160). (g) The synaptic currents in the four neuronal populations on the trial shown in f. (h) Rastergrams for the easy trial shown in f and g. 10 neurons from each of the four pools of neurons are shown. (After Rolls, Grabenhorst and Deco 2010b.) (See colour plates Appendix B.)

to trial, with on this trial prolonged competition between the D1 and D2 attractors until the D1 attractor finally won after approximately 1100 ms. Figure 6.2f shows firing rate plots for the four neuronal populations on an example of a single easy trial (ΔI=160), Fig. 6.2g shows the synaptic currents in the four neuronal populations on the same trial, and Fig. 6.2h shows rastergrams for the same trial.

Three important points are made by the results shown in Fig. 6.2. First, the network falls into its decision attractor faster on easy trials than on difficult trials. We would accordingly expect reaction times to be shorter on easy than on difficult trials. We might also expect the BOLD signal related to the activity of the network to be higher on easy than on difficult trials because it starts sooner on easy trials.

Second, the mean firing rate after the network has settled into the correct decision attractor is higher on easy than on difficult trials. We might therefore expect the BOLD signal related to the activity of the network to be higher on easy than on difficult trials because the maintained activity in the attractor is higher on easy trials. This shows that the exact firing rate in the attractor is a result not only of the internal recurrent collateral effect, but also of the external input to the neurons, which in Fig. 6.2a is 32 Hz to each neuron (summed across all synapses) of D1 and D2, but in Fig. 6.2e is increased by a further 80 Hz to D1, and decreased (from the 32 Hz added) by 80 Hz to D2 (i.e. the total external input to the network is the same, but ΔI=0 for Fig. 6.2a, and ΔI=160 for Fig. 6.2b).

Third, the variability of the firing rate is high, with the standard deviations of the mean firing rate calculated in 50 ms epochs indicated in order to quantify the variability. The large standard deviations on difficult trials for the first second after the decision cues are applied at t=2 s reflects the fact that on some trials the network has entered an attractor state after 1000 ms, but on other trials it has not yet reached the attractor, although it does so later. This trial by trial variability is indicated by the firing rates on individual trials and the rastergrams in the lower part of Fig. 6.2. The effects evident in Fig. 6.2 are quantified, and elucidated over a range of values for ΔI, next.

Fig. 6.3a shows the firing rates (mean $\pm$ sd) on correct trials when in the D1 attractor as a function of ΔI. ΔI=0 corresponds to the most difficult decision, and ΔI=160 corresponds to easy. The firing rates for both the winning population D1 and for the losing population D2 are shown. The firing rates were measured in the last 1 s of firing, i.e. between t=3 and t=4 s. It is clear that the mean firing rate of the winning population increases monotonically as ΔI increases, and interestingly, the increase is approximately linear (Pearson $r = 0.995$, p<10^{-6}). The higher mean firing rates as ΔI increases are due not only to higher peak firing, but also to the fact that the variability becomes less as ΔI increases ($r = -0.95$, p<10^{-4}), reflecting the fact that the system is more noisy and unstable with low ΔI, whereas the firing rate in the attractor is maintained more stably with smaller statistical fluctuations against the Poisson effects of the random spike timings at high ΔI. (The measure of variation indicated in the figure is the standard deviation, and this is shown here unless otherwise stated to quantify the degree of variation, which is a fundamental aspect of the operation of these neuronal decision-making networks.)

As shown in Fig. 6.3a, the firing rates of the losing population decrease as ΔI increases. The decrease of firing rate of the losing population is due in part to feedback inhibition through the inhibitory neurons by the winning population. Thus the difference of firing rates between the winning and losing populations, as well as the firing rate of the winning population D1, both clearly reflect ΔI, and in a sense the confidence in the decision.

The increase of the firing rate when in the D1 attractor (upper thick line) as ΔI increases thus can be related to the confidence in the decision, and, as will be shown next in Fig. 6.3b, the performance as shown by the percentage of correct choices. The firing rate of the losing attractor (D2, lower thick line) decreases as ΔI increases, due to feedback inhibition from

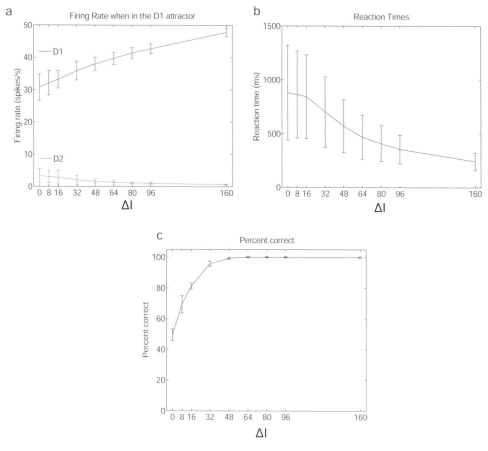

Fig. 6.3 (a) Firing rates (mean $\pm$ sd) on correct trials when in the D1 attractor as a function of ΔI. ΔI=0 corresponds to difficult, and ΔI=160 spikes/s corresponds to easy. The firing rates for both the winning population D1 and for the losing population D2 are shown for correct trials by thick lines. All the results are for 1000 simulation trials for each parameter value, and all the results shown are statistically highly significant. (b) Reaction times (mean $\pm$ sd) for the D1 population to win on correct trials as a function of the difference in inputs ΔI to D1 and D2. (c) Per cent correct performance, i.e. the percentage of trials on which the D1 population won, as a function of the difference in inputs ΔI to D1 and D2. The mean was calculated over 1000 trials, and the standard deviation was estimated by the variation in 10 groups each of 100 trials. (After Rolls, Grabenhorst and Deco, 2010b.)

the winning D1 attractor, and thus the difference in the firing rates of the two attractors also reflects well the decision confidence.

We emphasize from these findings (Rolls, Grabenhorst and Deco 2010b) that the firing rate of the winning attractor reflects ΔI, and thus the confidence in the decision which is closely related to ΔI.

6.3 Reaction times of the neuronal responses

The time for the network to reach the correct D1 attractor, i.e. the reaction time of the network, is shown as a function of ΔI in Fig. 6.3b (mean $\pm$ sd). Interestingly, the reaction time continues to decrease ($r = -0.95$, $p<10^{-4}$) over a wide range of ΔI, even when as shown in Fig. 6.3c the network is starting to perform at 100% correct. The decreasing reaction time as ΔI increases is attributable to the altered 'effective energy landscape' (see Section 2.3): a larger input to D1 tends to produce occasionally higher firing rates, and these statistically are more likely to induce a significant depression in the landscape towards which the network flows sooner than with low ΔI. Correspondingly, the variability (quantified by the standard deviation) of the reaction times is greatest at low ΔI, and decreases as ΔI increases ($r = -0.95$, $p<10^{-4}$). This variability would not be found with a deterministic system (i.e. the standard deviations would be 0 throughout, and such systems include those investigated with mean-field analyses), and is entirely due to the random statistical fluctuations caused by the random spiking of the neurons in the integrate-and-fire network.

6.4 Percentage correct

At $\Delta I=0$, there is no influence on the network to fall more into attractor D1 representing decision 1 than attractor D2 representing decision 2, and its decisions are at chance, with approximately 50% of decisions being for D1. As ΔI increases, the proportion of trials on which D1 is reached increases. The relation between ΔI and percentage correct is shown in Fig. 6.3c. Interestingly, the performance becomes 100% correct with $\Delta I=64$, whereas as shown in Figs. 6.3a and b the firing rates while in the D1 attractor (and therefore potentially the BOLD signal), continue to increase as ΔI increases further, and the reaction times continue to decrease as ΔI increases further. It is a clear prediction for neurophysiological and behavioural measures that the firing rates with decisions made by this attractor process continue to increase as ΔI is increased beyond the level for very good performance as indicated by the percentage of correct decisions, and the neuronal and behavioural reaction times continue to decrease as ΔI is increased beyond the level for very good performance. Fig. 6.3c also shows that the variability in the percentage correct (in this case measured over blocks of 100 trials) is large with $\Delta I=0$, and decreases as ΔI increases. This in consistent with unbiased effects of the noise producing very variable effects in the energy landscape at $\Delta I=0$, but in the external inputs biasing the energy landscape more and more as ΔI increases, so that the flow is much more likely to be towards the D1 attractor.

6.5 Simulation of fMRI signals: haemodynamic convolution of synaptic activity

The links between neural and synaptic activity, and functional magnetic resonance neuroimaging (fMRI) measurements, are still not fully understood. The fMRI signal is unfortunately strongly filtered and perturbed by the haemodynamic delay inherent in the blood oxygen level-dependent (BOLD) contrast mechanism (Buxton and Frank 1997). The fMRI signal is only a secondary consequence of neuronal activity, and yields therefore a blurred distortion of the temporal development of the underlying brain processes. Regionally, increased oxidative metabolism causes a transient decrease in oxyhaemoglobin and increase in deoxyhaemoglobin, as well as an increase in CO_2 and NO. This provokes over several seconds a local dilatation and increased blood flow in the affected regions that leads by overcompensation to a relative decrease in the concentration of deoxyhaemoglobin in the venules draining the activated region, and the alteration of deoxyhaemoglobin, which is paramagnetic, can be detected by changes in T2 or T2* in the MRI signal as a result of the decreased susceptibility and thus decreased local inhomogeneity which increases the MR intensity value (Glover 1999, Buxton and Frank 1997, Buxton, Wong and Frank 1998).

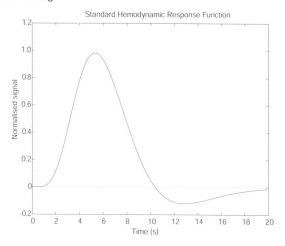

Fig. 6.4 The standard haemodynamic response function $h(t)$ (see text).

The fMRI BOLD (blood oxygen level-dependent) signal may reflect the total synaptic activity in an area (as ions need to be pumped back across the cell membrane) (perhaps, it has been suggested, better than the spiking neuronal activity (Logothetis et al. 2001)), and is spatially and temporally filtered. The filtering reflects the inherent spatial resolution with which the blood flow changes, as well as the resolution of the scanner, and spatial filtering which may be applied for statistical purposes, and the slow temporal response of the blood flow changes (Glover 1999, Buxton and Frank 1997, Buxton et al. 1998). Glover (1999) demonstrated that a good fit of the haemodynamical response $h(t)$ can be achieved by the following analytic function:

$$h(t) = c_1 t^{n_1} e^{-\frac{t}{t_1}} - a_2 c_2 t^{n_2} e^{-\frac{t}{t_2}}$$

$$c_i = \max(t^{n_i} e^{-\frac{t}{t_i}})$$

where t is the time, and c_1, c_2, a_2, n_1, and n_2 are parameters that are adjusted to fit the experimentally measured haemodynamical response. Figure 6.4 plots the haemodynamic standard response $h(t)$ for a biologically realistic set of parameters (see Deco, Rolls and Horwitz (2004)).

The temporal evolution of fMRI signals can be simulated from an integrate-and-fire population of neurons by convolving the total synaptic activity in the simulated population of neurons with the standard haemodynamic response formulation of Glover (1999) presented earlier (Deco, Rolls and Horwitz 2004, Horwitz and Tagamets 1999). The rationale for this is that the major metabolic expenditure in neural activity is the energy required to pump back against the electrochemical gradient the ions that have entered neurons as a result of the ion channels opened by synaptic activity, and that mechanisms have evolved to increase local blood flow in order to help this increased metabolic demand to be met. (In fact, the increased blood flow overcompensates, and the blood oxygenation level-dependent (BOLD) signal by reflecting the consequent alteration of deoxyhaemoglobin which is paramagnetic reflects this.)

The total synaptic current (I_{syn}) is given by the sum of the absolute values of the glutamatergic excitatory components (implemented through NMDA and AMPA receptors) and inhibitory components (GABA) (Tagamets and Horwitz 1998, Horwitz, Tagamets and McIntosh

1999, Rolls and Deco 2002, Deco, Rolls and Horwitz 2004). In our integrate-and-fire simulations the external excitatory contributions are produced through AMPA receptors ($I_{\mathrm{AMPA,ext}}$), while the excitatory recurrent synaptic currents are produced through AMPA and NMDA receptors ($I_{\mathrm{AMPA,rec}}$ and $I_{\mathrm{NMDA,rec}}$). The GABA inhibitory currents are denoted by I_{GABA}. Consequently, the simulated fMRI signal activity S_{fMRI} is calculated by the following convolution equation:

$$S_{\mathrm{fMRI}}(t) = \int_0^\infty h(t - t') I_{\mathrm{syn}}(t')\, dt'.$$

Deco, Rolls and Horwitz (2004) applied this approach to predicting fMRI BOLD signals based on activity simulated at the integrate-and-fire level of neuronal activity in the dorsolateral prefrontal cortex. They showed that differences in the fMRI BOLD signal from the dorsal as compared to the ventral prefrontal cortex in working memory tasks may reflect a higher level of inhibition in the dorsolateral prefrontal cortex. In their simulation the convolution was calculated numerically by sampling the total synaptic activity every 0.1 s and introducing a cut-off at a delay of 25 s. The parameters utilized for the haemodynamic standard response $h(t)$ were taken from the paper of Glover (1999), and were: $n_1 = 6.0$, $t_1 = 0.9$s, $n_2 = 12.0$, $t_2 = 0.9$s, and $a_2 = 0.2$.

6.6 Prediction of the BOLD signals on difficult vs easy decision-making trials

We now show how this model makes predictions for the fMRI BOLD signals that would occur in brain areas in which decision-making processing of the type described is taking place. The BOLD signals were predicted from the firing rates of the neurons in the network (or from the synaptic currents flowing in the neurons as described later) by convolving the neuronal activity with the haemodynamic response function in a realistic period, the two seconds after the decision cues are applied. This is a reasonable period to take, as decisions will be taken within this time, and the attractor state may not necessarily be maintained for longer than this. (The attractor states might be maintained for longer if the response that can be made is not known until later, as in some fMRI tasks with delays, and then the effects described might be expected to be larger, given the mean firing rate effects shown in Fig. 6.2.)

In more detail, the haemodynamic signal associated with the decision was calculated by convolving the neuronal activity or the synaptic currents of the neurons with the haemodynamic response function used with SPM5 (Statistical Parametric Mapping, Wellcome Department of Imaging Neuroscience, London) (as this was the function also used in the analyses by SPM of the experimental fMRI data) (Rolls, Grabenhorst and Deco 2010b). For the convolution, the predecision period of spontaneous activity was padded out so that it lasted for 30 s, and after the 2 s period of decision-making activity, each trial was padded out for a further 18 s with spontaneous activity, so that the effects found in a 2 s period of decision-making could be measured against a steady background. The predicted BOLD signals are shown with $t=0$ corresponding to the time when the decision stimuli were turned on, just as in related analyses of the experimental fMRI data, to enable direct comparisons (Rolls, Grabenhorst and Deco 2010b).

As shown in Fig. 6.5a, the predicted fMRI response is larger for easy ($\Delta I = 160$ spikes/s) than for difficult trials ($\Delta I=0$), with intermediate trials ($\Delta I=80$) producing an intermediate fMRI response. The difference in the peak response for $\Delta I=0$ and $\Delta I=160$ is highly significant ($p \ll 0.001$). Importantly, the BOLD response is inherently variable from brain regions associated with this type of decision-making process, and this is nothing to do with the noise

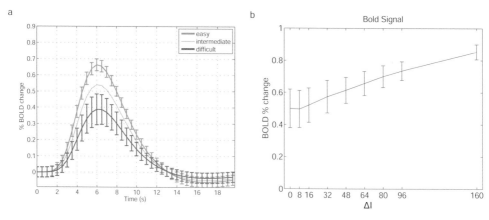

Fig. 6.5 (a) The percentage change in the simulated BOLD signal on easy trials (ΔI=160 spikes/s), on intermediate trials (ΔI=80), and on difficult trials (ΔI=0). The mean ± sd are shown for the easy and difficult trials. The percentage change in the BOLD signal was calculated from the firing rates of the D1 and D2 populations, and analogous effects were found with calculation from the synaptic currents averaged for example across all 4 populations of neurons. (b) The percentage change in the BOLD signal (peak mean ± sd) averaged across correct and incorrect trials as a function of ΔI. ΔI=0 corresponds to difficult, and ΔI=160 corresponds to easy. The percent change was measured as the change from the level of activity in a period of 1 s immediately before the decision cues were applied at t=0 s, and was calculated from the firing rates of the neurons in the D1 and D2 populations. The BOLD per cent change scaling is arbitrary, and is set so that the lowest value for the peak of a BOLD response is 0.5%. (After Rolls, Grabenhorst and Deco, 2010b.)

arising in the measurement of the BOLD response with a scanner. If the system were deterministic, the standard deviations, shown as a measure of the variability in Fig. 6.5a, would be 0. It is the statistical fluctuations caused by the noisy (random) spike timings of the neurons that account for the variability in the BOLD signals in Fig. 6.5a. Interestingly, the variability is larger on the difficult trials (ΔI=0) than on the easy trials (ΔI=160), as shown in Fig. 6.5a, and indeed this also can be taken as an indicator that attractor decision-making processes of the type described here are taking place in a brain region.

Fig. 6.5b shows that the percentage change in the BOLD signal (peak mean ± sd) averaged across correct and incorrect trials increases monotonically as a function of ΔI. This again can be taken as an indicator (provided that fMRI signal saturation effects are minimized) that attractor decision-making processes of the type described here are taking place in a brain region. The percentage change in Fig. 6.5b was calculated by convolution of the firing rates of the neurons in the D1 and D2 populations with the haemodynamic response function. Interestingly, the percentage change in the BOLD signal is approximately linearly related throughout this range to ΔI (r=0.995, p<10^{-7}). The effects shown in Figs. 6.5a and b can be related to the earlier onset of a high firing rate attractor state when ΔI is larger (see Figs. 6.2 and 6.3b), and to a higher firing rate when in the attractor state (as shown in Figs. 6.2 and 6.3a). As expected from the decrease in the variability of the neuronal activity as ΔI increases (Fig. 6.3a), the variability (standard deviation) in the predicted BOLD signal also decreases as ΔI increases, as shown in Fig. 6.5b (r=0.955, p<10^{-4}).

Similar effects, though smaller in degree, were found when the percentage change of the BOLD signal was calculated from the synaptic activity in all populations of neurons in the network (D1, D2, GABA, and non-specific, see Fig. 6.1), as shown in Fig. 6.6a. The percentage change in the BOLD signal is approximately linearly related throughout this range to ΔI (r=0.991, p<10^{-6}). Fig. 6.6b shows the relation in the model between the BOLD signal

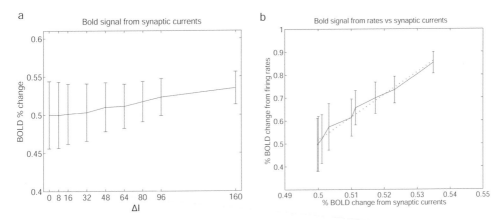

Fig. 6.6 (a) The percentage change of the BOLD signal calculated from the synaptic currents in all populations of neurons in the network (D1, D2, GABA, and non-specific, see Fig. 6.1). (Analogous results were found when the currents were calculated from the D1 and D2 populations; or from the D1, D2 and GABA populations.) (b) The relation between the BOLD current predicted from the firing rates and from the synaptic currents ($r=0.99$, $p<10^{-6}$) for values of ΔI between 0 and 160. The fitted linear regression line is shown. The BOLD per cent change scaling is arbitrary, and is set so that the lowest value is 0.5%. (After Rolls, Grabenhorst and Deco, 2010b.)

predicted from the firing rates and from the synaptic currents for values of ΔI between 0 and 160 Hz. The fitted linear regression line is shown ($r=0.99$, $p<10^{-6}$).

The findings shown in Fig. 6.6b are of interest when considering how the fMRI BOLD signal is generated in the cortex. The fMRI BOLD signal is produced by an alteration of blood flow in response to activity in a brain region. The BOLD signal, derived from the magnetic susceptibility of deoxyhaemoglobin, reflects an overcompensation in the blood flow, resulting in more deoxyhaemoglobin in active areas. The coupling of the activity in a brain region to the altered blood flow is complex (Logothetis et al. 2001, Logothetis 2008, Mangia, Giove, Tkac, Logothetis, Henry, Olman, Maraviglia, Di Salle and Ugurbil 2009), but may reflect the energy needed to pump ions that have crossed the cell membrane as a result of synaptic activity to be pumped back against the electrochemical gradient. In Fig. 6.5 predicted BOLD effects are shown related to the firing rate of the two populations D1 and D2 of neurons in the attractor network. In Fig. 6.6 are shown predicted BOLD effects related to the sum of the synaptic currents in all the neurons in the network, with analogous results found with the synaptic currents in only the D1 and D2 neurons, or in the D1, D2, and GABA neurons in the network. Although it has been suggested that synaptic activity is better coupled to the blood flow and hence the BOLD signal than the firing rates (Logothetis et al. 2001, Logothetis 2008, Mangia et al. 2009), in the neocortex these might be expected to be quite closely related, and indeed the integrate-and-fire model shows a linear relation between the BOLD signal predicted from the firing rates and from the synaptic currents ($r=0.99$, $p<10^{-6}$, Fig. 6.6b), providing an indication that at least in the cerebral cortex the BOLD signal may be related to the firing rates of the neurons as well as to the synaptic currents.

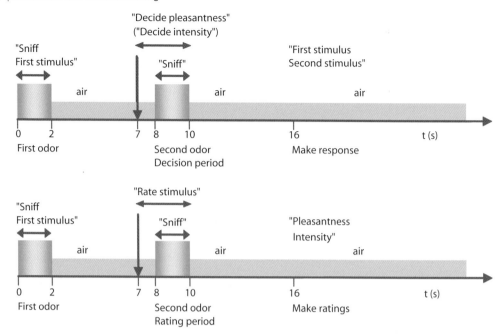

Fig. 6.7 Task design for trials of the olfactory task. On decision trials (upper), the task required a binary choice decision starting during the second odour about which of the two odours was more pleasant, or (on different trials, as indicated by the instruction at $t=7$ s) more intense. On rating trials (lower), identical stimuli were used, but no decision was required, and instead participants rated the second odour for pleasantness and intensity on continuous analog visual rating scales. The trial types were identical until $t=7$ s, when the instruction indicated whether the trial type was decide or rate. The second odour was delivered at $t=8$ s, the subjects were deciding or rating at that time, and the imaging was with respect to this period starting at $t=8$ s. No responses to indicate the decision made or the rating value could be made until t=16 s. (After Rolls, Grabenhorst and Parris, 2010b.)

6.7 Neuroimaging investigations of task difficulty, and confidence

Two functional neuroimaging investigations were performed to test the predictions of the model just described. Task difficulty was altered parametrically to determine whether there was a close relation between the BOLD signal and task difficulty (Rolls, Grabenhorst and Deco 2010b), and whether this was present especially in brain areas implicated in choice decision-making by other criteria (Grabenhorst, Rolls and Parris 2008, Rolls, Grabenhorst and Parris 2009c). The decisions were about the pleasantness of olfactory (Rolls, Grabenhorst and Parris 2009c) or thermal (Grabenhorst, Rolls and Parris 2008) stimuli.

6.7.1 Olfactory pleasantness decision task

An olfactory decision-making task was designed (Rolls, Grabenhorst and Parris 2009c) based on the vibrotactile decision-making studies of Romo and colleagues in which vibrotactile stimuli separated by a delay were presented, and a decision had to be made, when the second stimulus was presented, of whether the second frequency was higher than the first (Romo, Hernandez and Zainos 2004). In our design the decision was about whether the second odour was more pleasant than the first, or on other trials was more intense than the first (see Fig. 6.7). To allow a comparison with trials on which choices between stimuli were not made, there were also trials in which only ratings of the continuous affective value and intensity were made, without a choice between stimuli, as described in detail by Rolls et al. (2009c). The odours were delivered through a computer-controlled olfactometer. The pleasant odours were 1 M citral and 4 M vanillin. The unpleasant odours made up for the olfactometer were hexanoic acid (10% v/v) and isovaleric acid (15%). The experimental protocol consisted of an event-related interleaved fMRI design presenting in random permuted sequence the 3 experimental conditions and different pairs of olfactory stimuli for each condition. Each trial started at t=0 s with the first odour being delivered for 2 s accompanied by a visual label stating 'Sniff first stimulus'. There was then a 6 s period during which clean air was delivered. In this period at t=7 s a visual label was displayed stating either 'Decide pleasantness', 'Decide intensity' or 'Rate stimulus'. At t=8 s the second odour was presented for 2 s accompanied by a visual label stating either 'Sniff Decide pleasantness', 'Sniff Decide intensity' or 'Sniff Rate stimulus'. There was then a 6 s period during which clean air was delivered. Starting at t=16 s on decision trials, the words 'First stimulus' and 'Second stimulus' appeared on the screen for 2 s, and in this period the participant had to select which button key response to make (up button or down button) for the decision that had been taken at the time when the second odour was delivered. At t=18 s on decision trials, the word 'First stimulus' or 'Second stimulus' was then displayed to provide feedback to the participant that their choice was acknowledged.

On rating trials, starting at t=16 s the subjective ratings were made. The first rating was for the pleasantness of the second odour on a continuous visual scale from −2 (very unpleasant), through 0 (neutral), to +2 (very pleasant). The second rating was for the intensity of the second odour on a scale from 0 (very weak) to 4 (very intense). The ratings were made with a visual rating scale in which the subject moved the bar to the appropriate point on the scale using a button box. There was 4 s for each rating. Each of the trial types was presented in random permuted sequence 36 times. Which two of the four odours were presented on each trial, and the order in which they were presented, was determined by a random permuted sequence.

ΔI, the difference in pleasantness of the two stimuli between which a decision was being made, was obtained for each trial by the absolute value of the difference in the (average) rated pleasantness of that pair of stimuli for each subject. Thus, two odours of similar pleasantness would have a small ΔI, and two odours of different pleasantness would have a large ΔI. This measure thus reflects the difficulty of the decision, and is independent of whether the second odour happened to be pleasant or unpleasant. This value for ΔI on every trial was used to investigate whether at brain sites where there was more activation on easy vs difficult trials (as shown by a contrast analysis), the BOLD signal was related to ΔI. Because the stimuli were randomized, the analysis did not reflect the pleasantness or unpleasantness of the second odour, but only how different it was in pleasantness from the first odour, independently of the sign of the difference. The regressor was thus decision difficulty.

6.7.2 Temperature pleasantness decision task

Warm and cool thermal stimuli, and mixtures of them, were applied to the hand. In a previous investigation of the same dataset, we compared brain responses when participants were taking decisions about whether they would select a thermal stimulus (yes vs no), with activations to the same stimuli on different trials when only affective ratings were required and there was no decision about whether the participants would say yes or no to the stimuli if they were available in the future (Grabenhorst, Rolls and Parris 2008). In the investigation on task difficulty, we analysed data only on decision trials, and the analyses were about how activations with the stimuli were related to the difficulty of the decision (Rolls, Grabenhorst and Deco 2010b), which was not investigated previously. Both the decision and rating trials were identical from the start of each trial at t=0 s until t=5 s when a visual stimulus was shown for 1 s stating 'decide' or 'rate' the thermal stimulus being applied, and at t=6 s a green cross appeared until t=10 s. On decide trials from t=6 until t=10 s the participants had to decide whether yes or no was the decision on that trial. At t=10 s a visual stimulus with yes above no or vice versa in random order was shown for 2 s, and the participant had to press the upper or lower button on the button box as appropriate to indicate the response. On rating trials from t=6 until t=10 s the participants had to encode the pleasantness and intensity of the thermal stimulus being applied, so that the ratings could be made later. On rating trials at t=10 s the pleasantness rating could be made using the same button box, and then the intensity rating was made. The thermal stimuli were a warm pleasant stimulus (41°C) applied to the hand ('warm2'), a cool unpleasant stimulus (12°C) applied to the hand ('cold'), a combined warm and cold stimulus ('warm2+cold'), and a second combination designed to be less pleasant (39°C + 12°C) ('warm1+cold'), delivered with Peltier devices as described previously (Grabenhorst, Rolls and Parris 2008).

ΔI for the thermal stimuli was the absolute value of the pleasantness rating, based on the concept that it is more difficult to choose whether a stimulus should be repeated in future if it is close to neutral (0) in rated pleasantness, versus is rated as being pleasant (with the maximum pleasantness being +2), or as being unpleasant (with the most unpleasant being −2).

6.7.3 fMRI analyses

The criteria used to identify regions involved in choice decision-making in earlier investigations with the same data set were that a brain region should show more brain activity with identical stimuli on trials on which a choice decision was being made than when the continuous affective value of the stimuli were being rated, but no choice was being made between stimuli, or about whether the stimulus would be chosen again (Grabenhorst, Rolls and Parris 2008, Rolls, Grabenhorst and Parris 2009c, Rolls and Grabenhorst 2008). For the olfactory task, a contrast of decision vs rating trials showed activations in the medial prefrontal cortex medial area 10 at [2 50 −12] z=3.78 p<0.001 (Rolls, Grabenhorst and Parris 2009c). For the thermal task, a contrast of decision vs rating trials showed activations in the medial prefrontal cortex medial area 10, at [6 54 −8] z=3.24 p=0.022 (Grabenhorst, Rolls and Parris 2008).

6.7.4 Brain areas with activations related to easiness and confidence

Figure 6.8 shows experimental data with the fMRI BOLD signal measured on easy and difficult trials of the olfactory affective decision task (left) and the thermal affective decision task (right) (Rolls, Grabenhorst and Deco 2010b). The upper records are for prefrontal cortex medial area 10 in a region identified by the following criterion as being involved in choice decision-making. The criterion was that a brain region for identical stimuli should show more activity when a choice decision was being made than when a rating on a continuous scale of affective value was being made. Fig. 6.8 shows for medial prefrontal cortex area 10 that there is a larger BOLD signal on easy than on difficult trials. The top diagram shows the medial prefrontal area activated in this contrast for decisions about which olfactory stimulus was more pleasant (yellow), and for decisions about whether the thermal stimulus would be chosen in future based on whether it was pleasant or unpleasant (red).

In more detail, for the thermal stimuli, the contrast was the warm2 and the cold trials (which were both easy in that the percentage of the choices were far from the chance

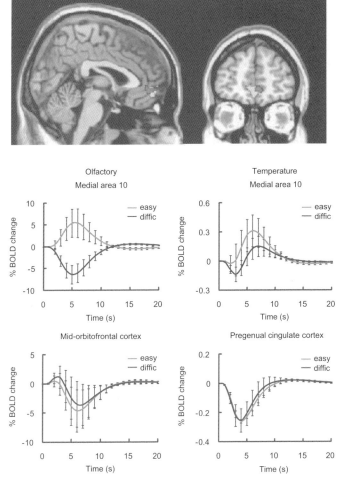

Fig. 6.8 Top: Medial prefrontal cortex area 10 activated on easy vs difficult trials in the olfactory pleasantness decision task (yellow) and the thermal pleasantness decision task (red). Middle: experimental data showing the BOLD signal in medial area 10 on easy and difficult trials of the olfactory affective decision task (left) and the thermal affective decision task (right). This medial area 10 was a region identified by other criteria (see text) as being involved in choice decision-making. Bottom: The BOLD signal for the same easy and difficult trials, but in parts of the pregenual cingulate and mid-orbitofrontal cortex implicated by other criteria (see text) in representing the subjective reward value of the stimuli on a continuous scale, but not in making choice decisions between the stimuli, or about whether to choose the stimulus in future. (After Rolls, Grabenhorst and Deco, 2010b.) (See colour plates Appendix B.)

value of 50%, and in particular were 96±1% (mean±sem) for the warm, and 18±6% for the cold), versus the mixed stimulus of warm2+cold (which was difficult in that the percentage of choices of 'Yes, it would be chosen in future' was 64±9%). For the temperature easy vs difficult decisions about pleasantness, the activation in medial area 10 had peaks at [4 42 −4] $z=3.59$ $p=0.020$ and [6 52 −4] $z=3.09$ $p=0.045$.

For the olfactory decision task, the activations in medial area 10 for easy vs difficult choices were at [−4 62 −2] $z=2.84$ $p=0.046$, confirmed in a finite impulse response (FIR) analysis with a peak at 6–8 s after the decision time at [−4 54 −6] $z=3.50$ $p=0.002$. (In the olfactory task, the easy trials were those in which one of the pair of odours was from the

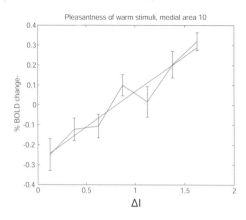
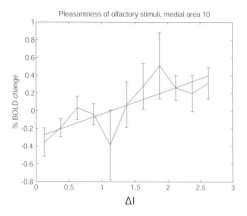

Fig. 6.9 Experimental fMRI data showing the change in the BOLD signal (mean±sem, with the fitted linear regression line shown) as a function of ΔI, the difference in pleasantness of warm stimuli or olfactory stimuli about which decision was being made, for medial prefrontal cortex area 10. (After Rolls, Grabenhorst and Deco, 2010b.)

pleasant set, and the other from the unpleasant set. The mean difference in pleasantness, corresponding to ΔI, was 1.76±0.25 (mean±sem). The difficult trials were those in which both odours on a trial were from the pleasant set, or from the unpleasant set. The mean difference in pleasantness, corresponding to ΔI, was 0.72±0.16. For easy trials, the percentage correct was 90±2, and for difficult trials was 59±8.) No other significant effects in the a priori regions of interest (Grabenhorst, Rolls and Parris 2008, Rolls, Grabenhorst and Parris 2009c) were found for the easy vs difficult trial contrast in either the thermal or olfactory reward decision task (Rolls, Grabenhorst and Deco 2010b).

The lower records in Fig. 6.8 are for the same easy and difficult trials, but in parts of the pregenual cingulate and mid-orbitofrontal cortex implicated by the same criteria in representing the subjective reward value of the stimuli, but not in making choice decisions between the stimuli. [For the pregenual cingulate cortex, there was a correlation of the activations with the subjective ratings of pleasantness of the thermal stimuli at [4 38 −2] z=4.24 p=0.001. For the mid-orbitofrontal cortex, there was a correlation of the activations with the subjective ratings of pleasantness of the thermal stimuli at [40 36 −12] z=3.13 p=0.024]. The BOLD signal was similar in these brain regions for easy and difficult trials, as shown in Fig. 6.8, and there was no effect in the contrast between easy and difficult trials.

Fig. 6.9 shows the experimental fMRI data with the change in the BOLD signal for medial prefrontal cortex area 10 indicated as a function of ΔI, the difference in pleasantness of warm stimuli or olfactory stimuli about which decision was being made, and thus the easiness of the decision. For the olfactory decision task, ΔI was the difference in pleasantness (for a given subject) between the mean pleasantness of the first odour and the mean pleasantness of the second odour between which decision was being taken, about which was more pleasant (Rolls et al. 2010b). It is shown in Fig. 6.9 (right) that there was a clear and approximately linear relation between the BOLD signal and ΔI for the olfactory pleasantness decision-making task ($r = 0.77, p = 0.005$). The coordinates for these data were as given for Fig. 6.8.

For the warm decision task, ΔI was the difference in mean pleasantness for a given subject from 0 for a given thermal stimulus about which a decision was being made about whether it should or should not be repeated in future (Grabenhorst et al. 2008). It is shown in Fig. 6.9 (left) that there was a clear and approximately linear relation between the BOLD signal and ΔI for the thermal pleasantness decision-making task ($r = 0.96, p < 0.001$). The coordinates for these data were as given for Fig. 6.8.

These experimental findings are thus consistent with the predictions made from the model, and provide strong support for the type of model of decision-making described in this book. Moreover, they show that decision confidence, which increases with ΔI, can be read off from fMRI BOLD signals in parts of the brain involved in making choices, as shown in this unifying theory.

6.8   Correct decisions vs errors, and confidence

If one makes a correct decision, consistent with the evidence, then one's confidence is higher than when one makes an incorrect decision (as shown by confidence ratings) (Vickers 1979, Vickers and Packer 1982). Consistent with this, the probability that a rat will abort a trial to try again is higher if a decision just made is incorrect (Kepecs et al. 2008).

Why does this occur, and in which brain regions is the underlying processing implemented? We show in this Section (6.8) that the integrate-and-fire attractor decision-making network described in this book predicts higher and shorter latency neuronal responses, and higher BOLD signals, in brain areas involved in making choices, on correct than on error trials, and how these changes vary with ΔI and thus compute decision confidence. The reason for this behaviour of the choice attractor network is that on correct trials, when the network influenced by the spiking noise has reached an attractor state supported by the recurrent connections between the neurons, the external inputs to the network that provide the evidence for the decision will have firing rates that are consistent with the decision, so the firing rate of the winning attractor will be higher when the correct attractor (given the evidence) wins the competition than when the incorrect attractor wins the competition. We then show from an experimental fMRI investigation that this BOLD signature of correct vs incorrect decisions, which reflects confidence in whether a decision just taken is correct, is found in the medial prefrontal cortex area 10, the posterior and subgenual cingulate cortex, and the dorsolateral prefrontal cortex, but not in the mid-orbitofrontal cortex, where activations instead reflect the pleasantness or subjective affective value of the stimuli used as inputs to the choice decision-making process (Rolls, Grabenhorst and Deco 2010c). This approach to decision-making, in contrast to mathematical models of decision-making as an accumulation or diffusion process (Vickers 1979, Vickers and Packer 1982, Ratcliff and Rouder 1998, Ratcliff, Zandt and McKoon 1999, Usher and McClelland 2001), thus makes testable predictions about how correct vs error performance is implemented in the brain, and these predictions are supported by experimental results showing that areas involved in choice decision-making (or that receive from them) have activity consistent with these predictions, and that other areas not involved in the choice-making part of the process do not.

6.8.1   Operation of the attractor network model on correct vs error trials

The attractor network model of decision-making is the same as that described in the first part of this Chapter, and the simulations are from the same test runs as those described earlier.

Figure 6.10a and d show the mean firing rates of the two neuronal populations D1 and D2 for correct trials (a) and incorrect trials (b) for an intermediate level of task difficulty (ΔI=32 Hz) (where ΔI is the difference in spikes/s summed across all synapses to each neuron between the two inputs, λ_1 to population D1, and λ_2 to population D2). In the correct trials, the population D1 won, and on incorrect trials, the population D2 won. The winning population

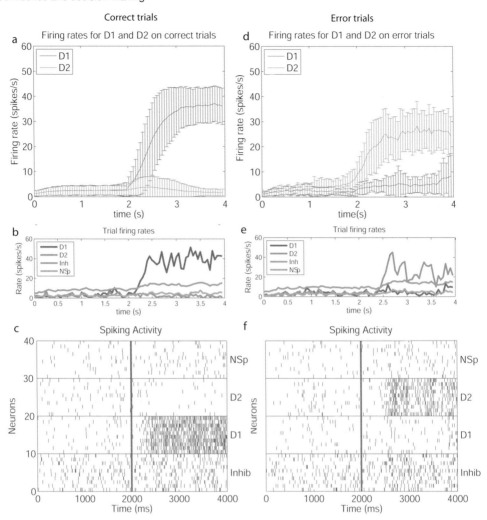

Fig. 6.10 (a and d) Firing rates (mean $\pm$ sd) for correct and error trials for an intermediate level of difficulty (ΔI=32). The period 0–2 s is the spontaneous firing, and the decision cues were turned on at time = 2 s. The means were calculated over 1000 trials. D1: firing rate of the D1 population of neurons, which is the correct population. D2: firing rate of the D2 population of neurons, which is the incorrect population. A correct trial was one in which in which the mean rate of the D1 attractor averaged $>$ 10 spikes/s for the last 1000 ms of the simulation runs. (Given the attractor nature of the network and the parameters used, the network reached one of the attractors on $>$90% of the 1000 trials, and this criterion clearly separated these trials, as indicated by the mean rates and standard deviations for the last 1 s of the simulation as shown.) (b and e) The firing rates of the four populations of neurons on a single trial for a correct (b) and incorrect (e) decision. Inh is the inhibitory population that uses GABA as a transmitter. NSp is the non-specific population of neurons. (c and f) Rastergrams for the trials shown in (b) and (e). 10 neurons from each of the four pools of neurons are shown. (After Rolls, Grabenhorst and Deco 2010c.) (See colour plates Appendix B.)

was defined as that which fired with a rate of $>$ 10 spikes/s for the last 1000 ms of the simulation runs. (This provided a clear criterion for which attractor won the competition as shown by the binary distribution of the rates of the two attractors found over 1000 simulation trials, and as exemplified by the single trials in Fig. 6.10.) Figures 6.10a and d show that the win-

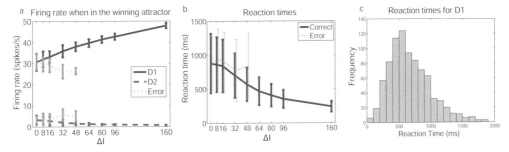

Fig. 6.11 (a) Firing rates (mean $\pm$ sd) on correct and error trials when in the winning attractor as a function of ΔI. ΔI=0 corresponds to difficult, and ΔI=160 corresponds to easy. The firing rates on correct trials for the winning population D1 are shown by solid lines, and for the losing population D2 by dashed lines. All the results are for 1000 simulation trials for each parameter value, and all the results shown are statistically highly significant. The results on error trials are shown by the dotted red lines, and in this case the D2 attractor wins, and the D1 attractor loses the competition. (There were no error trials for values of ΔI=64 Hz or above.) (b) Reaction times (mean $\pm$ sd) for the D1 population to win on correct trials (thick line), and for D2 to win on error trials (thin line), as a function of the difference in inputs ΔI to D1 and D2. The plot for the error trials has been offset by a small amount so that its values can be seen clearly. ΔI=0 corresponds to difficult, and ΔI=160 corresponds to easy. (c) The distribution of reaction times for the model for ΔI=32 illustrating the long tail of slow responses. Reaction times are shown for 837 correct trials, the level of performance was 95.7% correct, and the mean reaction time was 701 ms. (After Rolls, Grabenhorst and Deco, 2010c.) (See colour plates Appendix B.)

ning population had higher firing rates on correct than on incorrect trials. Figure 6.10b shows the firing rates of the four populations of neurons on a single trial for a correct decision, and Fig. 6.10c shows the rastergrams for the same trial. Figure 6.10e shows the firing rates of the four populations of neurons on a single trial for a incorrect decision, and Fig. 6.10f shows the rastergrams for the same trial. From Figs. 6.10a and d it is clear that the variability of the firing rate is high from trial to trial, with the standard deviations of the mean firing rate calculated in 50 ms epochs indicated in order to quantify the variability. This is due to the spiking noise in the network, which influences the decision taken, resulting in probabilistic decision-making, and incorrect choices on some trials. The effects evident in Fig. 6.10 are quantified, and elucidated over a range of values for ΔI, next.

Figure 6.11a shows the firing rates (mean $\pm$ sd) on correct and error trials of the winning and losing attractors as a function of ΔI. ΔI=0 corresponds to the most difficult decision, and ΔI=160 corresponds to easy. The firing rates were measured in the last 1 s of firing, i.e. between t=3 and t=4 s. It is clear that the mean firing rate of the winning attractor D1 on correct trials increases monotonically as ΔI increases, and interestingly, the increase is approximately linear (Pearson $r = 0.995, p < 10^{-6}$). The higher mean firing rates of the winning D1 attractor on correct trials as ΔI increases are due not only to higher peak firing, but also to the fact that the variability becomes less as ΔI increases ($r = -0.95, p < 10^{-4}$), reflecting the fact that the system is more noisy and unstable with low ΔI, whereas the firing rate in the correctly winning D1 attractor is maintained more stably with smaller statistical fluctuations against the Poisson effects of the random spike timings at high ΔI. (The measure of variation indicated in the Figure is the standard deviation, and this is shown throughout unless otherwise stated to quantify the degree of variation, which is a fundamental aspect of the operation of these neuronal decision-making networks.) The increase of the firing rate when in the D1 attractor on correct trials (solid black line) as ΔI increases thus reflects the confidence in the decision, and, as is shown in Fig. 6.3c, the performance as shown by the percentage of correct choices. On correct trials, the firing rate of the losing attractor (D2, dashed blue line) is of course low, and decreases further as ΔI increases on correct trials, due to feedback inhibition from the winning D1 attractor, and thus the difference in the firing rates of the two attractors also reflects well the decision confidence.

When we make an incorrect choice, our confidence in our decision is likely to be low (Vickers 1979, Vickers and Packer 1982). Exactly this is represented in the firing rates of the neurons on incorrect trials, for as shown in dotted red lines in Fig. 6.11a, when an incorrect choice is made (and D2 wins the competition because of noise), the firing rate of the D2 attractor when winning decreases as ΔI increases (upper dotted red line starting at 31 spikes/s). The reason for this is that the external inputs, the evidence for the decision, are now working against the internal recurrent attractor dynamics, which have reached the wrong decision because of the noise in the network. Thus on error trials, the confidence in a decision is also reflected in the firing rates of the attractors (Rolls, Grabenhorst and Deco 2010c).

Conversely, the firing rate of the losing attractor (D1) on incorrect trials, which is low with $\Delta I=0$ (approximately 3 spikes/s), increases by a few spikes/s as ΔI increases (lower dotted red line in Fig. 6.11a). The reason for this is that the noise has contributed to the D1 attractor losing, and there is more external input against this decision being applied to the D1 neurons as ΔI increases, tending to increase their firing rate from the low value. At the same time, the D1 neurons receive less feedback inhibition from the incorrectly winning population D2 via the inhibitory neurons as ΔI increases.

The time for the network to reach the correct D1 attractor, i.e. the reaction time of the network, is shown as a function of ΔI for correct and incorrect trials in Fig. 6.11b (mean $\pm$ sd). Interestingly, the reaction time continues to decrease ($r = -0.95, p < 10^{-4}$) over a wide range of ΔI, even when as shown in Fig. 6.11c the network is starting to perform at 100% correct. The decreasing reaction time as ΔI increases is attributable to the altered 'effective energy landscape' (see Section 2.3): a larger input to D1 tends to produce occasionally higher firing rates, and these statistically are more likely to induce a significant depression in the landscape towards which the network flows sooner than with low ΔI. Correspondingly, the variability (quantified by the standard deviation) of the reaction times is greatest at low ΔI, and decreases as ΔI increases ($r = -0.95, p < 10^{-4}$). This variability would not be found with a deterministic system (i.e. the standard deviations would be 0 throughout, and such systems include those investigated with mean-field analyses), and is entirely due to the random statistical fluctuations caused by the random spiking of the neurons in the integrate-and-fire network.

Very interestingly, the reaction times of the network are longer for incorrect (error) than for correct decisions, as shown by the thin graph in Fig. 6.11b (Rolls, Grabenhorst and Deco 2010c). This difference, though not large, was statistically significant. (The effect was found for the means of the 4 non-zero values of ΔI for which there were errors ($p < 0.05$), and for $\Delta I=8$ for example, the reaction time distribution was longer for error compared to correct trials, $p < 0.01$ and confirmed by the non-parametric Mann-Whitney test.) This models effects that are found in human performance (Vickers and Packer 1982), especially with difficult decisions (Luce 1986, Welford 1980), and that have also been found in lateral intraparietal (LIP) cortex neurons in a motion coherence discrimination task (Roitman and Shadlen 2002) and in a related attractor model of this (Wong et al. 2007). The actual mechanism for this in this biologically plausible attractor decision-making network is that on error trials, the noise is fighting the effects of the difference in the inputs λ_1 and λ_2 that bias the decision-making, and it takes on average relatively long for the noise, on error trials, to become by chance sufficiently large to overcome the effects of this input bias (Rolls, Grabenhorst and Deco 2010c). This effect will contribute to the smaller BOLD signal on error than on correct trials described below.

The distribution of reaction times for the model has a long tail of slow responses, as

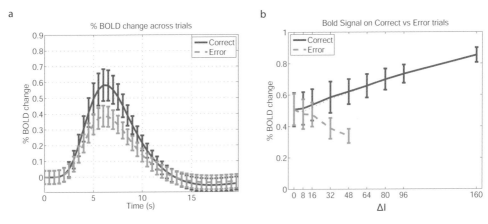

Fig. 6.12 (a) The predicted percentage change (mean $\pm$ sd) in the BOLD signal on correct and error trials for $\Delta I = 32$. The percentage change in the BOLD signal was calculated from the firing rates of the D1 and D2 populations, and analogous effects were found with calculation from the synaptic currents averaged for example across all 4 populations of neurons. (b) The percentage change in the BOLD signal (peak mean $\pm$ sd) computed separately for correct and error trials as a function of ΔI. $\Delta I = 0$ corresponds to difficult, and $\Delta I = 160$ corresponds to easy. The percentage change was measured as the change from the level of activity in the simulation period 1–2 s, before the decision cues were applied at t=2 s, and was calculated from the firing rates of the neurons in the D1 and D2 populations. The BOLD percentage change scaling is arbitrary, and is set so that the lowest value on correct trials is 0.5%. The thin dotted line just below the line for the predicted response on correct trials shows the predicted response from all trials, that is correct and error trials. (After Rolls, Grabenhorst and Deco, 2010c.)

shown in Fig. 6.11c for ΔI=32, capturing this characteristic of human reaction time distributions (Luce 1986, Welford 1980, Ratcliff and Rouder 1998, Ratcliff et al. 1999, Usher and McClelland 2001, Marti et al. 2008).

The relation between ΔI and percentage correct is shown in Fig. 6.3c on page 174, and described in Section 6.4.

6.8.2 Predictions of fMRI BOLD signals from the model

We now show how this model makes predictions for the fMRI BOLD signals that would occur on correct vs incorrect trials in brain areas in which decision-making processing of the type described is taking place. The BOLD signals were predicted from the firing rates of the neurons in the network (or from the synaptic currents flowing in the neurons as described later) by convolving the neuronal activity with the haemodynamic response function in a realistic period, the two seconds after the decision cues are applied. This is a reasonable period to take, as decisions will be taken within this time, and the attractor state may not necessarily be maintained for longer than this. (The attractor states might be maintained for longer if the response that can be made is not known until later, as in some fMRI tasks with delays including that described here, and then the effects described might be expected to be larger, given the mean firing rate effects shown in Fig. 6.10. Indeed, it is an advantage of the type of model described here that it is a short-term memory attractor network that can maintain its firing whether or not the decision cues remain, as this continuing firing enables the decision state to be maintained until a response based on it can be guided and made.) As shown in Fig. 6.12a, the predicted fMRI response is larger for correct vs incorrect trials in the decision-making network (Rolls, Grabenhorst and Deco 2010c). The results are shown for a level of difficulty in which there is a reasonable proportion of error trials (ΔI=32).

Fig. 6.12b shows that the difference between the BOLD signal on correct vs incorrect trials increases with ΔI. The reason for this is that on correct trials, with increasing ΔI the external evidence adds more strongly to the firing produced by the internal recurrent collateral attractor effect in the winning attractor (Fig. 6.11a), and thus the BOLD signal increases with ΔI (Fig. 6.12b). On the other hand, on incorrect trials, the external evidence adds less than on correct trials to the winning attractor, and also adds more than on correct trials to the firing rates in the losing attractor, the firing rate of which through the inhibitory interneurons decreases the firing in the winning attractor, as shown in Fig. 6.11a. Because on incorrect trials the firing of the winning attractor decreases with ΔI (Fig. 6.11a), the BOLD signal decreases with ΔI on incorrect trials (Fig. 6.12b). An important result of the simulations is that it is the firing rate of the winning attractor that dominates the predicted BOLD signals, not the smaller effects seen in the losing attractor (Rolls, Grabenhorst and Deco 2010c). Although it is largely the higher overall neural activity on correct trials that results in the larger predicted BOLD signal on correct than error trials (Fig. 6.11a), an additional contributor is the shorter decision latency for correct than error trials illustrated in Fig. 6.11b.

Fig. 6.12b also shows with the thin dotted line (just below the line for the predicted response on correct trials) the predicted response from all trials, that is correct and error trials. The prediction of the BOLD signal for all trials is relatively close to the prediction for the correct trials because with low ΔI the predictions for correct and error trials are close as indicated in Fig. 6.11b, and with larger values of ΔI there are fewer error trials.

6.8.3 fMRI BOLD signals that are larger on correct than error trials

The model predicts that the BOLD signal will be larger in at least some brain areas involved in choice decision-making on correct than on incorrect trials. We tested this by performing a contrast analysis on correct vs incorrect trials in the olfactory decision-making task described in Section 6.7.1, in which the participants had to decide whether the second odour was more pleasant than the first. Error trials were defined as choices that do not reflect the mean pleasantness ratings (Rolls, Grabenhorst and Deco 2010c). (The rationale for this was that noise in the system might influence the binary choice made on an individual trial, and each participant's estimate of what was the more pleasant of each pair was given by the mean pleasantness rating across all trials for each odour being compared on a particular trial.) For example, if citral was rated on average for pleasantness as +1.5 and vanillin as +1.0, and the subject chose vanillin over citral, this was classified as an error trial. At the single subject level, contrasts between correct vs error trials were computed within each odour pair. At the group level, these odour-specific contrast maps were combined across subjects.

Four brain regions had significant effects in the correct vs error contrast (Rolls, Grabenhorst and Deco 2010c). They were the medial prefrontal cortex area 10 [2 50 −12]; the posterior cingulate cortex ([−18 −32 34]; the subgenual cingulate cortex ([10 24 −8]; and the dorsolateral prefrontal cortex (DLPFC [24 14 34]). Figure 6.13 shows the timecourses of the activations on correct and on error trials for the dorsolateral prefrontal cortex (left) and subgenual cingulate cortex (right). The brain regions where these activations were found are close to those illustrated in Fig. 6.14 (top).

To investigate whether correct vs error effects are more likely to be found in areas that are implicated in decision-making by an earlier study that showed significant differences for trials on which decisions were made vs trials when only rating and no binary choices were made (Rolls, Grabenhorst and Parris 2009c), we performed ANOVAs on the parameter estimates for correct − incorrect decision contrasts vs the correlations with the pleasantness ratings. (In the previous investigation, brain areas implicated in decision-making included medial area 10, and in pleasantness ratings the orbitofrontal cortex.) We found that the posterior cingulate cor-

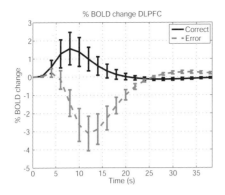
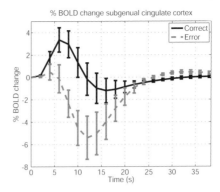

Fig. 6.13 The timecourses of the activations on correct and on error trials for the dorsolateral prefrontal cortex (left) and subgenual cingulate cortex (right), at sites where the contrast correct − error was significant. (After Rolls, Grabenhorst and Deco, 2010c.)

tex, subgenual cingulate cortex, dorsolateral prefrontal cortex, and medial prefrontal cortex area 10 had considerable parameter estimates from the general linear model implemented in SPM for correct − error trials, and low for correlations with pleasantness; whereas the orbitofrontal cortex showed high parameter estimates for correlations with pleasantness, but low for correct − incorrect contrasts (Rolls, Grabenhorst and Deco 2010c). ANOVAs on these effects showed very significant interactions which indicated that activations in the medial prefrontal cortex area 10, the posterior and subgenual cingulate cortex, and the dorsolateral prefrontal cortex were related more to whether a choice decision was correct or incorrect and less to pleasantness, whereas activations in a control region the mid-orbitofrontal were related more to pleasantness and less to whether the choice was correct or incorrect (Rolls, Grabenhorst and Deco 2010c).

6.8.4 fMRI signals linearly related to choice easiness with correct vs incorrect choices

Another prediction of the model is that in decision-making areas of the brain, the magnitude of the BOLD signal should increase with ΔI on correct trials, and decrease with ΔI on error trials (Fig. 6.12b). To investigate this, we performed separate SPM regression analyses for correct and error trials in which the regressor for this was ΔI, the mean absolute difference in pleasantness between the two odours presented on a given trial. The analyses showed that for correct trials, positive correlations of the BOLD signal with ΔI were found in the posterior cingulate cortex ([−8 −40 32], and on error trials, negative correlations of the BOLD signal with ΔI were found. These analyses also showed that on error trials, negative correlations of the BOLD signal with ΔI were found in the subgenual cingulate cortex ([10 16 −4]). This effect is illustrated in Fig. 6.14 which shows the BOLD signal as a function of (discretized values of) ΔI for correct and incorrect trials. The signs of the correlations ($r = 0.55$ on correct trials, and $r = -0.81$ on error trials) are as predicted for a decision-making area of the brain, and are significantly different (p<0.003). These analyses also showed that on error trials, negative correlations of the BOLD signal with ΔI were found in the dorsolateral prefrontal cortex ([30 18 28]), and again the signs of the correlations ($r = 0.65$ on correct trials, and $r = -0.95$ on error trials) are as predicted for a decision-making area of the brain, and are significantly different ($p < 0.001$) (Fig. 6.14). Similarly, in medial prefrontal cortex area 10 the signs of the correlations ($r = 0.53$ on correct trials, and $r = -0.62$ on error trials) are as predicted for a decision-making area of the brain, and are significantly different

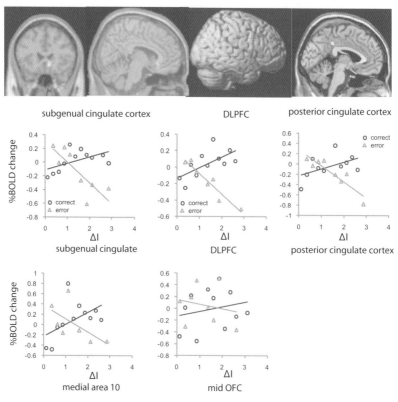

Fig. 6.14 Above: The subgenual cingulate cortex (with coronal and parasaggittal slices), dorsolateral prefrontal cortex (DLPFC) and posterior cingulate cortex regions with the percentage change in the BOLD signal positively correlated with ΔI on correct trials, and negatively correlated on error trials. Below: Separate regression plots for the relation between the BOLD signal and ΔI on correct and on error trials for these regions and for the medial prefrontal cortex area 10, and for a control region, the mid-orbitofrontal cortex (mid OFC) in which effects were not found. The regression plots show for discretized values of ΔI (averaged across subjects) the BOLD signal as a function of ΔI for correct and incorrect trials. For example, for the posterior cingulate cortex the signs of the correlations ($r = 0.51$ on correct trials, and $r = -0.91$ on error trials) are as predicted for a decision-making area of the brain, and are significantly different (p<0.001). The medial prefrontal cortex coordinates used for extracting the data were [−4 54 −6] with a 10 mm sphere around this location, the site in the same subjects for the peak of the contrast for easy − difficult trials (Rolls, Grabenhorst and Deco, 2010b). The same region [2 50 −12] showed more activity when binary decisions were made compared to when ratings were made in the same subjects (Rolls, Grabenhorst and Deco, 2010b). (After Rolls, Grabenhorst and Deco, 2010c.) (See colour plates Appendix B.)

($p = 0.032$) (Fig. 6.14) (Rolls, Grabenhorst and Deco 2010c).

For comparison, we also show in Fig. 6.14 the analyses of the percentage change in the BOLD signal for correct and for error trials for the mid-orbitofrontal cortex, chosen as a control brain region. Activations in this area did not have a significant relation to decision-making vs rating (Rolls, Grabenhorst and Parris 2009c), and were not correlated with trial easiness as indexed by ΔI (Rolls, Grabenhorst and Deco 2010b), but did have a correlation with the pleasantness of the odour that was presented (Rolls, Grabenhorst and Parris 2009c). In this mid-orbitofrontal cortex region the correlation between the BOLD signal and ΔI was $r = 0.22$ on correct trials, and $r = -0.19$ on error trials, and these are not significantly different ($p = 0.46$). Thus this brain area that is not implicated in the binary choice decision-making, but does represent information relevant to the decision, the pleasantness of the stimuli, did not have significantly different activations on correct − error trials, and did not have significantly different correlations with the BOLD signal on correct and incorrect trials. This is a useful control condition. Thus a difference in the correlations of the BOLD signal with ΔI for correct vs error trials appears to be a sensitive measure for a brain area involved in decision-making.

An implication is that a sensitive criterion for identifying areas of the brain involved in decision-making is a higher activation on easy than on difficult trials, and that this will be especially evident when only correct trials are included in the analysis. A further signature is that the activations are expected to increase with ΔI on correct trials, and to decrease on error trials with ΔI (Rolls, Grabenhorst and Deco 2010c). Part of the reason that activations are higher with larger ΔI with all trials included is that these trials will mostly be correct, whereas difficult trials, with lower values of ΔI, will include more error trials.

6.8.5 Evaluation of the model: a basis for understanding brain processes and confidence for correct vs incorrect decisions

The spiking network model predicts probabilistic decision-making with larger neuronal responses in the winning attractor on correct than on error trials, because the spiking noise-influenced decision attractor state of the network is consistent with the external evidence, and this neuronal activity in turn predicts larger BOLD responses on correct than error trials in brain regions involved in making choices. Moreover, the model predicts that the neuronal activity in the winning attractor and the BOLD response will become larger on correct trials as ΔI increases and confidence increases; and will become smaller in the winning attractor as confidence decreases on error trials as ΔI increases. These effects can be understood in that on an individual trial the spiking-related and other noise will influence the decision made and therefore which attractor is firing, that for the correct or the incorrect choice. If the winner is that for the correct choice, then the firing rate will be higher with increasing ΔI as then the external evidence, the incoming firing rates, will be higher for this attractor than for the losing attractor, and as ΔI increases, will increase the firing rates more and more for the winning attractor. On the other hand, if the incorrect attractor wins on a particular trial, then the incoming evidence will provide weak support for the winning (incorrect) attractor, and will instead tend to increase the relatively low firing rates of the losing (correct) attractor, which in turn through the inhibitory interneurons will tend to decrease the firing rates further as ΔI increases in the winning but incorrect attractor. These results of these effects are shown in Fig. 6.11a.

In this situation, the model is very useful, for because it includes synaptic currents and firing rates, it is possible to predict from these the BOLD signal by convolution with the haemodynamic response function. However, on correct trials as ΔI increases, the firing rate of the winning (correct) population increases, but of the losing (incorrect) population decreases, as shown in Fig. 6.11a, so the model is needed to show which effect dominates the BOLD response, and it is found quantitatively to be the increase in the high firing rate attractor state that dominates the BOLD signal, rather than the decrease in the low firing rate state in the losing attractor, as shown in Fig. 6.12b. Similarly, on incorrect trials as ΔI increases, the firing rate of the high firing rate incorrect attractor will decrease, and of the low firing rate correct attractor will increase, but the net effect predicted for the BOLD signal is a decrease as ΔI increases (see Fig. 6.12b). The model makes qualitatively similar predictions from the synaptic currents.

A fascinating and important property of this biologically based model of decision-making is that decision confidence on correct vs error trials is an *emergent property* of the neuronal decision-making mechanism, and is represented by the firing rates of the winning attractor on correct vs error trials, which also alter in the same way as decision confidence as a function of ΔI. This is shown for the first time. It is also an emergent property of the neuronal theory of decision-making, yet elegantly simple, that the firing rates of the neurons in the winning attractor on correct trials increase with ΔI, and that the firing rates of the neurons in the winning attractor on error trials decrease with ΔI (Rolls, Grabenhorst and Deco 2010c).

We note that the way we infer confidence from the ΔI of the ratings given is not only based on previous research in which confidence ratings and/or behaviour are related to ΔI (Vickers and Packer 1982, Jonsson et al. 2005, Kepecs et al. 2008), but is also not unlike the way in which 'decision utility' (Kahneman, Wakker and Sarin 1997), often referred to as 'subjective value' (Kable and Glimcher 2007), is usually inferred from choices or from neural activity (Kable and Glimcher 2007). In more detail, not only does decision confidence increase with ΔI, but also decision confidence is lower on error than on correct trials (Vickers and Packer 1982, Jonsson et al. 2005). The theory of decision-making described here indicates that in addition decision confidence on error trials decreases as ΔI increases, and shows how noise in the brain is related to these decision-related effects.

Kepecs, Uchida, Zariwala and Mainen (2008) have described a model of confidence estimation in which each stimulus, and a memory for the category boundary, is encoded as a distribution of values, and a choice is made by comparing each sample with the decision boundary, and a confidence value is estimated by calculating the absolute value of the distance from the decision boundary. This is unlike the present model, which instead is a plausible neuronal network model in which confidence emerges from the firing rates in the decision-making neuronal mechanism.

The fMRI results described here provide powerful support for this model of decision-making (Deco and Rolls 2006, Rolls, Grabenhorst and Deco 2010b, Rolls, Grabenhorst and Deco 2010c), in that not only does the BOLD signal in brain areas implicated in decision-making increase with ΔI when all trials are considered, but also the BOLD signal decreases on error trials as ΔI increases, and increases on correct trials as ΔI increases. In an attractor model of the activity of lateral intraparietal (LIP) cortex neurons in a motion coherence discrimination task (Roitman and Shadlen 2002), Wong et al. (2007) have also found longer reaction times on error than on correct trials, but have not related this to confidence or to ΔI, and also have not shown how the firing rates and the predicted BOLD signal from the process are influenced by ΔI and are related to decision confidence.

In another comparison, the accumulator, counter, or race models of decision-making in which the noisy evidence for different choices accumulates until some criterion is reached (Vickers 1979, Vickers and Packer 1982, Ratcliff and Rouder 1998, Ratcliff et al. 1999, Usher and McClelland 2001) do not describe a neurobiological mechanism, do not make specific predictions about how neuronal firing rates, synaptic currents, or BOLD signals will alter as decisions become easier, and do not make specific predictions about confidence in a decision, but do account for behavioural performance. In this context, it is thus of interest that the integrate-and-fire attractor approach as well as the other approaches can both be analyzed as diffusion processes (Roxin and Ledberg 2008), and that our approach is not inconsistent with these accumulator or with Bayesian (Ma, Beck, Latham and Pouget 2006, Beck, Ma, Kiani, Hanks, Churchland, Roitman, Shadlen, Latham and Pouget 2008, Ma, Beck and Pouget 2008) approaches to decision-making. However, while accumulator models implement a linear diffusion process operation with noisy inputs, the attractor model implements a non-linear diffusion process once it has left the boundary peak between two valleys in the energy landscape, which becomes faster due to the recurrent positive feedback in the network between the population of co-active neurons. Once the attractor has reached the new valley it may be relatively stable (see Fig. 6.1). Consistent with this analysis of a non-linear diffusion process for decision-making, it is harder to move the decision away from the way it is going by electrical stimulation if the stimulation is applied later rather than earlier in a trial in the lateral intraparietal area (LIP) during a motion discrimination task (Wong, Huk, Shadlen and Wang 2007).

In the fMRI study of the olfactory decision-making task, we found that neural activity was larger on correct than on error trials in the medial prefrontal cortex area 10, the posterior cingulate cortex, the subgenual cingulate cortex, and the dorsolateral prefrontal cortex. We also show that in the same brain regions the BOLD signal increases with ΔI on correct trials, and decreases with ΔI on error trials, showing that decision confidence is represented in neural activity in these areas. These effects are predicted from the model for brain areas involved in taking the decision, which in making the choice falls into a high firing rate attractor state. Medial prefrontal cortex area 10 is implicated in decision-making in that the BOLD signal is larger on trials on which choices must be made about which of the two stimuli presented is more pleasant, compared to trials on which the same stimuli are presented but a rating must be made on a continuous scale of the pleasantness of the (second) odour with no choice between the two odours (Rolls, Grabenhorst and Parris 2009c). Medial prefrontal cortex area 10 is also implicated in decision-making by the criterion used by others (Heekeren, Marrett, Bandettini and Ungerleider 2004, Heekeren, Marrett and Ungerleider 2008) that activation should increase with task easiness, that is with ΔI (Rolls, Grabenhorst and Deco 2010b). Medial area 10 is also implicated in decision-making in that a shopping decision-making task is impaired by medial prefrontal cortex damage (Burgess 2000).

The posterior cingulate cortex is implicated in decision-making in that some neurons there respond when risky, uncertain choices are made (McCoy and Platt 2005); and some neurons respond more when an expected large reward is not obtained, maintaining that firing until the next trial (Hayden, Nair, McCoy and Platt 2008) (probably reflecting input from orbitofrontal cortex error neurons that have attractor state-like persistent firing that encodes and maintains a negative reward prediction error signal (Thorpe, Rolls and Maddison 1983, Rolls and Grabenhorst 2008, Rolls 2009b)).

The subgenual cingulate cortex is connected with the ventromedial prefrontal cortical areas (Johansen-Berg, Gutman, Behrens, Matthews, Rushworth, Katz, Lozano and Mayberg 2008), and the activations we found there in the subgenual cingulate cortex may reflect inputs from the ventromedial prefrontal cortex. Further, neural activity in the subgenual cingulate cortex has been implicated in representing relative chosen value in an uncertain decision environment (Boorman, Behrens, Woolrich and Rushworth 2009).

The dorsolateral prefrontal cortex region activated in the present study has been implicated by earlier studies in decision-making, including decision-making about visual and vibrotactile stimuli (Heekeren et al. 2004, Kim and Shadlen 1999, Preuschhof, Heekeren, Taskin, Schubert and Villringer 2006).

An important control point is that activity in the mid-orbitofrontal cortex (which was not implicated in choices between the odours but that did have activations related to the pleasantness of the odours (Rolls, Grabenhorst and Parris 2009c)) did not show different activations on correct vs incorrect trials, and did not have activations that increased with ΔI on correct trials, and decreased with ΔI on error trials (Rolls, Grabenhorst and Deco 2010c). This is what is predicted for a brain area not involved in binary choice decision-making, but in representing pleasantness on a continuous scale, which is an important input to the decision-making system, providing the values of λ in Fig. 6.1. As shown earlier, there were significant dissociations between the decision-making areas and the mid-orbitofrontal cortex in terms of whether the activity was different on correct vs error trials, or was correlated on a continuous scale with subjective pleasantness, a representation that is likely to be a precursor and prerequisite for choice decisions about pleasantness (Rolls and Grabenhorst 2008, Grabenhorst, Rolls and Parris 2008, Rolls, Grabenhorst and Parris 2009c, Rolls, Grabenhorst and Deco 2010b, Rolls, Grabenhorst and Deco 2010c). We note that ΔI is not related in any simple way to the pleasantness of the second odour, presented at the onset time for the fMRI analyses described here. Instead, ΔI is the absolute value of the difference in the pleasantness of the first and second odours, and this could be high if the second odour is much less pleasant than the first.

It is a fundamentally important aspect of our theory of decision-making that the evidence for the decision-making, the values of λ in Fig. 6.1, are continuous-valued representations, and require a separate representation from the choice itself, which takes place in a separate network. It is this that enables us to simultaneously report the exact value on a continuous scale of two perhaps quite similar discriminanda, such as the pleasantness, motion or frequency of two stimuli, and at the same time make a binary choice between the stimuli using a different network in the brain.

6.9 Decisions based on confidence in one's decisions: self-monitoring

6.9.1 Decisions about confidence estimates

We have seen that after a binary decision there is nevertheless a continuous-valued representation of decision confidence encoded in the firing rates of the neurons in a decision-making integrate-and-fire attractor neuronal network (Fig. 6.11). What happens if instead of having to report or assess the continuous-valued representation of confidence in a decision one has taken, one needs to take a decision based on one's (partly) continuous-valued confidence estimate that one has just made a correct or incorrect decision? One might for example wait for a reward if one thinks one's decision was correct, or alternatively stop waiting on that trial and start another trial or action. We propose that in this case, one needs a second decision-making network, which takes decisions based on one's decision confidence (Insabato, Pannunzi, Rolls and Deco 2010).

6.9.2 A theory for decisions about confidence estimates

The architecture has a decision-making network, and a separate confidence decision network that receives inputs from the decision-making network, as shown in Fig. 6.15. The decision-making network has two main pools or populations of neurons, DA which becomes active for decision A, and DB which becomes active for decision B. Pool DA receives sensory information about stimulus A, and Pool DB receives sensory information about stimulus B. (We sometimes refer to stimulus A as odour A, and to stimulus B as odour B, as A and B were used in a neurophysiological investigation of odour decision-making to categorize an odour as A or B when there were different odour proportions in the mixture (Kepecs et al. 2008), as described in Section 6.9.3.) Each of these pools has strong recurrent collateral connections between its own neurons of strength w_+, so that each operates as an attractor population. There are inhibitory neurons with global connectivity to implement the competition between the attractor subpopulations. When stimulus A is applied, pool DA will usually win the competition and end up with high firing indicating that decision A has been reached. When stimulus B is applied, pool DB will usually win the competition and end up with high firing indicating that decision B has been reached. If a mixture of stimuli A and B is applied,

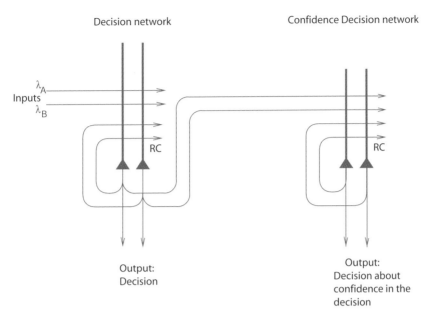

Fig. 6.15 Network architecture for decisions about confidence estimates. The first network is a decision–making network, and its outputs are sent to a second network that makes decisions based on the firing rates from the first network, which reflect the decision confidence. In the first network, high firing of neuronal population (or pool) DA represents decision A, and high firing of population DB represents decision B. Pools DA and DB receive a stimulus-related input (respectively λ_A and λ_B), the evidence for each of the decisions, and these bias the attractor networks, which have internal positive feedback produced by the recurrent excitatory connections (RC). Pools DA and DB compete through inhibitory interneurons. The neurons are integrate-and-fire spiking neurons with random spiking times (for a given mean firing rate) which introduce noise into the network and influence the decision-making, making it probabilistic. The second network is a confidence decision attractor network, and receives inputs from the first network. The confidence decision network has two selective pools of neurons, one of which (C) responds to represent confidence in the decision, and the other of which responds when there is little or a lack of confidence in the decision (LC). The C neurons receive the outputs from the selective pools of the (first) decision-making network, and the LC neurons receive $\lambda_{\text{Reference}}$ which is from the same source but saturates at 40 spikes/s, a rate that is close to the rates averaged across correct and error trials of the sum of the firing in the selective pools in the (first) decision-making network. (After Insabato, Pannunzi, Rolls and Deco, 2010.)

the decision-making network will probabilistically choose decision A or B influenced by the proportion of stimuli A and B in the mixture.

First we describe the operation of the decision-making (first) network, and show how the firing rates of its neurons reflect decision confidence. This part of the system operates in the same way as described earlier in this chapter. Fig. 6.16c shows the proportion of correct decisions as a function of the proportion of stimulus A and stimulus B in the mixture. The decision-making is probabilistic because of the spiking-related randomness in the network, as described in more detail elsewhere (Chapter 2 and Appendix A) (Wang 2002, Deco and Rolls 2006, Deco et al. 2009b, Deco and Marti 2007a, Marti et al. 2008). Fig. 6.16a shows that on trials when the DA neuronal population which represents decision A correctly wins and has a high firing rate, the firing rate increases further with the discriminability of the stimuli $\Delta\lambda$, and thus encodes increasing confidence. ($\Delta\lambda$ is defined as $(A-B)/((A+B)/2)$. $\Delta\lambda$ is thus large and positive if only A is present, is large and negative if only B is present, and is 0 if

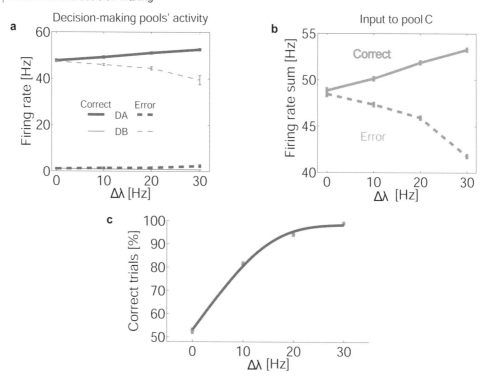

Fig. 6.16 Performance of the decision-making (first) network. (a) When pool DA correctly wins the competition, its firing rates are high and increase as a function of $\Delta\lambda$. When pool DB incorrectly wins making an error due to the noise and has a high firing rate, its firing rate decreases as a function of $\Delta\lambda$. The low firing rates when DB loses the competition on correct trials, and when DA loses the competition on error trials, are also shown. The error bars represent the sem. Confidence is thus encoded in the firing rates of the winning attractor, and is an emergent property of the decision-making network. (b) Sum of the firing rates from the DA and DB populations as a function of $\Delta\lambda$. This provides the input to the confidence (second) network selective pool C. The error bars show the sem. (c) The percentage correct performance of the decision-making network as a function of $\Delta\lambda$. (The error bars were estimated from the binomial distribution, and were small. The points are fitted by a Weibull function.) (After Insabato, Pannunzi, Rolls and Deco, 2010.)

A and B are present in equal proportions. $\Delta\lambda$ is another notation for ΔI.) The reason for the increase of firing rate with $\Delta\lambda$ on correct trials is that the external inputs from the stimuli A or B then support the (noise-influenced) winning attractor (pool DA) and add to the firing rates being produced by the recurrent collateral connections in the winning attractor. On the other hand, on error trials the firing rates of the winning pool (now DB, which represents decision B and wins despite the evidence because of noisy firing in the network) become lower as $\Delta\lambda$ increases, because then the external sensory inputs are inconsistent with the decision that has been taken, and do not support and increase the firing rate of the winning pool (Rolls, Grabenhorst and Deco 2010c). Confidence, which increases with $\Delta\lambda$ on correct trials and decreases with $\Delta\lambda$ on error trials (Vickers 1979, Vickers and Packer 1982, Jonsson, Olsson and Olsson 2005, Kepecs, Uchida, Zariwala and Mainen 2008, Rolls, Grabenhorst and Deco 2010c), is thus encoded in the firing rates of the winning attractor, and is an emergent property of the decision-making network. Moreover, the sum of the activity of the winning and losing populations also represents decision confidence on correct and error trials, as shown in Fig. 6.16b. It is this total firing from pools DA and DB of the first, decision-making, network,

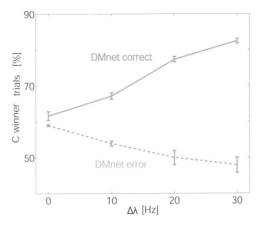

Fig. 6.17 Performance of the confidence decision (second) network. The percentage of trials on which in the second network the Confidence (C) population won the competition as a function of $\Delta\lambda$ for trials on which the decision-making (first) network (DMnet) was correct or incorrect. The performance of the LC population was the complement of this. (The parameters were set so that with $\Delta\lambda$ close to 0, approximately 60% of the trials were C trials, to be qualitatively in the same direction as in the experimental findings of Kepecs et al 2008.) (After Insabato, Pannunzi, Rolls and Deco, 2010.)

which reflects decision confidence, that is provided as the input to the confidence (second) network.

We now consider the operation of the confidence decision (second) network. The confidence network has two selective pools of neurons, one of which (C) responds to represent confidence in the decision, and the other of which responds when there is little or a lack of confidence in the decision (LC). (In the experiment of Kepecs et al. (2008), C corresponds to a decision to stay and wait for a reward, i.e. what they call the positive outcome population, though it really represents confidence or a prediction that the decision just taken will have a positive outcome. LC corresponds to a decision to abort a trial and not wait for a possible reward, i.e. what they call the negative outcome population, though it really represents lack of confidence that the decision just taken will have a positive outcome, equivalent to confidence that the decision just taken *will have* a negative outcome.) If the output firing of the winning attractor of the first, decision-making, network (whether DA or DB, reflected in their sum, as shown in Fig. 6.16b) is high because the decision just taken has sensory inputs consonant with the decision, then the confidence decision network acting as a second level network takes the decision, probabilistically as before, to have confidence in the decision, and the C population probabilistically wins the competition. If the output firing of DA and DB (reflected in their sum) is low because the decision just taken has sensory inputs that are not consonant with the decision, then with weaker driving inputs to the C network, it loses the competition with LC. The confidence network in this case takes the decision, probabilistically as before, to have a lack of confidence in the decision, in that the LC population wins the competition. The confidence decision network thus acts as a decision-making network to make confident decisions if the firing rates from the first, decision-making, network are high, and to make lack of confidence decisions if the firing rates from the first, decision-making, network are low.

As a result of the competition between the inputs to the C population and the inputs saturating at 40 Hz to the LC population, on trials when the (first) decision network is correct, the C population tends to win the competition more frequently as $\Delta\lambda$ increases, as shown in Fig. 6.17. Thus a decision to act confidently about one's first decision is more likely to be

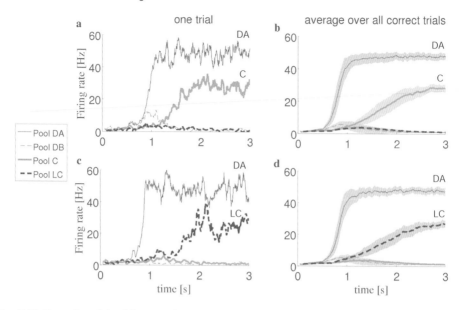

Fig. 6.18 Examples of the firing rate time courses of the selective pools in the decision-making (first) network and in the confidence decision (second) network. Each plot shows the activity in time of selective pools in the decision-making network in which pool DA won, and in the confidence network, for a difficult decision ($\Delta\lambda$=0). On 60% of trials the confidence network selective pool C won the competition as shown in the first row: (a) single trial, (b) average over all correct trials. On 40% of trials pool LC won the competition: (c) single trial, (d) average over all correct trials. The separation of the firing rates begins after the decision is taken and the general temporal structure of the network is in accordance with the experimental results of Kepecs et al (2008). (After Insabato, Pannunzi, Rolls and Deco, 2010.)

made as $\Delta\lambda$ increases on correct trials. On the other hand, when the first network makes an error, the C population tends to win the competition less frequently as $\Delta\lambda$ increases, as shown in Fig. 6.17, and correspondingly on error trials the proportion of trials on which the LC population wins increases with $\Delta\lambda$. (The percentage correct of the LC population is the complement of that shown in Fig. 6.17.) Thus a decision to lack confidence about one's first decision is more likely to be made as $\Delta\lambda$ increases on error trials, and this might make one abort such a trial, as in the experiment of Kepecs et al (Kepecs et al. 2008).

The general time structure of the neuronal activity in the model is in good accordance with the experimental results (Kepecs et al. 2008): as shown in Fig. 6.18, the confidence decision takes place after the first decision, and separation of the firing rates of the two selective populations C and LC occurs after the (first) decision-making network has reached a decision state, as in Fig. 3a-d of Kepecs et al. (2008).

It is important to examine the firing rates in the C and the LC attractor neuronal populations as a function of $\Delta\lambda$ on correct and incorrect trials, for they provide an account for neuronal responses recorded during decision-making (Kepecs et al. 2008), and those neurophysiological results in turn validate the model. We find for the confidence decision-making network that on correct trials with high $\Delta\lambda$=30 (easy decisions), C has a high firing rate, whereas it has a lower rate for $\Delta\lambda$=10, that is difficult decisions, as shown in Fig. 6.19. Conversely, on error trials when the firing rates in the first level, decision-making, network are lower, the confidence neurons C lose the competition and have relatively low firing rates, which decrease even further as the magnitude of $\Delta\lambda$ increases.

The firing rates (mean) in the confidence decision-making network of the C (confident,

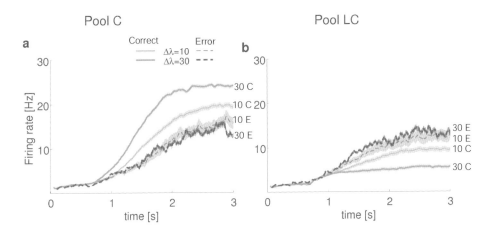

Fig. 6.19 Firing rates in the confidence decision-making network of the C (confident) and LC (lack of confidence) populations of neurons for trials when the first decision-making network is correct or incorrect for easy decisions ($\Delta\lambda$=30) and difficult ($\Delta\lambda$=10) decisions. Key: 30C–$\Delta\lambda$=30 first network correct; 10C–$\Delta\lambda$=10 first network correct; 10E–$\Delta\lambda$=10 first network error; 30E–$\Delta\lambda$=30 first network error. The firing rates are averaged across trials independently of which attractor the C / LC populations reached, and correspond to the firing rates in the thick box in Fig. 6.20. (The shaded areas represent the sem). The decision cues were turned on at t=500 ms. (After Insabato, Pannunzi, Rolls and Deco, 2010.)

'positive outcome') and LC (lack of confidence, 'negative outcome') populations of neurons for trials when the first decision-making network is correct or incorrect as a function of $\Delta\lambda$ are shown in Fig. 6.20. The thick lines show the mean firing rates for the C and LC pools for all trials on which the first network was correct, and separately for those in which the first network was in error. (We identify the LC population of neurons with the negative outcome population of neurons described by Kepecs et al (Kepecs et al. 2008), which have similar properties, as shown in Fig. 6.21. Further, we identify the C population of neurons with the positive outcome population of neurons described by Kepecs et al (Kepecs et al. 2008), which have similar properties, as shown in Fig. 6.21.)

However, as shown in Fig. 6.17, the confidence decision-making network was itself increasingly incorrect (i.e. took a confidence decision that was inconsistent with the decision taken by the (first) decision-making network) as $\Delta\lambda$ approached 0, and the firing rates in the thick lines of Fig. 6.20 reflect the fact that on some trials the C network won the competition, and on some trials it lost. This is effectively how Kepecs et al. (2008) presented their data (for they knew only whether the decision itself was correct, and did not measure while recording whether the rat took a confidence-related decision to stay with or abort a trial), and there is good correspondence, as can be seen by comparing Fig. 6.21 with Fig. 6.17.

If instead of taking the mean firing rate of the C neurons based only on whether the first decision was correct, we take just the trials on which the C (confidence) pool won the competition, the thin lines of Fig. 6.20 show for the C pool an average rate of close to 28 spikes/s that tends to increase with $\Delta\lambda$ when the first network is correct, and tends to decrease with $\Delta\lambda$ when the first network is in error. (This is supported by the data shown in Fig. 6.19.) If we take just the trials on which the C population lost the competition, the thin lines show for the C pool an average rate of close to 2 spikes/s. Conversely, for the LC pool of neurons, if we take just the trials on which the LC population won the competition, the thin lines show

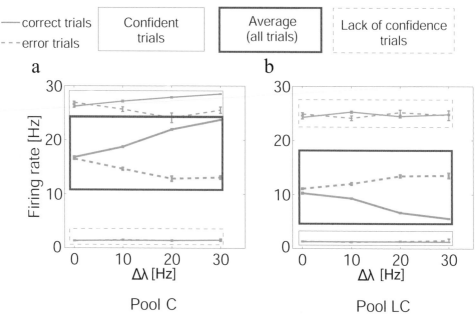

Fig. 6.20 Firing rates (mean ± s.e.m.) in the confidence decision-making network of the C (confident, 'positive outcome') (a), and LC (lack of confidence, 'negative outcome') (b) populations of neurons. In the solid thin boxes the firing rates of the C (left panel) and LC (right panel) pools are shown on trials on which the C (confidence) population won the competition. The lines in the solid thin boxes show for the C pool an average rate of close to 28 spikes/s which tends to increase with $\Delta\lambda$ when the first network is correct, and tends to decrease with $\Delta\lambda$ when the first network is in error. In the dashed thin boxes the firing rates of the C (left panel) and LC (right panel) pools are shown on trials on which the LC (lack of confidence) population won the competition. Again for the C population, if we take just the trials on which the C population lost the competition, the lines in the dashed thin boxes show for the C pool an average rate of close to 2 spikes/s. The activities of the C and LC populations of neurons, averaged over all trials, confident and lack of confidence, are shown in the thick boxes. As shown in Fig. 6.17, the confidence decision-making network was itself increasingly incorrect as $\Delta\lambda$ approached 0, and the firing rates in the thick lines reflect the fact that on some trials the C population won the competition, and on some trials it lost. (After Insabato, Pannunzi, Rolls and Deco, 2010.) (See colour plates Appendix B.)

for the LC pool an average rate of close to 26 spikes/s. If we take just the trials on which the LC population lost the competition, the thin lines show for this LC pool an average rate of close to 2 spikes/s. These firing rates shown in the thin lines in Fig. 6.20 are generally as expected, and the differences with $\Delta\lambda$ are due to whether the output of the decision-making (first) network shown in Fig. 6.16b are consistent or inconsistent with the decision taken by the confidence decision (second) network, which is of course influenced by the spiking noise in the confidence decision network, which can take the wrong decision given the evidence it receives from the decision-making network shown in Fig. 6.16b.

The fact that the changes of firing rates found in the rat by Kepecs et al. (2008) as shown in Fig. 6.21 as a function of $\Delta\lambda$ are comparable with those shown in the thick lines in Fig. 6.20 provides good support for the present model. However, Kepecs et al. (2008) did not distinguish trials in which a second-layer confidence decision network was in error or not as they did not record neuronal activity when they could examine whether the rat aborted a trial, and we suggest that it would now be interesting to do this. Further, it is notable that the change of firing rate with $\Delta\lambda$ found in the rat matches only that of the thick lines in Fig. 6.20 which includes all trials irrespective of the decision taken by the confidence decision (second) network, and not by the thin lines in Fig. 6.20 which reflect the decision taken by the confidence network. This leads to the novel prediction that different results will be found to those presented by Kepecs et al. (2008) if in a future experiment the responses of similar neurons are separated according to whether each trial is aborted or not. We predict in particular that the neurons will have activity like that shown in the thin lines in Fig. 6.20, and will be of two types (Insabato, Pannunzi, Rolls and Deco 2010).

One type will be similar to that of the C (confidence in the prior decision) neurons shown in Fig. 6.20 in which the firing rate is high on trials on which the confidence decision is to stay with the first decision, and low if the confidence (second) decision is to abort the trial. The prediction further is that the firing rates of these confidence neurons will change with $\Delta\lambda$ as shown by the thin lines in Fig. 6.20a, that is these high firing rates will tend to increase as a function of $\Delta\lambda$ if the first decision (taken by the decision-making, first, network) is consistent with the evidence (i.e. correct), as shown at the top of Fig. 6.20a, and to decrease as a function of $\Delta\lambda$ if the first decision (taken by the decision-making, first, network) is inconsistent with the evidence (i.e. is an error), as also shown at the top of Fig. 6.20a.

The second type of neuron will be similar to that of the LC (lack of confidence in the prior decision) neurons shown in Fig. 6.20b in which the firing rate is high on individual trials on which the confidence decision is to abort the trial, and low if the confidence (second) decision is to stay with the first decision. (The firing rates of the LC population do not change much with $\Delta\lambda$ as shown by the lines in the thin boxes of Fig. 6.20b because the input from the saturating neurons has a fixed firing rate.) It is only when we categorise the neurons according to whether the first decision was correct or not that curves similar to those shown by thick lines in Fig. 6.20 and as reported by Kepecs et al. (2008) will be found, and such curves and analyses do not capture fully the properties of the confidence decision-related neurons, which are we propose as shown in the thin lines in Fig. 6.20 (Insabato, Pannunzi, Rolls and Deco 2010).

These two predicted effects have not been sought yet in neurophysiological investigations (Kepecs et al. 2008), as all the correct trials were averaged, and all the incorrect trials were averaged. We thus make the novel prediction that when an animal incorrectly judges its choice, acting with confidence or lack of confidence that is inappropriate, new types of neuronal firing activity from the confidence and lack of confidence neurons will be found (Insabato, Pannunzi, Rolls and Deco 2010).

6.9.3 Decisions about confidence estimates: neurophysiological evidence

The model we describe accounts for results obtained by Kepecs, Uchida, Zariwala and Mainen (2008) from a binary decision task. The rats had to perform a binary categorization task with a mixture of two pure odourants (A, caproic acid; B, 1-hexanol), by entering one of two ports to indicate that the mixture was more like odour A or odour B (see Fig. 6.21). Correct choices were rewarded after a delay of 0.3–2 s. Varying the relative concentration of the odourants allowed the difficulty of the trial to be altered.

Neural activity related to decision confidence should occur just after the decision is taken and before the trial outcome. Kepecs et al. (2008) therefore analyzed recordings of neuronal activity during a delay period after a decision had been taken before a reward was given. The single neuron recordings were made in the rat orbitofrontal cortex. (Exactly what area in primates and humans corresponds to the area in which recordings were made is not yet clear.) The neurons were then divided into two groups based on whether they fired faster on correct or on error trials. Kepecs et al. (2008) found that the group of neurons ('negative outcome' neurons) with an increased firing rate on error trials had higher firing rates with

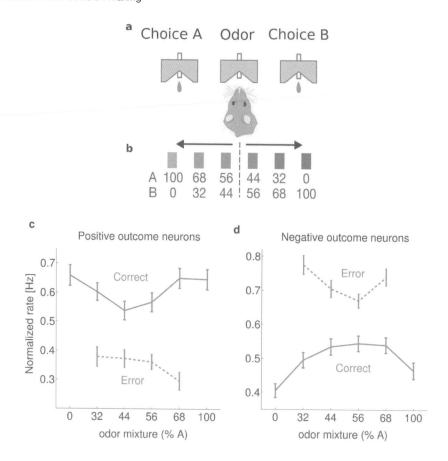

Fig. 6.21 (a) Odour mixture categorization task: when the rat enters the odour port, an odour mixture is delivered. Then the subject categorizes the mixture as A or B by moving left or right, according to the dominant odour. The stimulus is defined by the percentage of the two pure odourants A and B in the mixture as shown in (b). When odours are present in almost equal quantities, there is no bias to the decision process. Moving the stimulus towards pure odourants raises the bias towards one of the decisions and improves performance. (c) Mean normalized firing rate of the positive outcome selective neuronal population. These neurons fire faster on correct trials then on error trials, and increase their firing rate as the task becomes easier. On error trials, when the rat makes the wrong choice, these neurons fire more slowly, and decrease their firing rates more as the task becomes easier. If $\Delta\lambda$ is the discriminability of the stimuli, this is 0 for a 50% mixture of odours A and B, and $\Delta\lambda$ increases both to the left on the abscissa and to the right from 50% as the proportion of one of the odours in the mixture increases and the decision of which odour it is becomes easier. (d) Mean normalized firing rate of the negative outcome selective neuronal population in the orbitofrontal cortex as recorded by Kepecs et al (2008). These neurons have sustained activity on error trials, increasing their firing rate further when the decision is made easier. On correct trials they fire slower, and decrease their firing rates as the decision becomes easier. (Modified from Kepecs et al (2008)). (See colour plates Appendix B.)

easier stimuli. The same neurons fired at a substantially lower rate on correct trials, and on these trials the firing rates were lower when the decision was made easier (Fig. 6.21d). This produced the opposing V-shaped curves shown in Fig. 6.21d. These authors argued that this pattern of activity encoded decision confidence. The properties of these neurons, called 'negative outcome' neurons by Kepecs et al. (2008) (Fig. 6.21d), correspond to the Lack of Confidence (LC) neurons illustrated in Fig. 6.20b. We therefore interpret the 'negative outcome' neurons as being produced by a second network taking confidence decisions (about for example whether to abort a trial) based on the output of a first level decision-making network.

These 'negative outcome' neurons of Kepecs, Uchida, Zariwala and Mainen (2008) are very similar to the primate orbitofrontal cortex error neurons described by Thorpe, Rolls and Maddison (1983), which fire with a high rate which is prolonged for several seconds consistent with attractor dynamics on error trials but not on correct trials of a visual discrimination task (see Fig. 6.22). In more detail, these neurons fire when an expected reward is not received, for example during extinction or reversal of a visual discrimination task. Particularly when these neurons fire in extinction (when an expected reward is not delivered), their firing reflects low confidence in a decision just taken, in a way rather analogous to that described by Kepecs et al. (2008). Although the mechanism just described can take decisions about confidence without an outcome (reward or punishment) being received, the underlying mechanisms for the computation of error may we suggest be rather similar for the error neurons of Thorpe, Rolls and Maddison (1983), which compute negative reward prediction error (Rolls and Grabenhorst 2008, Rolls 2009b). Because the reversal of the behaviour described by Thorpe, Rolls and Maddison (1983) takes place in one trial, the actual mechanism proposed is the reversal of a rule attractor by the prolonged firing of error neurons, with the rule attractor biasing conditional reward neurons found in the orbitofrontal cortex (Rolls 2008d, Deco and Rolls 2005d, Rolls 2009b).

The 'positive outcome' neurons of Kepecs et al. (2008) fired at a higher rate on correct than error trials, and increased their firing rates as the decisions became easier on correct trials (Fig. 6.21c). On incorrect trials, these 'positive outcome' neurons fired at lower and lower rates as the decision was made easier, that is as ΔI increased. These neurons thus had the same properties as the Confidence population of neurons in the Confidence Decision network, as illustrated in Fig. 6.20a. We thus interpret the 'positive outcome' neurons as being produced by the computational mechanism just described, in a second level attractor network that effectively 'monitors' a first level decision-making network, and takes decisions about decision confidence.

Kepecs et al. (2008) performed a second experiment to investigate if rats were able to make use of this information encoded by the orbitofrontal cortex neurons they described. The delay period was prolonged up to 8 s in order to allow the rat to reinitiate the trial. The subject could decide to leave the choice port and reinitiate the trial, or could wait for the reward to be delivered at the choice port. It was found that when the likelihood of a reward (i.e. the confidence) was low, due to the decision difficulty and the choice just made, the rat reinitiated the trial by leaving the choice port and returning to the odour port to sample a new mixture on a new trial. The probability that the rat would restart a trial as a function of stimulus difficulty and the choice just made was consistent with the responses of the neurons with activity described as encoding decision confidence: in our terminology, in taking decisions based on one's confidence in an earlier decision.

This finding strengthens the interpretation we place on the function of the Confidence and Lack of Confidence neurons. The mechanism we describe is however very different from that suggested by Kepecs et al. (2008). The mechanism we describe is consistent with the stochastic operations performed by attractor networks in the unifying and simple approach consistent with the underlying cortical architecture and function that is described in this book.

The model we described of decisions about confidence (Insabato, Pannunzi, Rolls and Deco 2010) thus is confirmed by and provides a computational account of the findings of Kepecs, Uchida, Zariwala and Mainen (2008). Moreover, the model makes predictions, not yet tested, about the firing rates of the different populations of neurons on trials on which the confidence decision network makes an error due to noise in the system.

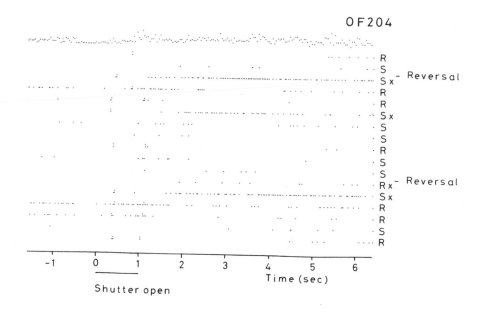

Fig. 6.22 Error neuron: responses of an orbitofrontal cortex neuron that responded only when the monkey licked to a visual stimulus during reversal, expecting to obtain fruit juice reward, but actually obtaining the taste of aversive saline because it was the first trial of reversal. Each single dot represents an action potential; each vertically arranged double dot represents a lick response. The visual stimulus was shown at time 0 for 1 s. The neuron did not respond on most reward (R) or saline (S) trials, but did respond on the trials marked x, which were the first trials after a reversal of the visual discrimination on which the monkey licked to obtain reward, but actually obtained aversive saline because the task had been reversed. This is thus an example of a 'negative reward prediction error neuron'. It is notable that after an expected reward was not obtained due to a reversal contingency being applied, on the very next trial the macaque selected the previously non-rewarded stimulus. This shows that rapid reversal can be performed by a non-associative process, and must be rule-based. (After Thorpe, Rolls and Maddison 1983.)

6.9.4 Decisions about decisions: self-monitoring

Our argument is as follows. Sometimes it may be necessary to take a categorical decision based on the continuous value representation of confidence encoded as just described in the decision-making network. To make the decision based on confidence, we therefore need a second decision-making network, the confidence network shown in Fig. 6.15. In the case of Kepecs et al. (2008), the animal had to decide whether to stay and wait because confidence was high for a reward after the choice had been made, or whether to abort the trial and start a new trial because confidence in the decision just taken was low. It is for this type of choice that the second network is needed, and the way it operates is exactly in the same way as the first network in terms of how it takes the decision. The difference is primarily that the input to the second (confidence decision) network is the output of the decision-making network, the first network, the firing rates of which reflect and effectively encode the continuous-valued representation of confidence.

If the decision-making network and the confidence decision network are localised separately in the brain, one could predict some interesting dissociations neuropsychologically. For example, if the functions of the confidence decision network were impaired, one would predict good decision-making, but little ability to later decide whether one should have confidence in the decision just taken. Given that the decision network logically must come before the confidence decision network, and the hierarchical nature of cortical processing, one might predict that in humans the confidence network for decisions about the pleasantness of stimuli might be at the anterior end of the medial prefrontal cortex area 10 region implicated in choice decision-making (Rolls and Grabenhorst 2008, Grabenhorst, Rolls and Parris 2008, Rolls, Grabenhorst and Parris 2009c, Shallice and Burgess 1991), or in another area connected to medial prefrontal cortex area 10.

6.10 Synthesis: decision confidence, noise, neuronal activity, the BOLD signal, and self-monitoring

6.10.1 Why there are larger BOLD signals for easy vs difficult decisions

The attractor neuronal network model of decision-making described here and in Chapter 5 provides a firm foundation for understanding fMRI BOLD signals related to decision-making (Rolls, Grabenhorst and Deco 2010b). The model provides two reasons for the fMRI signal being larger on easy than on difficult trials during decision-making.

The first reason is that the neurons have somewhat higher firing rates on easy trials (e.g. those with a large ΔI) than on difficult trials, as shown in Fig. 6.3a. The reason for the faster response is that the difference in the inputs to the two attractors modifies the flow landscape so that the activity has a smaller barrier over which to jump towards one of the attractors (see Fig. 6.1c and Section 2.3). The fMRI BOLD signal thus starts sooner, even if by only a time in the order of 0.5 s, as shown in Fig. 6.3b.

The second reason is that the firing rates become higher in the winning attractor as ΔI increases, as shown in Fig. 6.3a, and this also contributes to a larger BOLD signal for easy vs difficult decisions. The firing rate of the winning attractor is higher with a large ΔI because the external input adds to the internal recurrent collateral synaptic effect (see Fig. 6.1a) to lead to a higher firing rate. This effect is increased because the external input to the D2 neurons is low, and so it competes less with the D1 attractor, also leading to higher rates in the D1 attractor with high ΔI.

Thus this approach from integrate-and-fire networks provides a foundation for understanding the shorter neuronal response times (Kim and Shadlen 1999, Shadlen and Newsome 2001) and the larger BOLD signals on easy than on difficult trials found in brain regions (Heekeren et al. 2004, Heekeren et al. 2008) that include for decisions about pleasantness the medial prefrontal cortex area 10 (Fig. 6.8). Very interestingly, the BOLD signal is linearly related to task easiness, as quantified by ΔI. This is predicted from the model (Figs. 6.5 and 6.6), and found experimentally (Fig. 6.9) (Rolls, Grabenhorst and Deco 2010b, Rolls, Grabenhorst and Deco 2010c).

6.10.2 Validation of BOLD signal magnitude related to the easiness of a decision as a signature of neural decision-making

The findings (Rolls, Grabenhorst and Deco 2010b) validate a larger BOLD signal on easy than on difficult trials as an fMRI signature of a region involved in decision-making, in that the areas that show this effect are areas implicated by other criteria in decision-making, including their larger activation on trials on which choice decisions must be made, rather than continuous ratings of affective value (Grabenhorst, Rolls and Parris 2008, Rolls, Grabenhorst and Parris 2009c) (Fig. 6.8). The orbitofrontal and pregenual cortex, where continuous affective value but not choice is represented, do not show this easy vs difficult effect in the BOLD signal, as shown by the experimental fMRI results in Fig. 6.8.

6.10.3 Predictions of neuronal activity during decision-making

The model and its analysis described here also make predictions about the type of neuronal activity that will be found in brain areas involved in the choice part of decision-making, in which a categorical state must be reached. The predictions are that the neuronal activity will dynamically progress to a state in which there is a high firing of one population of neurons that occurs when one choice is made, with the other populations showing low activity, often below the spontaneous level, due to inhibition from the other attractor (especially when ΔI is high, as shown in Figs. 6.2 and 6.3a). This high vs low neuronal activity with almost a binary distribution of firing rates (as the decision is reached, either firing with a high firing rate, or a low firing rate, that predicts which decision is reached), is an important characteristic of a brain region performing the type of decision-making described here. However, although the firing rates are categorical in this sense, there is predicted to be some effect of ΔI on the firing rate when in the attractor, with as shown in Fig. 6.11a on correct trials an increase with ΔI of the winning population, with the additional prediction that on correct trials the firing rate of the losing populations will tend to become smaller as ΔI increases. Conversely, on error trials, as shown in Fig. 6.11a, there is predicted to be a decrease with ΔI of the winning population, with an increase of the (low) rate of the losing population as ΔI increases.

Another diagnostic property of the neuronal activity in this type of decision-making system is that the reaction times of the neurons (the time it takes to reach an attractor) will decrease as a function of ΔI, as shown in Fig. 6.3b.

Another fundamental property of this type of decision-making network is that it inherently shows statistical variation, with neuronal noise from the almost random, Poisson, spike times influencing the time when the system starts to fall into an attractor, and then a rapid change of firing rate, due to the supralinear positive feedback from the recurrent collateral synaptic connections (the recurrent collateral effect) starting to take off (see Section 1.10 and Chapter 5). It is to indicate and quantify this trial by trial statistical variability that the figures of the simulations in this chapter show the standard deviation of the responses. This is a case where it is not the peristimulus time histogram (i.e. an average across trials of neuronal activity) that is a main important measure of the neuronal functionality, but just as important and interesting is what happens on a trial by trial basis, and especially its variability, as the statistical fluctuations unfold. Some examples of the noisy temporal evolution of the firing rate on individual trials are shown in Figs. 6.2 and 6.10.

6.10.4 Multiple types of decision are made, each in its own brain region

It has been argued that this decision-making categorisation effect, where the attractor properties produced by for example the recurrent collateral effect make the decision on a given trial, is found in a number of different brain regions, each of which can be conceived of as making its own noise-influenced decision (Rolls 2008d) (see also Section 8.12). For example, the lateral intraparietal area can be thought of as reaching a decision about the state of global motion in a noisy display (Shadlen and Newsome 1996, Gold and Shadlen 2002, Gold and Shadlen 2007, Churchland, Kiani and Shadlen 2008); the medial prefrontal cortex area 10 as making choices about pleasantness between rewarding (affective) thermal (Grabenhorst, Rolls and Parris 2008) and olfactory (Rolls, Grabenhorst and Parris 2009c) stimuli; decisions about the intensity of odours as being related to larger activations in the dorsolateral prefrontal cortex and ventral premotor cortex (Rolls, Grabenhorst and Parris 2009c); and decisions about the frequency of vibrotactile stimuli to neuronal activity in the ventral premotor cortex (Romo and Salinas 2003, Deco, Rolls and Romo 2009b). Given that it is a property of a number of cortical regions that they can be involved in making choices between the inputs that are represented in those regions, we need not necessarily expect that at the neuronal level activity will always be related to the final behavioural choice of the subject, for there may be other decision-making processes in the final change to behavioural output, including those involved in response selection (Rolls 2008d).

How then is neuronal activity different in brain systems not involved in choice decision-making? The difference it is suggested is that in other brain areas, neuronal activity will be required to represent the exact details about the sensory stimulus or affective state that is produced, and not to be driven into a noise-influenced categorical decision state. For example, when we have to make a decision between two stimuli of similar pleasantness, we can at the same time rate on a continuous scale exactly what the pleasantness of each stimulus is, yet, presumably with a different brain system, come on an individual trial to a categorical choice of one rather than the other stimulus (Rolls and Grabenhorst 2008, Grabenhorst, Rolls and Parris 2008, Rolls, Grabenhorst and Parris 2009c). With respect to affect, representations correlated with continuous affective value (reward value) are found in brain regions such as the orbitofrontal cortex (O'Doherty, Kringelbach, Rolls, Hornak and Andrews 2001, Rolls 2005, Rolls 2007e, Rolls 2008d, Rolls and Grabenhorst 2008), and here activations are much less influenced than in medial prefrontal cortex by which choice is made (Rolls and Grabenhorst 2008, Grabenhorst, Rolls and Parris 2008, Rolls, Grabenhorst and Parris 2009c, Rolls, Grabenhorst and Deco 2010b).

It is suggested that the underlying computational basis for this is that in earlier cortical processing areas the recurrent collateral activity must not be allowed to be too strong, lest it produce a system that would fall into attractors where the internal recurrent collateral effect dominated the external input. This would distort the sensory signal, which should not occur in early sensory areas, where some approach to linearity can be useful so that undistorted signal is available for later processing. It is suggested that in contrast in later cortical processing areas where decisions are being taken or categorical states are being reached, the external inputs to the network must influence where the system reaches, but the recurrent collateral effects can be stronger so that an attractor state with categorical properties is reached. It could be a simple difference, genetically guided, between cortical areas that accounts for this difference, with the relative strength of the recurrent collateral synaptic connections being stronger in later cortical areas such as the prefrontal cortex, where this also serves the same purpose of enabling short-term memory to be maintained by the attractor even when no stimuli remain (Rolls 2008d). It could also be that transmitter effects, such as those produced by acetylcholine (Giocomo and Hasselmo 2007), could alter the relative balance between the external inputs to a network and the internal recurrent collateral effects, and thus alter dynamically how a network in the brain is being used for decision-making vs signal processing (Rolls 2008d).

6.10.5   The encoding of decision confidence in the brain

An interesting property of this network model of decision-making revealed by these analyses is that the degree of confidence in the decision is encoded in the firing rate of the attractor networks that take the decision, as shown in Fig. 6.3a (with the per cent correct, which is closely correlated with subjective confidence (Vickers and Packer 1982, Jonsson, Olsson and Olsson 2005), shown in Fig. 6.3c). (The higher firing rates when in the attractor on easy vs difficult trials are also clearly illustrated in Fig. 6.2.) In essence, the higher the firing rate of a particular attractor, the more confident or certain one is that the decision is correct. In fact, as shown in Fig. 6.3a, the difference of firing rates between the winning and losing populations also clearly reflects the confidence in the decision. In Section 6.8 we show how similar analyses follow through for confidence on correct vs error trials, and how these confidence estimates depend on ΔI. Thus confidence may not be a special property of a group of neurons (Kepecs, Uchida, Zariwala and Mainen 2008), but instead emerges as a property of the type of integrate-and-fire attractor network model of decision-making described here.

Consistent with this, activations in the medial prefrontal cortex, area 10, an area implicated by other criteria in decision-making, are correlated with the confidence in the decision, as indexed by increasing values of ΔI, as shown in Section 6.7 and Figs. 6.8 and 6.9.

Moreover, the theoretical approach described here is strongly supported by the evidence that neurons in the macaque parietal cortex involved in perceptual decision-making about motion stimuli have firing rates that are higher for decisions in which there is confidence (Kiani and Shadlen 2009).

The performance of the attractor network on error trials is also of interest in relation to the observation that when we make an incorrect choice, our confidence in our decision (or certainty) is likely to be low. As shown earlier, exactly this is represented in the firing rates of the neurons on incorrect trials, for as shown by the thin dotted curves in Fig. 6.11a, when an incorrect choice is made (and D2 wins the competition because of noise), the firing rate of the D2 attractor when winning decreases as ΔI increases (upper thin dotted line). The reason for this is that the external inputs, the evidence for the decision, are now working against the internal recurrent attractor dynamics, which have reached the wrong decision just because of the noise in the network. Thus on error trials, the confidence or certainty of a decision is also reflected in the firing rates of the attractors (Rolls, Grabenhorst and Deco 2010b).

We note that many further factors influence decision-making, including sensitivity to the non-reward that occurs on error trials, and impulsiveness (Rolls, Hornak, Wade and McGrath 1994a, Hornak, Bramham, Rolls, Morris, O'Doherty, Bullock and Polkey 2003, Hornak, O'Doherty, Bramham, Rolls, Morris, Bullock and Polkey 2004, Clark, Cools and Robbins 2004, Rolls and Grabenhorst 2008), both of which show important differences between individuals (Berlin, Rolls and Kischka 2004, Berlin and Rolls 2004, Berlin, Rolls and Iversen 2005), so that the inherent properties of the decision-making circuitry described here have an overlay that contributes to the final behaviour (see chapter 10 in Rolls (2008d)).

Another interesting property of the decision-making framework described and analyzed in this book, with quite clear neurophysiological and behavioural predictions, is that the percentage correct performance asymptotes to perfect at 100% with quite moderate levels of ΔI (Fig. 6.3c), whereas the neuronal responses continue to become higher as ΔI increases further (Fig. 6.3a), the BOLD signals become larger (Fig. 6.5b), and the reaction times decrease further (Fig. 6.3b).

Further, the inherent trial-by-trial variability in human choice reaction times, which has always been an enigma, can now be understood in terms of statistical fluctuations in decision-making networks in the brain. Further, a firm foundation for longer reaction times on error than on correct trials, and the relation to ΔI, has now been developed (see Fig. 6.11b).

What has been described in this chapter demonstrates that a major part of the strength of the current approach is that the framework for decision-making incorporates effects from many levels, including the synaptic and biophysical levels, and the abstract level of statistical fluctuations in dynamical systems, and is able to make predictions all the way from synaptic currents, to neuronal activity, to fMRI signals, and to behavioural and subjective performance and confidence, and offers a quantitative and unifying approach to our understanding at all these levels (Rolls 2008d).

6.10.6   Self-monitoring: correction of previous decisions

The approach described in Section 6.9 indicates that an integrate-and-fire attractor neuronal decision-making network encodes confidence in its firing rates, and that adding a second attractor network allows decisions to be made about whether to change the decision made by the first network, and for example abort the trial or strategy (see Fig. 6.15). The second network, the confidence decision network, is in effect monitoring the decisions taken by the first network, and can cause a change of strategy or behaviour if the assessment of the decision taken by the first network does not seem a confident decision.

Now this is the type of description, and language used, to describe 'monitoring' functions, taken to be high level cognitive processes, possibly related to consciousness (Lycan 1997, Block 1995). For example, in an experiment performed by Hampton (2001) (experiment 3), a monkey had to remember a picture over a delay. He was then given a choice of a 'test flag', in which case he would be allowed to choose from one of four pictures the one seen before the delay, and if correct earn a large reward (a peanut). If he was not sure that he remembered the first picture, he could choose an 'escape flag', to start another trial. With longer delays, when memory strength might be lower partly due to noise in the system, and confidence therefore in the memory on some trials might be lower, the monkey was more likely to choose the escape flag. The experiment is described as showing that the monkey is thinking about his own memory, that is, is a case of meta-memory, which may be related to consciousness (Heyes 2008). However, the decision about whether to escape from a trial can be taken just by adding a second decision network to the first decision network. Thus we can account for what seem like complex cognitive phenomena with a simple system of two attractor decision-making networks (Fig. 6.15). The design of Kepecs et al. (2008) was analogous, in that the rat could choose to abort a trial if decision confidence was low, and again this functionality can be implemented by two attractor decision-making networks, as described in Section 6.9.

The implication is that some types of 'self-monitoring' can be accounted for by simple, two-attractor network, computational processes. But what of more complex 'self-monitoring', such as is described as occurring in a commentary that might be based on reflection on previous events, and appears to be closely related to consciousness (Weiskrantz 1997). This approach has been developed into a higher order syntactic theory (HOST) of consciousness (Rolls 1999, Rolls 2004b, Rolls 2005, Rolls 2007a, Rolls 2007b, Rolls 2008a), in which there is a credit assignment problem if a multi-step reasoned plan fails, and it may be unclear which step failed. Such plans are described as syntactic as there are symbols at each stage that must be linked together with the syntactic relationships between the symbols specified, but kept separate across stages of the plan. It is suggested that in this situation being able to have higher order syntactic thoughts will enable one to think and reason about the first order plan, and detect which steps are likely to be at fault. Now this type of 'self-monitoring' is much more complex, as it requires syntax. The thrust of the argument is that some types of 'self-monitoring' are computationally simple, for example in decisions made based on confidence in a first decision, and may have little to do with consciousness; whereas higher order thought processes are very different in terms of the type of syntactic computation required, and may be more closely related to consciousness (Rolls 1999, Rolls 2004b, Rolls 2005, Rolls 2007a, Rolls 2007b, Rolls 2008a).

7 Perceptual detection and stochastic dynamics

7.1 Introduction

Pioneering investigations in several sensory systems have shown how neural activity represents the physical parameters of sensory stimuli in the peripheral nervous system and in the brain (Hubel and Wiesel 1968, Mountcastle, Talbot, Darian-Smith and Kornhuber 1967, Talbot et al. 1968). These investigations have paved the way for new questions that are more closely related to cognitive processing. For example, where and how in the brain do the neuronal responses that encode the sensory stimuli translate into responses that encode a perceptual decision (Romo and Salinas 2001, Schall 2001)? What components of the neuronal activity evoked by a sensory stimulus are directly related to perception (Romo, Hernandez, Zainos and Salinas 1998, Newsome, Britten and Movshon 1989)? Where and how in the brain is sensory information stored in memory (Romo, Brody, Hernandez and Lemus 1999)? In the next Section, 7.2, neurophysiological approaches to these issues are described, and in Section 7.3 computational approaches to understanding the processing and how noise plays an important role are described, building on the foundation in Chapter 5.

7.2 Psychophysics and neurophysiology of perceptual detection

The detection of sensory stimuli is among the simplest perceptual experiences and is a prerequisite for any further sensory processing. A fundamental problem posed by sensory detection tasks is that repeated presentation of a near-threshold stimulus might unpredictably fail or succeed in producing a sensory percept. Where in the brain are the neuronal correlates of these varying perceptual judgments? This problem has been addressed by de Lafuente and Romo (2005) and de Lafuente and Romo (2006). These authors trained monkeys to perform a detection task (Fig. 7.1). On each trial, the animal had to report whether the tip of a mechanical stimulator vibrated or not on the skin of a finger tip (Fig. 7.1B). Crucially, the stimuli varied in amplitude across trials. Stimulus-present trials were interleaved with an equal number of stimulus-absent trials in which no mechanical vibrations were delivered (Fig. 7.1B). Depending on the monkey's responses, trials could be classified into four response types: hits and misses in the stimulus-present condition, and correct rejections and false alarms in the stimulus-absent condition. Stimulus detection thresholds were calculated from the behavioural responses (Fig. 7.1C). Previous studies seeking the neuronal correlates of sensory detection showed that, in the case of the vibrotactile stimuli, the responses of neurons of the primary somatosensory cortex (S1) account for the measured psychophysical accuracy (Mountcastle, Talbot, Sakata and Hyvarinen 1969). However, imaging and physiological studies show that, in addition to the sensory cortices, areas of the frontal lobes are also active during sensory detection and discrimination (Shulman, Olliger, Linenweber, Petersen and Corbetta 2001, Hernandez, Zainos and Romo 2002, Romo, Hernandez and

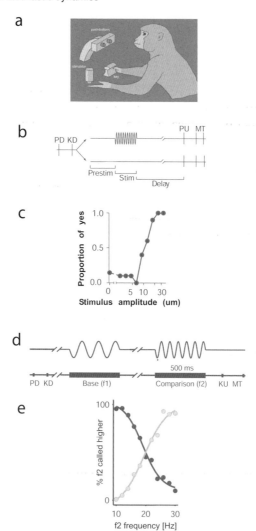

Fig. 7.1 (a) Drawing of a monkey working in a detection or discrimination task. (b) The sequence of events during the detection trials. Trials began when the stimulation probe indented the skin of one fingertip of the left, restrained hand (probe down, PD). The monkey then placed its right, free hand on an immovable key (key down, KD). On half of the randomly selected trials, after a variable pre-stimulus period (Prestim, 1.5 to 3.5 s), a vibratory stimulus (Stim, 20 Hz, 0.5 s) was presented. Then, after a fixed delay period (Delay, 3 s), the stimulator probe moved up (probe up, UP), indicating to the monkey that it could make the response movement (MT) to one of the two buttons. The button pressed indicated whether or not the monkey felt the stimulus (henceforth referred to as 'yes' and 'no' responses, respectively). Depending on whether the stimulus was present or absent and on the behavioural response, the trial outcome was classified as a hit, miss, false alarm or correct reject. Trials were pseudo-randomly chosen: 90 trials were with the stimulus absent (amplitude 0), and 90 trials were with the stimulus present with varying amplitudes (9 amplitudes with 10 repetitions each). (c) Classical psychometric detection curve obtained by plotting the proportion of 'yes' responses as a function of the stimulus amplitude. (d) Sequence of events during the discrimination trials. Trials began exactly as in the detection task but the probe oscillates at the base frequency (f1), and after a delay (3 s), a second mechanical vibration is delivered at the comparison frequency (f2). The monkey then releases the key (KU) and presses either the medial or lateral push-button to indicate whether the comparison frequency was lower or higher than the base frequency. (e) Discrimination performance curve, plotted as the animal's capacity to judge whether f2 is higher (grey line) or lower (black line) in frequency than f1=20 Hz. (After de Lafuente and Romo, 2005, and Hernandez et al., 1997.)

Zainos 2004, Romo, Hernandez, Zainos and Salinas 2003). This evidence raises an important question: what are the specific functional roles of primary sensory cortices and association areas of the frontal lobes in perceptual detection?

To further test the contributions of these cortical areas in perceptual detection, de Lafuente and Romo (2005) and de Lafuente and Romo (2006) recorded the activity of S1 and medial premotor cortex (MPC, a frontal lobe area involved in decision-making and motor planning (Hernandez, Zainos and Romo 2002, Tanji 2001)) neurons, while monkeys performed the task (Fig. 7.1). They basically found that the responses of S1 neurons varied as a function of the stimulus strength but did not predict the behavioural responses. Conversely, the responses of MPC neurons did not vary as a function of the stimulus strength, but predicted the behavioural responses on a trial-by-trial basis. These results further support a detection model in which, to judge the stimulus presence or absence, a central area (or areas) with internal fluctuations must track the activity of S1. These results suggest that perceptual judgments emerge from the activity of frontal lobe neurons but not in sensory cortices. Importantly, the internal fluctuations of frontal lobe neurons are closely related to the animal's behaviour, as shown by the data of de Lafuente and Romo (2005) and de Lafuente and Romo (2006).

7.3   Computational models of probabilistic signal detection

An aim of this chapter is to show how stochastic dynamics helps to understand the computational mechanisms involved in perceptual detection. The computational analysis of detection focuses on the paradigm and experimental results of de Lafuente and Romo (2005) described earlier. [To summarize, the firing rates of S1 neurons did not covary with the animals' perceptual reports, but their firing rate did show a monotonically increasing graded dependence on the stimulus amplitude (Figure 3c and 3f in de Lafuente and Romo (2005)). On the other hand, the activity of Medial Premotor Cortex neurons was only weakly modulated by the stimulus amplitude, but were related to the monkeys' trial-by-trial reports of whether they detected a stimulus.] The fact that the firing rates of MPC neurons correlate with whether the stimulus is detected, with a high firing rate for a 'yes' report and a low firing rate for a 'no' report, suggests not only that this area is involved in the detection, but also that the detection is implemented by an attractor network with bistable dynamics.

Deco, Perez-Sanagustin, de Lafuente and Romo (2007a) investigated two integrate-and-fire attractor network models of this stimulus detection. One (Fig. 7.2 top left) has one population or pool of neurons that when the stimulus λ is presented can either remain in the spontaneous state which represents that no stimulus has been detected (a 'No' response), or can transition to a high firing rate state indicating that the stimulus has been detected (a 'Yes' response). This was called a NCYN model (Non-Competing Yes Neurons). The arrows indicate the recurrent synaptic connections between the different neurons in a pool. The strengths of these synaptic connections $\omega+$ were obtained by a mean field analysis to produce two stable states, one where the activity of the Yes neurons was low (spontaneous firing rate, representing a 'No' decision), and the other where the Yes neurons fire fast, representing a 'Yes' decision.

The performance of this model is shown in Fig. 7.2 (Modelling Results, left), where it can be seen that the network probabilistically enters a 'Yes' state with a probability influenced by the stimulus strength λ. Once the network is in the 'Yes' state the firing rate is high, independently of the stimulus strength (see Deco et al. (2007a)). It is the spiking noise in the network that makes the detection probabilistic. The model captures the probabilistic detection of the monkeys (Fig. 7.2, middle left), and also the neurophysiological data that the MPC neurons after the decision has been made have either a low firing rate (usually when the

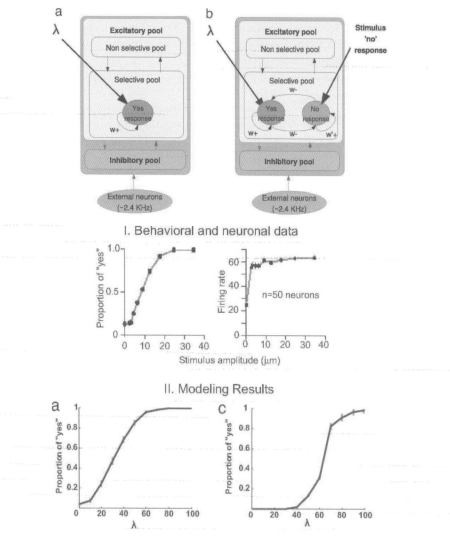

Fig. 7.2 Stimulus detection. Top. (a) Left: A stimulus detection model with one population or pool of neurons that when the stimulus λ is presented can either remain in the spontaneous state which represents that no stimulus has been detected (a 'No' response), or can transition to a high firing rate state indicating that the stimulus has been detected (a 'Yes' response). This was called a NCYN model (non-competing yes neurons). (b) Right: A stimulus detection model with two populations of competing neurons. The 'No' population starts at a high rate, and remains in this attractor state if the stimulus is not detected. The 'Yes' population receives inputs λ from the sensory stimulus, and if this population wins the competition, this corresponds to detection and a 'Yes' response. This was called a CYNN model (competing yes-no neurons). The arrows indicate the recurrent synaptic connections between the different neurons in a pool, and the synaptic weights ω as determined from a mean field analysis are indicated as being stronger $\omega+$ or weaker $\omega-$ than the other connection strengths in the network. Neurons in the networks are connected via three types of receptors that mediate the synaptic currents flowing into them: AMPA and NMDA glutamate receptors, and GABA receptors. Below. I. Behavioural and neuronal data. (Left) Behavioural detection curve of the macaques. Proportion of 'yes' responses as a function of the amplitude of a 20 Hz vibratory stimulus applied to a fingertip. (Right) Mean response of Medial Premotor Cortex neurons as a function of stimulus amplitude. In contrast to what has been observed in primary somatosensory areas, stimulus amplitude has little effect on the firing rate of MPC neurons once the decision has been taken: they either fire very little, or at a high rate. II. Integrate-and-fire modelling results. The probability of a high firing rate 'yes'-response (hit) as a function of the stimulus strength λ for (left) the NCYN model and (right) the CYNN model. (After Deco, Perez-Sanagustin, de Lafuente and Romo (2007).

stimulus is weak, see Fig. 7.2, middle right), or have a high firing rate (usually when the stimulus is strong, see Fig. 7.2, middle right).

The second integrate-and-fire model (Fig. 7.2 top right) has two populations of neurons that compete with each other, one which starts with a relatively high firing rate at the beginning of a trial and remains in this attractor for a 'No' decision, and a second population that receives the stimulus λ and can remain in its spontaneous state, or can transition to a high firing rate state indicating that the stimulus has been detected (a 'Yes' response). This was called a CYNN model (Competing Yes/No Neurons). The arrows indicate the recurrent synaptic connections between the different neurons in a pool. The strengths of these synaptic connections were set to be stronger $\omega+$ or weaker $\omega-$ than the other connections in the network following a mean field analysis to produce two stable states, one where the activity of the Yes neurons was low (spontaneous firing rate, representing a 'No' decision), and the other where the Yes neurons fire fast, representing a 'Yes' decision. The firing of the 'No' population is approximately the reciprocal of the 'Yes' population. The performance of this model is shown in Fig. 7.2 (Modeling Results, right), where it can be seen that the network probabilistically enters a 'Yes' state with a probability influenced by the stimulus strength λ. Once the network is in the 'Yes' state the firing rate is high, independently of the stimulus strength (see Deco et al. (2007a)). It is again the spiking noise in the network that makes the detection probabilistic.

By analysing the temporal evolution of the firing rate activity of neurons on trials associated with the two different behavioural responses of 'Yes' or 'No', Deco et al. (2007a) produced evidence in favour of the CYNN (2 population competing) model compared to the non-competing one population model (NCYN). Specifically, the CYNN model predicts the existence of some neurons that encode the 'no' response, and other neurons that encode the 'yes' response. The first set of neurons slightly decrease their activity at the end of the trial, whereas the second group of neurons increase their firing activity when a stimulus is presented, which is what is found neurophysiologically. Thus in this case, the simulations indicate that the CYNN model fits the experimental data better than the NCYN model.

The important aspects of both models are as follows. First, the spiking noise in the network is part of the mechanism by which the detection outcome for a given stimulus strength is probabilistic. Second, once the decision has been taken, the firing rates of the population(s) reach a binary state, of being high or low, and do not reflect the exact value of the stimulus strength. This property, and the fact that a binary decision state is reached, is a property of attractor networks. The competition implemented by the inhibitory neurons helps a binary decision to be reached, and this is facilitated by the non-linear positive feedback implemented by the recurrent collateral excitatory connections within a population of neurons. Third, the mean-field analysis is important because it enables the synaptic connection strengths that can produce the required stable states to be determined; and the integrate-and-fire simulations with those weights are important in allowing the probabilistic dynamical transition into a decision state to be investigated, and to be compared with the responses of neurons recorded in the brain. Fourth, these integrate-and-fire models incorporate synaptically activated ion channels and neurons with realistic dynamics, and so pave the way for relating detailed experimental neuroscience measurements and manipulations of the overall decision-making at the global level of behaviour.

7.4 Stochastic resonance

If we had a deterministic neuron without noise and a fixed threshold above which spikes were emitted, then if the signal was below the threshold there would be no output, and if the signal was above threshold the neuron would emit a spike, and indeed continuous spike trains if the signal remained above threshold. In particular, if the signal was just below the threshold of the neuron, there would be no evidence that a signal close to threshold was present. However, if noise is present in the system (due for example to the afferent neurons having probabilistic spiking activity similar to that of a Poisson process), then occasionally with a signal close to threshold a spike would occur due to the summation of the signal and the noise. If the signal was a bit weaker, then the neuron might still occasionally spike, but at a lower average rate. If the signal was a bit closer to threshold, then the neuron would emit spikes at a higher average rate. Thus in this way some evidence about the presence of a subthreshold signal can be made evident in the spike trains emitted by a neuron if there is noise in the inputs to the neuron. The noise in this case is useful, and may have an adaptive function (cf. Faisal et al. (2008)). This process is known as *stochastic resonance*, and is a well known example of how noise can have beneficial effects in signal detection systems operating close to a threshold (Longtin 1993, Weisenfeld 1993, Stocks 2000, Riani and Simonotto 1994, Shang, Claridge-Chang, Sjulson, Pypaert and Miesenböck 2007, Faisal, Selen and Wolpert 2008, Goldbach, Loh, Deco and Garcia-Ojalvo 2008).

7.5 Synthesis

We are led to the conclusion that in a perceptual signal detection task, noise, of which one type is the noise contributed by statistical fluctuations related to spiking dynamics, can help, as in stochastic resonance. The noise makes the detection process probabilistic across trials. However, the attractor dynamics allows the process to come to a 'Yes' or 'No' categorical decision on each trial.

The source of the noise in the detection mechanism in the brain, and in fact the noise in signal detection theory (Green and Swets 1966) may be at least in part due to the statistical fluctuations caused by the probabilistic spiking of neurons in networks (Section 2.4).

We may note that decisions about whether a signal has been detected are not typically taken at the periphery, in that the distribution of false positive decisions etc does not necessarily accurately reflect on a trial-by-trial basis variations at the periphery, but instead fluctuations in more central brain areas (de Lafuente and Romo 2005). A good example of this is that the explicit, conscious, recognition of which face was seen is set with a threshold which is higher than that at which information is present in the inferior temporal visual cortex, and at which guessing can be much better than chance (Rolls 2007b).

8 Applications of this stochastic dynamical theory to brain function

8.1 Introduction

The theory described in this book as to how networks take decisions stochastically has implications throughout the brain, some of which are described in this chapter.

8.2 Memory recall

The theory is effectively a model of the stochastic dynamics of the recall of a memory in response to a recall cue. The memory might be a long-term memory, but the theory applies to the retrieval of any stored representation in the brain. The way in which the attractor is reached depends on the strength of the recall cue, and inherent noise in the attractor network performing the recall because of the spiking activity in a finite size system. The recall will take longer if the recall cue is weak. Spontaneous stochastic effects may suddenly lead to the memory being recalled, and this may be related to the sudden recovery of a memory which one tried to remember some time previously. These processes are considered further by Rolls (2008d).

The theory applies to a situation where the representation may be being 'recalled' by a single input, which is perceptual detection as described in Chapter 7.

The theory also applies to a situation where the representation may be being 'recalled' by two or more competing inputs λ, which is decision-making as described in Chapters 5 and 6.

The theory also applies to short-term memory, in which the continuation of the recalled state as a persistent attractor is subject to stochastic noise effects, which may knock the system out of the short-term memory attractor, as described in Chapter 3.

The theory also applies to attention, in which the continuation of the recalled state as a persistent attractor is subject to stochastic noise effects, which may knock the system out of the short-term memory attractor that is normally stable because of the non-linear positive feedback implemented in the attractor network by the recurrent collateral connections, as described in Chapter 4.

8.3 Decision-making with multiple alternatives

This framework can also be extended very naturally to account for the probabilistic decision taken when there are multiple, that is more than two, choices. One such extension models choices between continuous variables in a continuous or line attractor network (Furman and Wang 2008, Liu and Wang 2008) to account for the responses of lateral intraparietal cortex neurons in a 4-choice random dot motion decision task (Churchland, Kiani and Shadlen 2008). In another approach, a network with multiple discrete attractors (Albantakis and Deco 2009) can account well for the same data.

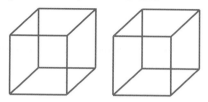

Fig. 8.1 Two Necker cubes. (It may be helpful to increase the viewing distance.)

A clear example of a discrete multiple choice decision is a decision about which of a set of objects, perhaps with different similarity to each other, has been shown on each trial, and where the decisions are only probabilistically correct. When a decision is made between different numbers of alternatives, a classical result is Hick's law, that the reaction time increases linearly with $\log_2$ of the number of alternatives from which a choice is being made. This has been interpreted as supporting a series of binary decisions each one taking a unit amount of time (Welford 1980). As the integrate-and-fire model we describe works completely in parallel, it will be very interesting to investigate whether Hick's law is a property of the network. If so, this could be related in part to the fact that the activity of the inhibitory interneurons is likely to increase linearly with the number of alternatives between which a decision is being made (as each one adds additional bias to the system through a λ input), and that the GABA inhibitory interneurons implement a shunting, that is divisive, operation (Rolls 2008d).

8.4 Perceptual decision-making and rivalry

Another application is to changes in perception. Perceptions can change 'spontaneously' from one to another interpretation of the world, even when the visual input is constant, and a good example is the Necker cube, in which visual perception flips occasionally to make a different edge of the cube appear nearer to the observer (Fig. 8.1). We hypothesize that the switching between these multistable states is due in part to the statistical fluctuations in the network due to the Poisson-like spike firing that is a form of noise in the system. It will be possible to test this hypothesis in integrate-and-fire simulations. (This may or may not be supplemented by adaptation effects (of the synapses or neurons) in integrate-and-fire networks.) (You may observe interesting effects in Fig. 8.1, in which when one cube flips which face appears closer, the other cube performs a similar flip, so that the two cubes remain for most of the time in the same configuration. This effect can be accounted for by short-range cortico-cortical excitatory connections between corresponding depth feature cue combination neurons that normally help to produce a consistent interpretation of 3D objects. The underlying mechanism is that

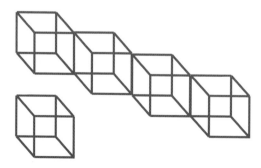

Fig. 8.2 Linked Necker cubes, and a companion.

of attractor dynamics linking in this case corresponding features in different objects. When the noise in one of the attractors make the attractor flip, this in turn applies a bias to the other attractor, making it very likely that the attractor for the second cube will flip soon under the influence of its internal spiking-related noise. The whole configuration provides an interesting perceptual demonstration of the important role of noise in influencing attractor dynamics, and of the cross-linking between related attractors which helps the whole system to move towards energy minima, under the influence of noise. This example of the noisy brain in action appears on the cover of this book. Another interesting example is shown in Fig. 8.2.)

The same approach should provide a model of pattern and binocular rivalry, where one image is seen at a time even though two images are presented simultaneously, and indeed an attractor-based noise-driven model of perceptual alternations has been described (Moreno-Botel, Rinzel and Rubin 2007). When these are images of objects or faces, the system that is especially important in the selection is the inferior temporal visual cortex (Blake and Logothetis 2002, Maier, Logothetis and Leopold 2005), for it is here that representations of whole objects are present (Rolls 2008d, Rolls and Stringer 2006, Rolls 2009c), and the global interpretation of one object can compete with the global interpretation of another object. These simulation models are highly feasible, in that the effects in integrate-and-fire simulations to influence switching between stable states not only of noise, but also of synaptic adaptation and neuronal adaptation which may contribute, have already been investigated (Deco and Rolls 2005d, Deco and Rolls 2005c, Moreno-Botel et al. 2007).

8.5 The matching law

Another potential application of this model of decision-making is to probabilistic decision tasks. In such tasks, the proportion of choices reflects, and indeed may be proportional to, the expected value of the different choices. This pattern of choices is known as the matching law (Sugrue, Corrado and Newsome 2005). An example of a probabilistic decision task in which the choices of the human participants in the probabilistic decision task clearly reflected the expected value of the choices is described by Rolls, McCabe and Redoute (2008c).

A network of the type described here in which the biasing inputs λ_1 and λ_2 to the model are the expected values of the different choices alters the proportion of the decisions it makes as a function of the relative expected values in a way similar to that shown in Fig. 5.7, and provides a model of this type of probabilistic reward-based decision-making (Marti, Deco, Del Giudice and Mattia 2006). It was shown for example that the proportion of trials on which one stimulus was shown over the other was approximately proportional to the difference of the two values between which choices were being made. In setting the connection weights to the two attractors that represent the choices, the returns (the average reward per choice), rather than the incomes (the average reward per trial) of the two targets, are relevant (Soltani and Wang 2006).

This type of model also accounts for the observation that matching is not perfect, and the relative probability of choosing the more rewarding option is often slightly smaller than the relative reward rate ('undermatching'). If there were no neural variability, decision behaviour would tend to get stuck with the more rewarding alternative; stochastic spiking activity renders the network more exploratory and produces undermatching as a consequence (Soltani and Wang 2006).

8.6   Symmetry-breaking

It is of interest that the noise that contributes to the stochastic dynamics of the brain through the spiking fluctuations may be behaviourally adaptive, and that the noise should not be considered only as a problem in terms of how the brain works. This is the issue raised for example by the donkey in the medieval Duns Scotus paradox, in which a donkey situated between two equidistant food rewards might never make a decision and might starve.

The problem raised is that with a deterministic system, there is nothing to break the symmetry, and the system can become deadlocked. In this situation, the addition of noise can produce probabilistic choice, which is advantageous. We have shown here that stochastic neurodynamics caused for example by the relatively random spiking times of neurons in a finite sized cortical attractor network can lead to probabilistic decision-making, so that in this case the stochastic noise is a positive advantage.

8.7   The evolutionary utility of probabilistic choice

Probabilistic decision-making can be evolutionarily advantageous in another sense, in which sometimes taking a decision that is not optimal based on previous history may provide information that is useful, and which may contribute to learning. Consider for example a probabilistic decision task in which choice 1 provides rewards on 80% of the occasions, and choice 2 on 20% of the occasions. A deterministic system with knowledge of the previous reinforcement history would always make choice 1. But this is not how animals including humans behave. Instead (especially when the overall probabilities are low and the situation involves random probabilistic baiting, and there is a penalty for changing the choice), the proportion of choices made approximately matches the outcomes that are available, in what is called the matching law (Sugrue, Corrado and Newsome 2005, Corrado, Sugrue, Seung and Newsome 2005, Rolls, McCabe and Redoute 2008c) (Section 8.5). By making the less favoured choice sometimes, the organism can keep obtaining evidence on whether the environment is changing (for example on whether the probability of a reward for choice 2 has increased), and by doing this approximately according to the matching law minimizes the cost of the disadvantageous choices in obtaining information about the environment.

This probabilistic exploration of the environment is very important in trial-and-error learning, and indeed has been incorporated into a simple reinforcement algorithm in which noise is added to the system, and if this improves outcomes above the expected value, then changes are made to the synaptic weights in the correct direction (in the associative reward-penalty algorithm) (Sutton and Barto 1981, Barto 1985, Rolls 2008d).

In perceptual learning, probabilistic exploratory behaviour may be part of the mechanism by which perceptual representations can be shaped to have appropriate selectivities for the behavioural categorization being performed (Sigala and Logothetis 2002, Szabo, Deco, Fusi, Del Giudice, Mattia and Stetter 2006).

Another example is in food foraging, which probabilistically may reflect the outcomes (Krebs and Davies 1991, Kacelnik and Brito e Abreu 1998), and is a way optimally in terms of costs and benefits to keep sampling and exploring the space of possible choices.

Another sense in which probabilistic decision-making may be evolutionarily advantageous is with respect to detecting signals that are close to threshold, as described in Chapter 7.

Intrinsic indeterminacy may be essential for unpredictable behaviour (Glimcher 2005). For example, in interactive games like matching pennies or rock–paper–scissors, any trend that deviates from random choice by an agent could be exploited to his or her opponent's advantage.

8.8   Selection between conscious vs unconscious decision-making, and free will

Another application of this type of model is to taking decisions between the implicit (unconscious) and explicit (conscious) systems in emotional decision-making (see Rolls (2005) and Rolls (2008d)), where again the two different systems could provide the biasing inputs λ_1 and λ_2 to the model. An implication is that noise will influence with probabilistic outcomes which system, the implicit or the conscious reasoning system, takes a decision.

When decisions are taken, sometimes confabulation may occur, in that a verbal account of why the action was performed may be given, and this may not be related at all to the environmental event that actually triggered the action (Gazzaniga and LeDoux 1978, Gazzaniga 1988, Gazzaniga 1995, Rolls 2005, LeDoux 2008). It is accordingly possible that sometimes in normal humans when actions are initiated as a result of processing in a specialized brain region such as those involved in some types of rewarded behaviour, the language system may subsequently elaborate a coherent account of why that action was performed (i.e. confabulate). This would be consistent with a general view of brain evolution in which, as areas of the cortex evolve, they are laid on top of existing circuitry connecting inputs to outputs, and in which each level in this hierarchy of separate input–output pathways may control behaviour according to the specialized function it can perform.

This raises the issue of free will in decision-making.

First, we can note that in so far as the brain operates with some degree of randomness due to the statistical fluctuations produced by the random spiking times of neurons, brain function is to some extent non-deterministic, as defined in terms of these statistical fluctuations. That is, the behaviour of the system, and of the individual, can vary from trial to trial based on these statistical fluctuations, in ways that are described in this book. Indeed, given that each neuron has this randomness, and that there are sufficiently small numbers of synapses on the neurons in each network (between a few thousand and 20,000) that these statistical fluctuations are not smoothed out, and that there are a number of different networks involved in typical thoughts and actions each one of which may behave probabilistically, and with 10^{11} neurons in the brain each with this number of synapses, the system has so many degrees of freedom that it operates effectively as a non-deterministic system. (Philosophers may wish to argue about different senses of the term deterministic, but it is being used here in a precise, scientific, and quantitative way, which has been clearly defined.)

Second, do we have free will when both the implicit and the explicit systems have made the choice? Free will would in Rolls' view (Rolls 2005, Rolls 2008a, Rolls 2008d, Rolls 2010c, Rolls 2010a) involve the use of language to check many moves ahead on a number of possible series of actions and their outcomes, and then with this information to make a choice from the likely outcomes of different possible series of actions. (If, in contrast, choices were made only on the basis of the reinforcement value of immediately available stimuli, without the arbitrary syntactic symbol manipulation made possible by language, then the choice strategy would be much more limited, and we might not want to use the term free will, as all the consequences of those actions would not have been computed.) It is suggested that when this type of reflective, conscious, information processing is occurring and leading to action, the system performing this processing and producing the action would have to believe that it could cause the action, for otherwise inconsistencies would arise, and the system might no longer try to initiate action. This belief held by the system may partly underlie the feeling of free will. At other times, when other brain modules are initiating actions (in the implicit systems), the conscious processor (the explicit system) may confabulate and believe that it caused the action, or at least give an account (possibly wrong) of why the action was initiated. The fact that the conscious processor may have the belief even in these circumstances that it initiated the action may arise as a property of it being inconsistent for a system that can take overall control using conscious verbal processing to believe that it was overridden by another system. This may be the underlying computational reason why confabulation occurs.

The interesting view we are led to is thus that when probabilistic choices influenced by stochastic dynamics are made between the implicit and explicit systems, we may not be aware of which system made the choice. Further, when the stochastic noise has made us choose with the implicit system, we may confabulate and say that we made the choice of our own free will, and provide a guess at why the decision was taken. In this scenario, the stochastic dynamics of the brain plays a role even in how we understand free will (Rolls 2010c).

8.9   Creative thought

Another way in which probabilistic decision-making may be evolutionarily advantageous is in creative thought, which is influenced in part by associations between one memory, representation, or thought, and another. If the system were deterministic, i.e. for the present purposes without noise, then the trajectory through a set of thoughts would be deterministic and would tend to follow the same furrow each time. However, if the recall of one memory or thought from another were influenced by the statistical noise due to the random spiking of neurons, then the trajectory through the state space would be different on different occasions, and we might be led in different directions on different occasions, facilitating creative thought (Rolls 2008d).

Of course, if the basins of attraction of each thought were too shallow, then the statistical noise might lead one to have very unstable thoughts that were too loosely and even bizarrely associated to each other, and to have a short-term memory and attentional system that is unstable and distractible, and indeed this is an account that we have proposed for some of the symptoms of schizophrenia (Rolls 2005, Rolls 2008d, Loh, Rolls and Deco 2007a, Loh, Rolls and Deco 2007b, Rolls, Loh, Deco and Winterer 2008b) (see Section 8.14).

The stochastic noise caused by the probabilistic neuronal spiking plays an important role in these hypotheses, because it is the noise that destabilizes the attractors when the depth of the basins of attraction is reduced. If the basins of attraction were too deep, then the noise might be insufficient to destabilize attractors, and this leads to an approach to understanding obsessive-compulsive disorders (Rolls, Loh and Deco 2008a) (see Section 8.15).

8.10   Unpredictable behaviour

An area where the spiking-related noise in the decision-making process may be evolutionarily advantageous is in the generation of unpredictable behaviour, which can be advantageous in a number of situations, for example when a prey is trying to escape from a predator, and perhaps in some social and economic situations in which organisms may not wish to reveal their intentions (Maynard Smith 1982, Maynard Smith 1984, Dawkins 1995). We note that such probabilistic decisions may have long-term consequences. For example, a probabilistic decision in a 'war of attrition' such as staring down a competitor e.g. in dominance hierarchy formation, may fix the relative status of the two individual animals involved, who then tend to maintain that relationship stably for a considerable period of weeks or more (Maynard Smith 1982, Maynard Smith 1984, Dawkins 1995).

Thus intrinsic indeterminacy may be essential for unpredictable behaviour. We have noted the example provided by Glimcher (2005): in interactive games like matching pennies or rock–paper–scissors, any trend that deviates from random choice by an agent could be exploited to his or her opponent's advantage.

8.11 Dreams

Similar noise-driven processes may lead to dreams, where the content of the dream is not closely tied to the external world because the role of sensory inputs is reduced in paradoxical (desynchronized, high frequency, fast wave, dreaming) sleep, and the cortical networks, which are active in fast-wave sleep (Kandel, Schwartz and Jessell 2000, Carlson 2006, Horne 2006), may move under the influence of noise somewhat freely on from states that may have been present during the day. Thus the content of dreams may be seen as a noisy trajectory through state space, with starting point states that have been active during the day, and passing through states that will reflect the depth of the basins of attraction (which might well reflect ongoing concerns including anxieties and desires), and will be strongly influenced by noise. Moreover, the top-down attentional and monitoring control from for example the prefrontal cortex appears to be less effective during sleep than during waking, allowing the system to pass into states that may be bizarre.

I suggest that this provides a firm theoretical foundation for understanding the interpretation of dreams, which may be compared with that of Freud (1900).

In this context, the following thoughts follow. Dreams, or at least the electrical activity of paradoxical sleep, may occur in cortical areas that are concerned with conscious experience, such as those involved in higher order thoughts, and in others where processing is unconscious, and cannot be reported, such as those in the dorsal visual stream concerned with the control of actions (Rolls 2004b, Goodale 2004, Rolls 2005, Rolls 2007a, Rolls 2008a, Rolls 2008d). Thus it may be remarked that dreams may occur in conscious and unconscious processing systems. Dreams appear to be remembered that occur just before we wake up, and consistent with this, memory storage (implemented by synaptic long-term potentiation, (Rolls 2008d)) appears to be turned off during sleep. This may be adaptive, for then we do not use up memory capacity (Rolls 2008d) on noise-related representations. However, insofar as we can memorize and later remember dreams that are rehearsed while we wake up, it could be that bizarre thoughts, possibly unpleasant, could become consolidated. This consolidation could lead to the relevant attractor basins becoming deeper, and returning to the same set of memories on subsequent nights. This could be a mechanism for the formation of nightmares. A remedy is likely to be not rehearsing these unpleasant dreams while waking, and indeed to deliberately move to more pleasant thoughts, which would then be consolidated, and increase the probability of dreams on those more pleasant subjects, instead of nightmares, on later nights, given the memory attractor landscape over which noise would move one's thoughts.

In slow-wave sleep, and more generally in resting states, the activity of neurons in many cortical areas is on average low (Carlson 2006), and stochastic spiking-related noise may contribute strongly to the states that are found.

8.12 Multiple decision-making systems in the brain

Each cortical area can be conceived as performing a local type of decision-making using attractor dynamics of the type described (Rolls 2008d). Even memory recall is in effect the same local 'decision-making' process.

The orbitofrontal cortex for example is involved in providing evidence for decisions about which visual stimulus is currently associated with reward, in for example a visual discrimination reversal task. Its computations are about stimuli, primary reinforcers, and secondary reinforcers (Rolls 2005). The orbitofrontal cortex appears to represent reward value on a continuous scale, with binary choice decisions being made in the immediately adjoining area anteriorly, the medial prefrontal cortex area 10, as described in Chapter 6 (Rolls and Grabenhorst 2008, Grabenhorst, Rolls and Parris 2008, Rolls, Grabenhorst and Parris 2009c, Rolls, Grabenhorst and Deco 2010b, Rolls, Grabenhorst and Deco 2010c).

The dorsolateral prefrontal cortex takes an executive role in decision-making in a working memory task, in which information must be held available across intervening stimuli (Rolls 2008d). The dorsal and posterior part of the dorsolateral prefrontal cortex may be involved in short-term memory-related decisions about where to move the eyes (Rolls 2008d).

The parietal cortex is involved in decision-making when the stimuli are for example optic flow patterns (Glimcher 2003, Gold and Shadlen 2007).

The hippocampus is involved in (providing evidence for) decision-making when the allocentric places of stimuli must be associated with rewards or objects (Rolls and Kesner 2006, Rolls 2008d).

The somatosensory cortex and ventral premotor cortex are involved in decision-making when different vibrotactile frequencies must be compared (as described in Chapter 5).

The cingulate cortex may be involved when action–outcome decisions must be taken (Rushworth, Walton, Kennerley and Bannerman 2004, Rolls 2005, Rolls 2008d, Rolls 2009a).

In each of these cases, local cortical processing that is related to the type of decision being made takes place, and all cortical areas are not involved in any one decision. The style of the decision-making-related computation in each cortical area appears to be of the form described here, in which the local recurrent collateral connections enable the decision-making process to accumulate evidence across time, falling gradually into an attractor that represents the decision made in the network. Because there is an attractor state into which the network falls, this can be described statistically as a non-linear diffusion process, the noise for the diffusion being the stochastic spiking of the neurons, and the driving force being the biasing inputs.

If decision-making in the cortex is largely local and typically specialized, it leaves open the question of how one stream for behavioural output is selected. This type of 'global decision-making' is considered in *Memory, Attention, and Decision-Making* (Rolls 2008d).

8.13 Stochastic noise, attractor dynamics, and aging

In Section 8.14 we show how some cognitive symptoms such as poor short-term memory and attention could arise due to reduced depth in the basins of attraction of prefrontal cortical networks, and the effects of noise. The hypothesis is that the reduced depth in the basins of attraction would make short-term memory unstable, so that sometimes the continuing firing of neurons that implement short-term memory would cease, and the system under the influence of noise would fall back out of the short-term memory state into spontaneous firing.

Given that top-down attention requires a short-term memory to hold the object of attention in mind, and that this is the source of the top-down attentional bias that influences competition

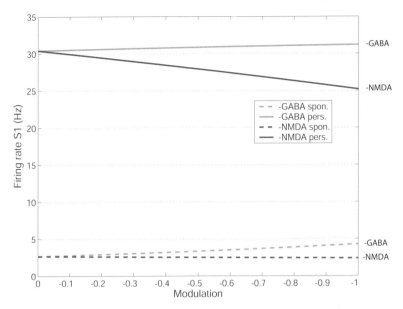

Fig. 8.3 The firing rate of an attractor network as a function of a modulation of the NMDA receptor activated ion channel conductances. On the abscissa, 0 corresponds to no reduction, and -1.0 to a reduction by 4.5%. The change in firing rate for the high firing rate short-term memory state of persistent firing, and for the spontaneous rate, are shown. The values were obtained from the mean-field analysis of the network described in Section 8.14, and thus for the persistent state represent the firing rate when the system is in its high firing rate state. Curves are also shown for a reduction of up to 9% in the GABA receptor activated ion channel conductances. (Data of Loh, Rolls and Deco.)

in other networks that are receiving incoming signals, then disruption of short-term memory is also predicted to impair the stability of attention.

These ideas are elaborated in Section 8.14, where the reduced depth of the basins of attraction in schizophrenia is related to down-regulation of NMDA receptors, or to factors that influence NMDA receptor generated ion channel currents such as dopamine D1 receptors.

Could similar processes in which the stochasticity of the dynamics is increased because of a reduced depth in the basins of attraction contribute to the changes in short-term memory and attention that are common in aging? Reduced short-term memory and less good attention are common in aging, as are impairments in episodic memory (Grady 2008). What changes in aging might contribute to a reduced depth in the basins of attraction? We describe recent hypotheses (Rolls, Deco and Loh 2010a).

8.13.1 NMDA receptor hypofunction

One factor is that NMDA receptor functionality tends to decrease with aging (Kelly, Nadon, Morrison, Thibault, Barnes and Blalock 2006). This would act, as investigated in detail in Sections 3.8.3 and 8.14, to reduce the depth of the basins of attraction, both by reducing the firing rate of the neurons in the active attractor, and effectively by decreasing the strength of the potentiated synaptic connections that support each attractor as the currents passing through these potentiated synapses would decrease. These two actions are clarified by considering Equation 1.21 on page 34, in which the Energy E reflects the depth of the basin of attraction.

An example of the reduction of firing rate in an attractor network produced by even a

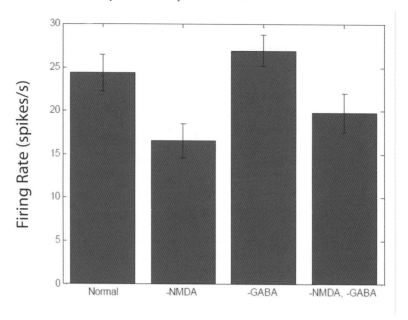

Fig. 8.4 The firing rate (mean ± sd across neurons in the attractor and simulation runs) of an attractor network in the baseline (Normal) condition, and when the NMDA receptor activated ion channel conductances in integrate-and-fire simulations are reduced by 4.5% (−NMDA). The firing rate for the high firing rate short-term memory state of persistent firing is shown. The values were obtained from integrate-and–fire simulations, and only for trials in which the network was in a high firing rate attractor state using the criterion that at the end of the 3 s simulation period the neurons in the attractor were firing at 10 spikes/s or higher. The firing rates are also shown for a reduction of 9% in the GABA receptor activated ion channel conductances (−GABA), and for a reduction in both the NMDA and GABA receptor activated ion channel conductances (−NMDA, −GABA). (Data of Loh, Rolls and Deco.)

small downregulation (by 4.5%) of the NMDA receptor activated ion channel conductances is shown in Fig. 8.3, based on a mean-field analysis. The effect of this reduction would be to decrease the depth of the basins of attraction, both by reducing the firing rate, and by producing an effect similar to weakened synaptic strengths, as shown in Equation 1.21.

In integrate-and-fire simulations, the effect on the firing rates of reduction in the NMDA activated channel conductances by 4.5% and in the GABA activated channel conductances by 9% are shown in Fig. 8.4.

The reduced depth in the basins of attraction could have a number of effects that are relevant to cognitive changes in aging.

First, the stability of short-term memory networks would be impaired, and it might be difficult to hold items in short-term memory for long, as the noise might push the network easily out of its shallow attractor, as illustrated in Section 3.8.3 and Fig. 3.11.

Second, top-down attention would be impaired, in two ways. First, the short-term memory network holding the object of attention in mind would be less stable, so that the source of the top-down bias for the biased competition in other cortical areas might disappear. Second, and very interestingly, even when the short-term memory for attention is still in its persistent attractor state, it would be less effective as a source of the top-down bias, because the firing rates would be lower, as shown in Figs. 8.3 and 8.4.

Third, the recall of information from episodic memory systems in the temporal lobe (Rolls 2008d, Dere, Easton, Nadel and Huston 2008, Rolls 2008e) would be impaired. This would arise because the positive feedback from the recurrent collateral synapses that helps the system to fall into a basin of attraction (see Fig. 5.5), representing in this case the recalled memory, would be less effective, and so the recall process would be more noisy overall.

Fourth, any reduction of the firing rate of the pyramidal cells caused by NMDA receptor hypofunction (Figs. 8.3 and 8.4) would itself be likely to impair new learning involving LTP.

In addition, if the NMDA receptor hypofunction were expressed not only in the prefrontal cortex where it would affect short-term memory, and in the temporal lobes where it would affect episodic memory, but also in the orbitofrontal cortex, then we would predict some reduction in emotion and motivation with aging, as these functions rely on the orbitofrontal cortex (see Rolls (2005) and Section 8.14).

Although NMDA hypofunction may contribute to cognitive effects such as poor short-term memory and attention in aging and in schizophrenia, the two states are clearly very different. Part of the difference lies in the positive symptoms of schizophrenia (the psychotic symptoms, such as thought disorder, delusions, and hallucinations) which may be related to the additional downregulation of GABA in the temporal lobes, which would promote too little stability of the spontaneous firing rate state of temporal lobe attractor networks, so that the networks would have too great a tendency to enter states even in the absence of inputs, and to not be controlled normally by input signals (Loh, Rolls and Deco 2007a, Loh, Rolls and Deco 2007b, Rolls, Loh, Deco and Winterer 2008b) (see Section 8.14). However, in relation to the cognitive symptoms of schizophrenia, there has always been the fact that schizophrenia is a condition that often has its onset in the late teens or twenties, and I suggest that there could be a link here to changes in NMDA and related receptor functions that are related to aging. In particular, short-term memory is at its peak when young, and it may be the case that by the late teens or early twenties NMDA and related receptor systems (including dopamine) may be less efficacious than when younger, so that the cognitive symptoms of schizophrenia are more likely to occur at this age than earlier.

8.13.2 Dopamine

D1 receptor blockade in the prefrontal cortex can impair short-term memory (Sawaguchi and Goldman-Rakic 1991, Sawaguchi and Goldman-Rakic 1994, Goldman-Rakic 1999, Castner, Williams and Goldman-Rakic 2000). Part of the reason for this may be that D1 receptor blockade can decrease NMDA receptor activated ion channel conductances, among other effects (Seamans and Yang 2004, Durstewitz, Kelc and Gunturkun 1999, Durstewitz, Seamans and Sejnowski 2000a, Brunel and Wang 2001, Durstewitz and Seamans 2002) (see further Section 8.14). Thus part of the role of dopamine in the prefrontal cortex in short-term memory can be accounted for by a decreased depth in the basins of attraction of prefrontal attractor networks (Loh, Rolls and Deco 2007a, Loh, Rolls and Deco 2007b, Rolls, Loh, Deco and Winterer 2008b). The decreased depth would be due to both the decreased firing rate of the neurons, and the reduced efficacy of the modified synapses as their ion channels would be less conductive (see Equation 1.21). The reduced depth of the basins of attraction can be thought of as decreasing the signal-to-noise ratio (Loh, Rolls and Deco 2007b, Rolls, Loh, Deco and Winterer 2008b). Given that dopaminergic function in the prefrontal cortex may decline with aging (Sikstrom 2007), and in conditions in which there are cognitive impairments such as Parkinson's disease, the decrease in dopamine could contribute to the reduced short-term memory and attention in aging.

In attention deficit hyperactivity disorder (ADHD), in which there are attentional deficits including too much distractibility, catecholamine function more generally (dopamine and noradrenaline (i.e. norepinephrine)) may be reduced (Arnsten and Li 2005), and I suggest that these reductions could produce less stability of short-term memory and thereby attentional states by reducing the depth of the basins of attraction.

8.13.3 Impaired synaptic modification

Another factor that may contribute to the cognitive changes in aging is that long-lasting associative synaptic modification as assessed by long-term potentiation (LTP) is more difficult to achieve in older animals and decays more quickly (Barnes 2003, Burke and Barnes 2006, Kelly et al. 2006). This would tend to make the synaptic strengths that would support an attractor weaker, and weaken further over the course of time, and thus directly reduce the depth of the attractor basins. This would impact episodic memory, the memory for particular past episodes, such as where one was at breakfast on a particular day, who was present, and what was eaten (Rolls 2008d, Rolls 2008e, Dere et al. 2008). The reduction of synaptic strength over time could also affect short-term memory, which requires that the synapses that support a short-term memory attractor be modified in the first place using LTP, before the attractor is used (Kesner and Rolls 2001).

In view of these changes, boosting glutamatergic transmission is being explored as a means of enhancing cognition and minimizing its decline in aging. Several classes of AMPA receptor potentiators have been described in the last decade. These molecules bind to allosteric sites on AMPA receptors, slow desensitization, and thereby enhance signalling through the receptors. Some AMPA receptor potentiator agents have been explored in rodent models and are now entering clinical trials (Lynch and Gall 2006, O'Neill and Dix 2007). These treatments might increase the depth of the basins of attraction. Agents that activate the glycine or serine modulatory sites on the NMDA receptor (Coyle 2006) would also be predicted to be useful.

Another factor is that Ca^{2+}-dependent processes affect Ca^{2+} signaling pathways and impair synaptic function in an aging-dependent manner, consistent with the Ca^{2+} hypothesis of brain aging and dementia (Thibault, Porter, Chen, Blalock, Kaminker, Clodfelter, Brewer and Landfield 1998, Kelly et al. 2006). In particular, an increase in Ca^{2+} conductance can occur in aged neurons, and CA1 pyramidal cells in the aged hippocampus have an increased density of L-type Ca^{2+} channels that might lead to disruptions in Ca^{2+} homeostasis, contributing to the plasticity deficits that occur during aging (Burke and Barnes 2006).

8.13.4 Cholinergic function

Another factor is acetylcholine. Acetylcholine in the neocortex has its origin largely in the cholinergic neurons in the basal magnocellular forebrain nuclei of Meynert. The correlation of clinical dementia ratings with the reductions in a number of cortical cholinergic markers such as choline acetyltransferase, muscarinic and nicotinic acetylcholine receptor binding, as well as levels of acetylcholine, suggested an association of cholinergic hypofunction with cognitive deficits, which led to the formulation of the cholinergic hypothesis of memory dysfunction in senescence and in Alzheimer's disease (Bartus 2000, Schliebs and Arendt 2006). Could the cholinergic system alter the function of the cerebral cortex in ways that can be illuminated by stochastic neurodynamics?

The cells in the basal magnocellular forebrain nuclei of Meynert lie just lateral to the lateral hypothalamus in the substantia innominata, and extend forward through the preoptic area into the diagonal band of Broca (Mesulam 1990). These cells, many of which are cholinergic, project directly to the cerebral cortex (Divac 1975, Kievit and Kuypers 1975, Mesulam 1990). These cells provide the major cholinergic input to the cerebral cortex, in that if they are lesioned the cortex is depleted of acetylcholine (Mesulam 1990). Loss of these cells does occur in Alzheimer's disease, and there is consequently a reduction in cortical acetylcholine in this disease (Mesulam 1990, Schliebs and Arendt 2006). This loss of cortical acetylcholine may contribute to the memory loss in Alzheimer's disease, although it may not be the primary factor in the aetiology.

In order to investigate the role of the basal forebrain nuclei in memory, Aigner, Mitchell, Aggleton, DeLong, Struble, Price, Wenk, Pettigrew and Mishkin (1991) made neurotoxic lesions of these nuclei in monkeys. Some impairments on a simple test of recognition memory, delayed non-match-to-sample, were found. Analysis of the effects of similar lesions in rats showed that performance on memory tasks was impaired, perhaps because of failure to attend properly (Muir, Everitt and Robbins 1994). Damage to the cholinergic neurons in this region in monkeys with a selective neurotoxin was also shown to impair memory (Easton and Gaffan 2000, Easton, Ridley, Baker and Gaffan 2002).

There are quite limited numbers of these basal forebrain neurons (in the order of thousands). Given that there are relatively few of these neurons, it is not likely that they carry the information to be stored in cortical memory circuits, for the number of different patterns that could be represented and stored is so small. (The number of different patterns that could be stored is dependent in a leading way on the number of input connections on to each neuron in a pattern associator, see e.g. Rolls (2008d)). With these few neurons distributed throughout the cerebral cortex, the memory capacity of the whole system would be impractically small. This argument alone indicates that these cholinergic neurons are unlikely to carry the information to be stored in cortical memory systems. Instead, they could modulate storage in the cortex of information derived from what provides the numerically major input to cortical neurons, the glutamatergic terminals of other cortical neurons. This modulation may operate by setting thresholds for cortical cells to the appropriate value, or by more directly influencing the cascade of processes involved in long-term potentiation (Rolls 2008d). There is indeed evidence that acetylcholine is necessary for cortical synaptic modifiability, as shown by studies in which depletion of acetylcholine and noradrenaline impaired cortical LTP/synaptic modifiability (Bear and Singer 1986). However, non-specific effects of damage to the basal forebrain cholinergic neurons are also likely, with cortical neurons becoming much more sluggish in their responses, and showing much more adaptation, in the absence of cholinergic inputs (Markram and Tsodyks 1996, Abbott, Varela, Sen and Nelson 1997) (see later).

The question then arises of whether the basal forebrain cholinergic neurons tonically release acetylcholine, or whether they release it particularly in response to some external influence. To examine this, recordings have been made from basal forebrain neurons, at least some of which project to the cortex (see Rolls (2005)) and will have been the cholinergic neurons just described.

It has been found that some of these basal forebrain neurons respond to visual stimuli associated with rewards such as food (Rolls 1975, Rolls 1981b, Rolls 1981a, Rolls 1982, Rolls 1986c, Rolls 1986a, Rolls 1986b, Rolls 1990b, Rolls 1993, Rolls 1999, Rolls, Burton and Mora 1976, Burton, Rolls and Mora 1976, Mora, Rolls and Burton 1976, Wilson and Rolls 1990b, Wilson and Rolls 1990a), or with punishment (Rolls, Sanghera and Roper-Hall 1979), that others respond to novel visual stimuli (Wilson and Rolls 1990c), and that others respond to a range of visual stimuli. For example, in one set of recordings, one group of these neurons (1.5%) responded to novel visual stimuli while monkeys performed recognition or visual discrimination tasks (Wilson and Rolls 1990c).

A complementary group of neurons more anteriorly responded to familiar visual stimuli in the same tasks (Rolls, Perrett, Caan and Wilson 1982, Wilson and Rolls 1990c).

A third group of neurons (5.7%) responded to positively reinforcing visual stimuli in visual discrimination and in recognition memory tasks (Wilson and Rolls 1990b, Wilson and Rolls 1990a).

In addition, a considerable proportion of these neurons (21.8%) responded to any visual stimuli shown in the tasks, and some (13.1%) responded to the tone cue that preceded the presentation of the visual stimuli in the task, and was provided to enable the monkey to alert to the visual stimuli (Wilson and Rolls 1990c). None of these neurons responded to touch to the leg which induced arousal, so their responses did not simply reflect arousal.

Neurons in this region receive inputs from the amygdala (Mesulam 1990, Amaral et al. 1992, Russchen, Amaral and Price 1985) and orbitofrontal cortex, and it is probably via the amygdala (and orbitofrontal cortex) that the information described here reaches the basal forebrain neurons, for neurons with similar response properties have been found in the amygdala, and the amygdala appears to be involved in decoding visual stimuli that are associated with reinforcers, or are novel (Rolls 1990b, Rolls 1992, Davis 1992, Wilson and Rolls 1993, Rolls 2000b, LeDoux 1995, Wilson and Rolls 2005, Rolls 2005, Rolls 2008d).

It is therefore suggested that the normal physiological function of these basal forebrain neurons is to send a general activation signal to the cortex when certain classes of environmental stimulus occur. These stimuli are often stimuli to which behavioural activation is appropriate or required, such as positively or negatively reinforcing visual stimuli, or novel visual stimuli. The effect of the firing of these neurons on the cortex is excitatory, and in this way produces activation. This cortical activation may produce behavioural arousal, and may thus facilitate concentration and attention, which are both impaired in Alzheimer's disease. The reduced arousal and concentration may themselves contribute to the memory disorders. But the acetylcholine released from these basal magnocellular neurons may in addition be more directly necessary for memory formation, for Bear and Singer (1986) showed that long-term potentiation, used as an indicator of the synaptic modification which underlies learning, requires the presence in the cortex of acetylcholine as well as noradrenaline. For comparison, acetylcholine in the hippocampus makes it more likely that LTP will occur, probably through activation of an inositol phosphate second messenger cascade (Markram and Siegel 1992, Seigel and Auerbach 1996, Hasselmo and Bower 1993, Hasselmo, Schnell and Barkai 1995).

The adaptive value of the cortical strobe provided by the basal magnocellular neurons may thus be that it facilitates memory storage especially when significant (e.g. reinforcing) environmental stimuli are detected. This means that memory storage is likely to be conserved (new memories are less likely to be laid down) when significant environmental stimuli are not present. In that the basal forebrain projection spreads widely to many areas of the cerebral cortex, and in that there are relatively few basal forebrain neurons (in the order of thousands), the basal forebrain neurons do not determine the actual memories that are stored. Instead the actual memories stored are determined by the active subset of the thousands of cortical afferents on to a strongly activated cortical neuron (Treves and Rolls 1994, Rolls and Treves 1998, Rolls 2008d). The basal forebrain magnocellular neurons would then according to this analysis when activated increase the probability that a memory would be stored. Impairment of the normal operation of the basal forebrain magnocellular neurons would be expected to interfere with normal memory by interfering with this function, and this interference could contribute in this way to the memory disorder in Alzheimer's disease.

Thus one way in which impaired cholinergic neuron function is likely to impair memory is by reducing the depth of the basins of attraction of cortical networks, in that these networks store less strongly the representations that are needed for episodic memory and for short-term memory, thus making the recall of long-term episodic memories less reliable in the face of stochastic noise, and the maintenance of short-term memory less reliable in the face of stochastic noise. Such changes would thereby also impair attention.

Another property of cortical neurons is that they tend to adapt with repeated input (Abbott, Varela, Sen and Nelson 1997, Fuhrmann, Markram and Tsodyks 2002a). However, this adap-

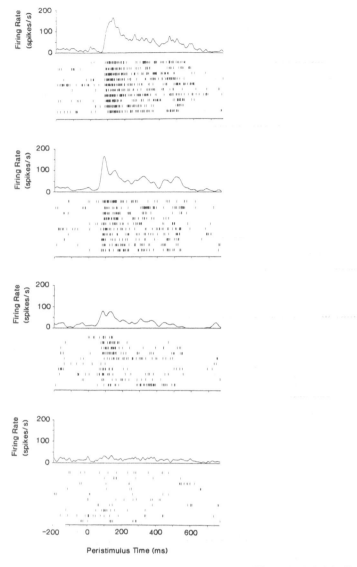

Fig. 8.5 Peristimulus time histograms and rastergrams showing the responses on different trials (originally in random order) of a face-selective neuron in the inferior temporal visual cortex to four different faces. (In the rastergrams each vertical line represents one spike from the neuron, and each row is a separate trial. Each block of the figure is for a different face.) The stimulus onset was at 0 ms, and the duration was 500 ms. (The neuron stopped firing after the 500 ms stimulus disappeared towards the end of the period shown on each trial.) These four faces were presented in random permuted sequence and repeated very many times in the experiment. (From Tovee, Rolls, Treves and Bellis 1993.)

tation is most marked in slices, in which there is no acetylcholine. One effect of acetylcholine is to reduce this adaptation (Power and Sah 2008). The mechanism is understood as follows. The afterpolarization (AHP) that follows the generation of a spike in a neuron is primarily mediated by two calcium-activated potassium currents, I_{AHP} and sI_{AHP} (Sah and Faber 2002), which are activated by calcium influx during action potentials. The I_{AHP} current is mediated by small conductance calcium-activated potassium (SK) channels, and its time course primarily follows cytosolic calcium, rising rapidly after action potentials and decaying with a time constant of 50 to several hundred milliseconds (Sah and Faber 2002). In contrast, the kinetics of the sI_{AHP} are slower, exhibiting a distinct rising phase and decaying with a time constant of 1–2 s (Sah 1996). A variety of neuromodulators, including acetylcholine (ACh), noradrenaline, and glutamate acting via G-protein-coupled receptors, suppress the sI_{AHP} and thus reduce spike-frequency adaptation (Nicoll 1988).

When recordings are made from single neurons operating in physiological conditions in the awake behaving monkey, peristimulus time histograms of inferior temporal cortex neurons to visual stimuli show only limited adaptation. There is typically an onset of the neuronal response at 80–100 ms after the stimulus, followed within 50 ms by the highest firing rate. There is after that some reduction in the firing rate, but the firing rate is still typically more than half-maximal 500 ms later (see example in Fig. 8.5). Thus under normal physiological conditions, firing rate adaptation can occur, but does not involve a major adaptation, even when cells are responding fast (at e.g. 100 spikes/s) to a visual stimulus. One of the factors that keeps the response relatively maintained may however be the presence of acetylcholine. The depletion of acetylcholine in aging and some disease states (Schliebs and Arendt 2006) could lead to less sustained neuronal responses (i.e. more adaptation), and this may contribute to the symptoms found. In particular, the reduced firing rate that may occur as a function of time if acetylcholine is low would gradually over a few seconds reduce the depth of the basin of attraction, and thus destabilize short-term memory when noise is present, by reducing the firing rate component shown in Equation 1.21. I suggest that such changes would thereby impair short-term memory, and thus also top-down attention.

The effects of this adaptation can be studied by including a time-varying intrinsic (potassium like) conductance in the cell membrane, as described in Section 2.2.2. A specific implementation of the spike-frequency adaptation mechanism using Ca^{++}-activated K^+ hyperpolarizing currents (Liu and Wang 2001) is described next, and was used by Deco and Rolls (2005c) (who also describe two other approaches to adaptation). We assume that the intrinsic gating of K^+ after-hyper-polarizing current (I_{AHP}) is fast, and therefore its slow activation is due to the kinetics of the cytoplasmic Ca^{2+} concentration. This can be introduced in the model by adding an extra current term in the integrate-and-fire model, i.e. by adding I_{AHP} to the right hand of Equation 8.1, which describes the evolution of the subthreshold membrane potential $V(t)$ of each neuron:

$$C_m \frac{dV(t)}{dt} = -g_m(V(t) - V_L) - I_{syn}(t) \tag{8.1}$$

where $I_{syn}(t)$ is the total synaptic current flow into the cell, V_L is the resting potential, C_m is the membrane capacitance, and g_m is the membrane conductance. The extra current term that is introduced into this equation is as follows:

$$I_{AHP} = -g_{AHP}[Ca^{2+}](V(t) - V_K) \tag{8.2}$$

where V_K is the reversal potential of the potassium channel. Further, each action potential generates a small amount (α) of calcium influx, so that I_{AHP} is incremented accordingly. Between spikes the $[Ca^{2+}]$ dynamics is modelled as a leaky integrator with a decay constant τ_{Ca}. Hence, the calcium dynamics can be described by the following system of equations:

$$\frac{d[Ca^{2+}]}{dt} = -\frac{[Ca^{2+}]}{\tau_{Ca}} \tag{8.3}$$

If $V(t) = \theta$, then $[Ca^{2+}] = [Ca^{2+}] + \alpha$ and $V = V_{reset}$, and these are coupled to the previously-mentioned modified equations. The $[Ca^{2+}]$ is initially set to be 0 μM, τ_{Ca}= 600 ms, $\alpha = 0.005$, V_K=-80 mV and g_{AHP}=7.5 nS.

It is predicted that enhancing cholinergic function will help to reduce the instability of attractor networks involved in short-term memory and attention that may occur in aging.

Part of the interest of this stochastic dynamics approach to aging is that it provides a way to test combinations of pharmacological treatments, that may together help to minimize the cognitive symptoms of aging. Indeed, the approach facilitates the investigation of drug combinations that may together be effective in doses lower than when only one drug is given. Further, this approach may lead to predictions for effective treatments that need not necessarily restore the particular change in the brain that caused the symptoms, but may find alternative routes to restore the stability of the dynamics.

8.14   Stochastic noise, attractor dynamics, and schizophrenia

8.14.1   Introduction

Schizophrenia is a major mental illness, which has a great impact on patients and their environment. One of the difficulties in proposing models for schizophrenia is the complexity and heterogeneity of the illness. We propose that part of the reason for the inconsistent symptoms may be a reduced signal-to-noise ratio and increased statistical fluctuations in different cortical brain networks. The novelty of the approach described here is that instead of basing our hypothesis purely on biological mechanisms, we develop a top-down approach based on the different types of symptoms and relate them to instabilities in attractor neural networks (Rolls 2005, Loh, Rolls and Deco 2007a, Loh, Rolls and Deco 2007b, Rolls, Loh, Deco and Winterer 2008b, Rolls 2008d, Deco, Rolls and Romo 2009b).

The main assumption of our hypothesis is that attractor dynamics are important in cognitive processes. Our hypothesis is based on the concept of attractor dynamics in a network of interconnected neurons which in their associatively modified synaptic connections store a set of patterns, which could be memories, perceptual representations, or thoughts (Hopfield 1982, Amit 1989, Rolls and Deco 2002). The attractor states are important in cognitive processes such as short-term memory, attention, and action selection (Deco and Rolls 2005a). The network may be in a state of spontaneous activity, or one set of neurons may have a high firing rate, each set representing a different memory state, normally recalled in response to a retrieval stimulus. Each of the states is an attractor in the sense that retrieval stimuli cause the network to fall into the closest attractor state, and thus to recall a complete memory in response to a partial or incorrect cue. Each attractor state can produce stable and continuing or persistent firing of the relevant neurons. The concept of an energy landscape (Hopfield 1982) is that each pattern has a basin of attraction, and each is stable if the basins are far apart, and also if each basin is deep, caused for example by high firing rates and strong synaptic connections between the neurons representing each pattern, which together make the attractor state resistant to distraction by a different stimulus. The spontaneous firing state, before a retrieval cue is applied, should also be stable. Noise in the network caused by statistical fluctuations in the stochastic spiking of different neurons can contribute to making the network transition from one state to another, and we take this into account by performing integrate-and-fire simulations with spiking activity, and relate this to the concept of an altered signal-to-noise ratio in schizophrenia (Winterer et al. 2000, Winterer et al. 2004, Winterer et al. 2006).

Schizophrenia is characterized by three main types of symptom: cognitive dysfunction, negative symptoms, and positive symptoms (Liddle 1987, Baxter and Liddle 1998, Mueser and McGurk 2004). We consider how the basic characteristics of these three categories might be produced in a neurodynamical system, as follows.

Dysfunction of working memory, the core of the cognitive symptoms, may be related to instabilities of persistent attractor states (Durstewitz, Seamans and Sejnowski 2000b, Wang 2001) which we show can be produced by reduced firing rates in attractor networks, in brain regions such as the prefrontal cortex.

The negative symptoms such as flattening of affect or reduction of emotions may be caused by a consistent reduction in firing rates of neurons in regions associated with emotion such as the orbitofrontal cortex (Rolls 2005, Rolls 2008d). These hypotheses are supported by the frequently observed hypofrontality, a reduced activity in frontal brain regions in schizophrenic patients during cognitive tasks (Ingvar and Franzen 1974, Kircher and Thienel 2005, Scheuerecker, Ufer, Zipse, Frodl, Koutsouleris, Zetzsche, Wiesmann, Albrecht, Bruckmann, Schmitt, Moller and Meisenzahl 2008).

The positive symptoms are characterized by phenomenologically overactive perceptions or thoughts such as hallucinations or delusions which are reflected for example by higher activity in the temporal lobes (Shergill, Brammer, Williams, Murray and McGuire 2000, Scheuerecker et al. 2008). We relate this category of symptoms to a spontaneous appearance of activity in attractor networks in the brain and more generally to instability of both the spontaneous and persistent attractor states (Loh, Rolls and Deco 2007a).

In particular we were interested in how these symptoms are related. Schizophrenia is a neurodevelopmental disease, and negative and cognitive symptoms typically precede the first psychotic episode (Lieberman, Perkins, Belger, Chakos, Jarskog, Boteva and Gilmore 2001, Hafner, Maurer, Loffler, an der Heiden, Hambrecht and Schultze-Lutter 2003). Positive symptoms can be treated in most cases with neuroleptics whereas negative and cognitive symptoms persist, at least for typical neuroleptics. Can a mapping onto a dynamical system help to understand these relations? After proposing a dynamical systems hypothesis for the different symptoms of schizophrenia, we study a standard neural network model (Brunel and Wang 2001) of cortical dynamics specifically in relation to our hypothesis (Rolls 2005, Loh, Rolls and Deco 2007a, Rolls, Loh, Deco and Winterer 2008b). We were especially interested in how excitation and inhibition implemented by NMDA and GABA synapses affect the network dynamics. Alterations in the efficacies of the NMDA and GABA channels have been identified in the pathology of schizophrenia (Coyle, Tsai and Goff 2003, Lewis, Hashimoto and Volk 2005), and transmitters such as dopamine influence the currents in these receptor-activated channels (Seamans and Yang 2004). Do NMDA and GABA currents have antagonistic effects or do they have a special role in the network dynamics? How could this be related to our hypothesis of schizophrenia? Building upon the current body of neural network research, we describe neural network simulations to substantiate our dynamical systems hypothesis of schizophrenia. While focussing on NMDA and GABA synapses in the simulations, we also consider how altered transmission at D1 and D2 receptors by modulating NMDA and GABA conductances could not only influence working memory, which has been investigated previously (Durstewitz, Kelc and Gunturkun 1999, Durstewitz, Seamans and Sejnowski 2000a, Brunel and Wang 2001, Durstewitz and Seamans 2002), but could in particular influence the different symptoms of schizophrenia.

8.14.2 A dynamical systems hypothesis of the symptoms of schizophrenia

We relate the three types of symptoms of schizophrenia to the dynamical systems attractor framework described earlier as follows (Rolls 2005).

The *cognitive symptoms* of schizophrenia include distractibility, poor attention, and the dysexecutive syndrome (Liddle 1987, Green 1996, Mueser and McGurk 2004). The core of the cognitive symptoms is a working memory deficit in which there is a difficulty in maintaining items in short-term memory (Goldman-Rakic 1994, Goldman-Rakic 1999), which could directly or indirectly account for a wide range of the cognitive symptoms. We propose that these symptoms may be related to instabilities of persistent states in attractor neural networks, consistent with the body of theoretical research on network models of working memory (Durstewitz, Seamans and Sejnowski 2000b). The neurons are firing at a lower rate, leading to shallower basins of attraction of the persistent states, and thus a difficulty in maintaining a stable short-term memory, normally the source of the bias in biased competition models of attention (Rolls and Deco 2002, Deco and Rolls 2005a). The shallower basins of attraction would thus result in working memory deficits, poor attention, distractibility, and problems with executive function and action selection (Deco and Rolls 2003, Deco and Rolls 2005a).

The *negative symptoms* refer to the flattening of affect and a reduction in emotion. Behavioural indicators are blunted affect, emotional and passive withdrawal, poor rapport, lack of spontaneity, motor retardation, and disturbance of volition (Liddle 1987, Mueser and McGurk 2004). We propose that these symptoms are related to decreases in firing rates in the orbitofrontal cortex and/or anterior cingulate cortex (Rolls 2005), where neuronal firing rates and activations in fMRI investigations are correlated with reward value and pleasure (Rolls 2005, Rolls 2006b, Rolls 2007e, Rolls and Grabenhorst 2008, Rolls 2008c). Consistent with this, imaging studies have identified a relationship between negative symptoms and prefrontal hypometabolism, i.e. a reduced activation of frontal areas (Wolkin, Sanfilipo, Wolf, Angrist, Brodie and Rotrosen 1992, Aleman and Kahn 2005).

The *positive symptoms* of schizophrenia include bizarre (psychotic) trains of thoughts, hallucinations, and (paranoid) delusions (Liddle 1987, Mueser and McGurk 2004). We propose that these symptoms are related to shallow basins of attraction of both the spontaneous and persistent states in the temporal lobe semantic memory networks and to the statistical fluctuations caused by the probabilistic spiking of the neurons. This could result in activations arising spontaneously, and thoughts moving too freely round the energy landscape, loosely from thought to weakly associated thought, leading to bizarre thoughts and associations, which may eventually over time be associated together in semantic memory to lead to false beliefs and delusions. Consistent with this, neuroimaging studies suggest higher activation especially in areas of the temporal lobe (Weiss and Heckers 1999, Shergill et al. 2000, Scheuerecker et al. 2008).

To further investigate our hypothesis, we use an attractor network, as this is likely to be implemented in many parts of the cerebral cortex by the recurrent collateral connections between pyramidal cells, and has short-term memory properties with basins of attraction which allow systematic investigation of stability and distractibility. The particular neural network implementation we adopt includes channels activated by AMPA, NMDA and GABA$_A$ receptors and allows not only the spiking activity to be simulated, but also a consistent mean-field approach to be used (Brunel and Wang 2001). The network, and some of the simulations performed with it, are described in Chapters 3 and 4.

8.14.3   The depth of the basins of attraction: mean-field flow analysis

First we introduced an analytical approach to the concepts of how changes in transmitters could affect the depth of the basins of attraction in networks in ways that may be related to the symptoms of schizophrenia. Figure 3.10 on page 107 shows the flow between the spontaneous and persistent state in a network featuring one selective pool. A reduction of NMDA receptor activated ion channel currents ($-$NMDA) produces a stronger flow than the unchanged condition at low firing rates towards the spontaneous attractor (at about 2 Hz). The absolute values of the function are higher compared to the normal condition until the first unstable fixed point (at around 6–7 Hz). The basin of attraction towards the persistent attractor at high firing rates yields the reverse picture. Here the (−NMDA) curve is clearly below the unchanged condition and the flow towards the high firing rate attractor is smaller. Overall, the basin of attraction is deeper for the spontaneous state and shallower for the persistent state compared to the unchanged condition. This pattern fits to the cognitive symptoms of schizophrenia as proposed in our hypothesis. We also note that the firing rate of the persistent fixed point is reduced in the (−NMDA) condition (crossing with the flow=0 axis), which is consistent with the hypothesis for the negative symptoms.

A reduction of the GABA receptor activated ion channel conductances (−GABA) yields the opposite pattern to that in the reduced NMDA condition. Here the basin of attraction of the persistent high firing rate state is deeper. This is not a condition that we suggest is related to the symptoms of schizophrenia.

However, in the condition in which both the NMDA and the GABA conductances are reduced (−NMDA, −GABA), the persistent high firing rate state basin of attraction is shallower, and the spontaneous state basin is a little shallower. This condition corresponds to the proposed landscape for the positive symptoms as considered earlier. In particular, in the (−NMDA, −GABA) condition, the system would be less stable in the persistent state, tending to move to another attractor easily, and less stable in the spontaneous state, so tending to move too readily into an attractor from spontaneous activity.

Overall, the mean-field flow analysis suggests that both the cognitive and negative symptoms could be related to a decrease in the NMDA conductances. This is consistent with the fact that these two symptoms usually appear together. The flow analysis suggests that the positive symptoms are related to a reduction in both NMDA and GABA. Thus, the transition from the cognitive and negative symptoms to the positive, psychotic, symptoms might be caused by an additional decrease in the GABA conductance.

Very importantly, it is notable that excitation and inhibition do not cancel each other out as assumed by many models, but have distinct influences on the stochastic dynamics of the network.

8.14.4 Decreased stability produced by reductions of NMDA receptor activated synaptic conductances

We assessed how the stability of both the spontaneous and persistent states changes when NMDA and GABA efficacies are modulated. Specifically we ran multiple trial integrate-and-fire network simulations and counted how often the system maintained the spontaneous or persistent state, assessed by the firing rate in the last second of the simulation (2–3 s) of each 3 s trial. Figure 3.11 (on page 108) shows the stability of the spontaneous and persistent attractor relative to the unmodulated reference state (Normal). A negative percentage means that the system was less stable than in the unmodulated state.

A reduction of the NMDA receptor activated synaptic conductances (−NMDA) reduces the stability of the persistent state drastically, while slightly increasing the stability of the spontaneous state (see Fig. 3.11). We hypothesized that this type of change might be related to the cognitive symptoms, since it shows a reduced stability of the working memory properties. A reduction of GABA shows the opposite pattern: a slight reduction in the stability of the spontaneous state, and an increased stability of the persistent (i.e. attractor) state (see Fig. 3.11).

When both NMDA and GABA are reduced one might think that these two counterbalancing effects (excitatory and inhibitory) would either cancel each other out or yield a tradeoff between the stability of the spontaneous and persistent state. However, this is not the case. The stability of both the spontaneous state and the persistent state is reduced (see Figure 3.11). We relate this pattern to the positive symptoms of schizophrenia, in which both the spontaneous and attractor states are shallow, and the system merely jumps by the influence of statistical fluctuations between the different (spontaneous and high firing rate attractor) states.

To investigate more directly the wandering between spontaneous and several different persistent attractor states, we simulated the condition with decreased NMDA and GABA conductances over a long time period in which no cue stimulus input was given. Figure 3.12 shows the firing rates of the two selective pools S1 and S2. The high activity switches between the two attractors due to the influence of fluctuations, which corresponds to spontaneous wandering in a shallow energy landscape, corresponding for example to sudden jumps between unrelated cognitive processes. These results are consistent with the flow analysis and demonstrate that the changes in the attractor landscape influence the behaviour at the stochastic level.

8.14.5 Increased distractibility produced by reductions of NMDA receptor activated synaptic conductances

As distractibility is directly related to the symptoms of schizophrenia, we ran simulations specifically to assess this property using persistent and distractor simulations (see Fig. 3.8 on page 104). A distractor strength of 0 Hz corresponds to the persistent condition described in Section 3.8.2. Figure 4.10 on page 136 shows the stability and distractibility for reductions of NMDA and GABA currents. The reference state is labelled 'Normal'. In this state, pool S1 continued to maintain its attractor firing without any distractor (distractor strength = 0 Hz) throughout the delay period on almost 90% of the trials. In both conditions which reduce the NMDA current (labelled −NMDA) the network was less and less able to maintain the S1 attractor firing as the distractor stimulus strength was increased through the range 0–80 Hz. The stability of the persistent state was reduced, and also the distractibility was increased, as shown by the fact that increasing distractor currents applied to S2 could move the attractor away from S1.

The implication therefore is that a reduction of the NMDA currents could cause the cognitive symptoms of schizophrenia, by making short-term memory networks less stable and more distractible, thereby reducing the ability to maintain attention. Reducing only the GABA currents (−GABA) reduces the distractibility for low distractor strengths and coincides with the reference (Normal) condition at high values of the distractor strengths.

8.14.6 Signal-to-noise ratio in schizophrenia

We further investigated the signal-to-noise ratio in relation to the changes in synaptic conductances, as described in Section 3.8.3.3. In an attractor network, a high signal-to-noise ratio indicates that the network will maintain the attractor stably, as it will be unlikely to be disrupted by spiking-related statistical fluctuations that are the source of the noise in the network. Figure 3.13 on page 111 shows the signal-to-noise ratio of a measure related to the fMRI BOLD signal. [This measure described in the legend to Fig. 3.13 and below was used because the experimental data with which we wish to compare the simulation results use fMRI measures (Winterer et al. 2000, Winterer et al. 2004, Winterer et al. 2006). The index we used of the activity of the network was the total synaptic current of selective pool 1 averaged over the whole simulation time of 3 s, to take the temporal filtering properties of the BOLD signal into account, given the typical time course which lasts for several seconds of the

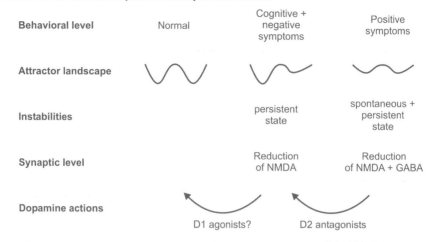

Fig. 8.6 Summary of attractor hypothesis of schizophrenic symptoms and simulation results (see text). The first basin (from the left) in each energy landscape is the spontaneous state, and the second basin is the persistent attractor state. The vertical axis of each landscape is the energy potential. (After Loh, Rolls and Deco 2007b.)

haemodynamic response function (Deco, Rolls and Horwitz 2004) (see Section 6.5). Further, we subtracted the averages of the spontaneous trial simulations which represent the baseline activity from the persistent trial simulation values. The signal-to-noise ratio was calculated from the mean of this index across trials divided by the standard deviation of the index, both measured using 1000 simulation trials. If the network sometimes had high activity, and sometimes low, then the signal-to-noise measure gave a low value. If the network reliably stayed in the high persistent firing states, then the signal-to-noise ratio measure was high.] As shown in Fig. 3.13, we found that in all the cases in which the NMDA, or the GABA, conductance, or both, are reduced, the signal-to-noise ratio, computed by the mean divided by the standard deviation, is also reduced. This relates to recent experimental observations which show a decreased signal-to-noise ratio in schizophrenic patients (Winterer et al. 2000, Winterer et al. 2004, Winterer et al. 2006, Rolls et al. 2008b). Here we directly relate a decrease in the signal-to-noise ratio to changes (in this case decreases) in receptor activated synaptic ion channel conductances.

8.14.7   Synthesis: network instability and schizophrenia

We have proposed a hypothesis that relates the cognitive, negative, and positive symptoms of schizophrenia (Liddle 1987, Mueser and McGurk 2004) to the depth of basins of attraction and to the stability properties of attractor networks caused by statistical fluctuations of spiking neurons (Rolls 2005, Loh, Rolls and Deco 2007a, Loh, Rolls and Deco 2007b, Rolls 2008d, Rolls, Loh, Deco and Winterer 2008b). This assumes that some cognitive processes can be understood as dynamical attractor systems, which is an established hypothesis in areas such as working memory, but has also been used many other areas (Rolls and Deco 2002, O'Reilly 2006). Our approach applies this concept to mental illnesses (Bender, Albus, Moller and Tretter 2006). Due to the diversity of the symptoms of schizophrenia, our general hypothesis is meant to serve as a heuristic for how the different kinds of symptoms might arise and are related. We investigated the hypothesis empirically in a computational attractor framework to capture an important aspect of cortical functionality. Figure 8.6 summarizes our hypothesis and its relation to the investigations of stochastic neurodynamics.

The middle column in Fig. 8.6 shows the overview for the cognitive and negative symptoms. The core of the cognitive symptoms is a failure of working memory and attentional mechanisms. Working memory activity is related to the ongoing (i.e. persistent) firing of neurons during the delay period of cognitive tasks (Goldman-Rakic 1994, Goldman-Rakic 1999). This could be implemented by associatively modifiable synapses between the recurrent collateral synapses of cortical pyramidal cells (Rolls and Treves 1998, Durstewitz et al. 2000b, Renart et al. 2001, Wang 2001). We propose that the cognitive symptoms of schizophrenia could arise because the basins of attraction of the persistent states in the prefrontal cortex become too shallow. This leads in combination with the statistical fluctuations due to randomness of the spiking activity to either a fall out of an active working memory state or to a shift to a different attractor state, leading to a failure to maintain attention and thereby impairing executive function. The hypofrontality in schizophrenia, that is less activation in frontal brain regions during working memory tasks (Ingvar and Franzen 1974, Carter, Perlstein, Ganguli, Brar, Mintun and Cohen 1998), is in line with our hypothesis, since the firing rates of the persistent state are lower in the reduced NMDA condition (Fig. 3.10), and the system spends on average less time in the persistent state, since it is less stable than in the normal condition (Figure 3.11). In addition, a reduced signal-to-noise ratio as shown in our simulations (Fig. 3.13) has also been identified in imaging studies (Winterer et al. 2000, Winterer et al. 2004, Winterer et al. 2006). Our simulations suggest that a reduction in NMDA conductance at the synaptic level (see Figs. 8.6, 8.3 and 8.4) can account for this phenomenon. This is in line with previous work on the stability of working memory networks (Wang 1999, Durstewitz et al. 2000a, Wang 2001).

A reduction of the NMDA conductance also results in a reduction of the firing rates of the neurons in the persistent state (see Figs. 3.10, 8.3 and 8.4, and Brunel and Wang (2001)). We relate this, following Rolls (2005), to the negative symptoms which include flattening of affect, a reduction in emotion, emotional and social withdrawal, poor rapport, passive withdrawal, lack of spontaneity, motor retardation, apathy, and disturbance of motivation. These symptoms are related to decreases in activity in the orbitofrontal cortex and/or anterior cingulate cortex (Wolkin et al. 1992, Aleman and Kahn 2005), both of which are implicated in emotion (Rolls 2005, Rolls 2006b, Rolls and Grabenhorst 2008, Rolls 2009a). The emotional states represented in the orbitofrontal cortex and anterior cingulate cortex include states elicited both by rewards and punishers. Our hypothesis is that both would be reduced by the mechanism described. Correspondingly, motivation would be reduced in the same way, in that motivation is a state in which we work to obtain goals (rewards) or avoid punishers (Rolls 2005).

Both the negative and cognitive symptoms thus could be caused by a reduction of the NMDA conductance in attractor networks. The proposed mechanism links the cognitive and negative symptoms of schizophrenia in an attractor framework and is consistent with a close relation between the cognitive and negative symptoms: blockade of NMDA receptors by dissociative anesthetics such as ketamine produces in normal subjects schizophrenic symptoms including both negative and cognitive impairments (Malhotra, Pinals, Weingartner, Sirocco, Missar, Pickar and Breier 1996, Newcomer, Farber, Jevtovic-Todorovic, Selke, Melson, Hershey, Craft and Olney 1999); agents that enhance NMDA receptor function reduce the negative symptoms and improve the cognitive abilities of schizophrenic patients (Goff and Coyle 2001); and the cognitive and negative symptoms occur early in the illness and precede the first episode of positive symptoms (Lieberman et al. 2001, Hafner et al. 2003, Mueser and McGurk 2004). Consistent with this hypothesized role of a reduction in NMDA conductances being involved in schizophrenia, postmortem studies of schizophrenia have identified abnormalities in glutamate receptor density in regions such as the prefrontal cortex, thalamus and the temporal lobe (Goff and Coyle 2001, Coyle et al. 2003), brain areas that are active during the performance of cognitive tasks.

The dopamine D1 receptor has been shown to modulate the performance of working memory tasks (Sawaguchi and Goldman-Rakic 1991, Sawaguchi and Goldman-Rakic 1994, Goldman-Rakic 1999, Castner et al. 2000). An increase in D1 receptor activation has been shown to increase the NMDA current (Durstewitz and Seamans 2002, Seamans and Yang 2004), and modelling studies have shown that this increase is related to the stability of working memory states (Durstewitz et al. 1999, Durstewitz et al. 2000a, Brunel and Wang 2001). Imaging data also support the importance of the D1 receptor in schizophrenia (Okubo, Suhara, Sudo and Toru 1997a, Okubo, Suhara, Suzuki, Kobayashi, Inoue, Terasaki, Someya, Sassa, Sudo, Matsushima, Iyo, Tateno and Toru 1997b). We therefore suggest that an increased activation of D1 receptors might alleviate both the cognitive and the negative symptoms of schizophrenia (Goldman-Rakic, Castner, Svensson, Siever and Williams 2004, Miyamoto, Duncan, Marx and Lieberman 2005), by increasing NMDA receptor mediated synaptic currents (Fig. 8.6). Atypical neuroleptics might use this mechanism by not only blocking D2 receptors, but also by increasing the presynaptic release of dopamine which in turn would increase the activation of the extrasynaptic D1 receptors (Castner et al. 2000, Moller 2005).

Taken together, we suggest that the cognitive and negative symptoms could be caused by the same synaptic mechanism, namely a reduction in the NMDA conductance, which reduces the stability and increases the distractibility of the persistent attractors, and reduces the activity (firing rates) of neurons (Fig. 8.6, middle column). The reduced depth of the basins of attraction can be understood in the following way. Hopfield (1982) showed that the recall state in an attractor network can be thought of as the local minimum in an energy landscape, where the energy would be defined as shown in Equation 1.21 on page 34. In general neuronal systems do not admit an energy function. Nevertheless, we can assume an effective energy function: in fact, the flow picture shown in Fig. 3.10 resulting from the mean-field reduction associated with the spiking network analyzed here, can be viewed as an indirect description of an underlying effective energy function. From Equation 1.21, it follows that the depth of a basin of attraction is deeper if the firing rates are higher and if the synaptic strengths that couple the neurons that are part of the same attractor are strong. (The negative sign results in a low energy, and thus a stable state, if the firing rates of the neurons in the same attractor and their synaptic coupling weights are high.) If we reduce the NMDA receptor activated channel conductances, then the depth of the basins of attraction will be reduced both because the firing rates are reduced by reducing excitatory inputs to the neurons, and because the synaptic coupling weights are effectively reduced because the synapses can pass only reduced currents.

The positive symptoms (Fig. 8.6, right column) of schizophrenia include delusions, hallucinations, thought disorder, and bizarre behaviour. Examples of delusions are beliefs that others are trying to harm the person, impressions that others control the person's thoughts, and delusions of grandeur. Hallucinations are perceptual experiences, which are not shared by others, and are frequently auditory but can affect any sensory modality. These symptoms may be related to activity in the temporal lobes (Liddle 1987, Epstein, Stern and Silbersweig 1999, Mueser and McGurk 2004). The attractor framework approach taken here hypothesizes that the basins of attraction of both spontaneous and persistent states are shallow (Fig. 8.6). Due to the shallowness of the spontaneous state, the system can jump spontaneously up to a high activity state causing hallucinations to arise and leading to bizarre thoughts and associations. This might be the cause for the higher activations in schizophrenics in temporal lobe areas which are identified in imaging experiments (Shergill et al. 2000, Scheuerecker et al. 2008).

We relate the positive symptoms to not only a reduction in NMDA conductance, but also to a reduction in GABA conductance. This is consistent with the fact that the positive symptoms usually follow the cognitive and negative ones and represent a qualitative worsening of the illness (Mueser and McGurk 2004). Alterations in GABA receptors have been identified in schizophrenia (Wang, Tegner, Constantinidis and Goldman-Rakic 2004, Lewis et al. 2005).

D2 receptor antagonism remains a main target for antipsychotics (Seeman and Kapur 2000, Leuner and Muller 2006). Dopamine receptor D2 antagonists mainly alleviate the positive symptoms of schizophrenia, whereas the cognitive and negative symptoms persist, especially for the typical neuroleptics (Mueser and McGurk 2004). Together with the simulations, our hypothesis suggests that an increase in the GABA current in the state corresponding to the positive symptoms ($-$NMDA, $-$GABA) might have the same effect as D2 antagonists. The therapeutic effect of D2 antagonists might thus be caused by an increase in GABA currents. Indeed, it has been found that D2 receptors decrease the efficacy of the GABA system (Seamans, Gorelova, Durstewitz and Yang 2001, Trantham-Davidson, Neely, Lavin and Seamans 2004). (For example, Seamans et al. (2001) found that the application of D2 antagonists prevented a decrease in eIPSC amplitude produced by dopamine.) Thus D2 antagonists would, in a hypersensitive D2 receptor state (Seeman, Weinshenker, Quirion, Srivastava, Bhardwaj, Grandy, Premont, Sotnikova, Boksa, El-Ghundi, O'Dowd, George, Perreault, Mannisto, Robinson, Palmiter and Tallerico 2005, Seeman, Schwarz, Chen, Szechtman, Perreault, McKnight, Roder, Quirion, Boksa, Srivastava, Yanai, Weinshenker and Sumiyoshi 2006), increase GABA inhibition in the network, and this we suggest could increase the stability of attractor networks involved in the positive symptoms of schizophrenia, and thus ameliorate the positive symptoms. Since the concentration of dopamine in the cortex depends on cortical-subcortical interactions (Carlsson 2006), the causes of the described changes could also result from subcortical deficits. A detailed analysis of these feedback loops would require specific modelling.

Earlier accounts of the relation of dopamine and schizophrenia in the cortex (Seamans et al. 2001, Seamans and Yang 2004) have suggested two distinct states of dopamine modulation. One is a D2-receptor-dominated state in which there is weak gating and information can easily affect network activity. The other is a D1-receptor-dominated state in which network activity is stable and maintained. We have proposed a more detailed account for stability and discussed this separately for the spontaneous and persistent attractor states. This allows us to account for the dichotomy between the cognitive/negative and positive symptoms. We emphasize that, in biophysically realistic network simulations, excitation and inhibition are not merely antagonistic but implement different functions in the network dynamics. Thereby our stochastic dynamics modeling approach (Loh, Rolls and Deco 2007a, Loh, Rolls and Deco 2007b, Rolls, Loh, Deco and Winterer 2008b) provides a missing link between the symptoms of schizophrenia and network models of working memory and dopamine (Durstewitz et al. 1999, Durstewitz et al. 2000a, Brunel and Wang 2001).

We considered earlier a possible cause for the proposed alterations of the attractor landscape related to schizophrenia, namely changes in NMDA and GABA conductance as these are directly related to schizophrenia (Coyle et al. 2003, Lewis et al. 2005). We did not investigate changes in AMPA conductance. In this particular model the contribution of the AMPA current is relatively small (Brunel and Wang 2001). A more detailed investigation could also include AMPA conductance, especially because it is known to be influenced by NMDA synaptic plasticity (Bagal, Kao, Tang and Thompson 2005). Indeed, if reduced NMDA currents led in turn by synaptic plasticity to reduced AMPA currents, this would amplify the effects we describe.

The proposed alterations in the attractor landscape could have a variety of causes at the neurobiological level: abnormalities in glutamate and GABA receptors and signalling, modulations in synaptic plasticity, aberrant dopamine signalling, reduced neuropil, genetic mechanisms, and brain volume reduction (Goldman-Rakic 1999, Winterer and Weinberger 2004, Mueser and McGurk 2004, Stephan, Baldeweg and Friston 2006). Besides cortical mechanisms, cortical-subcortical dynamics could also cause the proposed alterations in the cortical attractor landscape for example via neuromodulatory influences such as dopamine or serotonin or cortical-subcortical feedback loops (Capuano, Crosby and Lloyd 2002, Carlsson 2006). Our general hypothesis regarding the attractor landscape is meant to describe the aberrant dynamics in cortical regions which could be caused by several pathways. Future work could analyze further how changes of different factors such as regional differences, subcortical-cortical networks or even more detailed neural and synaptic models might influence the stability of the type of neurodynamical system described here. Moreover, future work could investigate within the framework we describe the effect of treatment with particular combinations of drugs designed to facilitate both glutamate transmission (including modulation with glycine or serine of the NMDA receptor) and GABA effects. Drug combinations identified in this way could be useful to explore, as combinations might work where treatment with a single drug on its own may not be effective, and possibilities can be systematically explored with the approach described here.

8.15 Stochastic noise, attractor dynamics, and obsessive-compulsive disorder

8.15.1 Introduction

Obsessive-compulsive disorder (OCD) is a chronically debilitating disorder with a lifetime prevalence of 2–3% (Robins, Helzer, Weissman, Orvaschel, Gruenberg, Burke and Regier 1984, Karno, Golding, Sorenson and Burnam 1988, Weissman, Bland, Canino, Greenwald, Hwu, Lee, Newman, Oakley-Browne, Rubio-Stipec, Wickramaratne et al. 1994). It is characterized by two sets of symptoms, obsessive and compulsive. Obsessions are unwanted, intrusive, recurrent thoughts or impulses that are often concerned with themes of contamination and 'germs', checking household items in case of fire or burglary, order and symmetry of objects, or fears of harming oneself or others. Compulsions are ritualistic, repetitive behaviours or mental acts carried out in relation to these obsessions e.g. washing, household safety checks, counting, rearrangement of objects in symmetrical array or constant checking of oneself and others to ensure no harm has occurred (Menzies, Chamberlain, Laird, Thelen, Sahakian and Bullmore 2008). Patients with OCD experience the persistent intrusion of thoughts that they generally perceive as foreign and irrational but which cannot be dismissed. The anxiety associated with these unwanted and disturbing thoughts can be extremely intense; it is often described as a feeling that something is incomplete or wrong, or that terrible consequences will ensue if specific actions are not taken. Many patients engage in repetitive, compulsive behaviours that aim to discharge the anxieties associated with these obsessional thoughts. Severely affected patients can spend many hours each day in their obsessional thinking and resultant compulsive behaviours, leading to marked disability (Pittenger, Krystal and Coric 2006). While OCD patients exhibit a wide variety of obsessions and compulsions, the symptoms tend to fall into specific clusters. Common patterns include obsessions of contamination, with accompanying cleaning compulsions; obsessions with symmetry or order, with accompanying ordering behaviours; obsessions of saving, with accompanying hoarding; somatic obsessions; aggressive obsessions with checking compulsions; and sexual and religious obsessions (Pittenger, Krystal and Coric 2006).

In this section we describe a theory of how obsessive-compulsive disorders arise, and of the different symptoms. The theory (Rolls, Loh and Deco 2008a) is based on the top-down proposal that there is overstability of attractor neuronal networks in cortical and related areas in obsessive-compulsive disorders. The approach is top-down in that it starts with the set of symptoms and maps them onto the dynamical systems framework, and only after this considers detailed underlying biological mechanisms, of which there could be many, that might produce the effects. (In contrast, a complementary bottom-up approach starts from detailed neurobiological mechanisms, and aims to interpret their implications with a brain-like model for higher level phenomena.) We show by integrate-and-fire neuronal network simulations that the overstability could arise by for example overactivity in glutamatergic excitatory neurotransmitter synapses, which produces an increased depth of the basins of attraction, in the presence of which neuronal spiking-related and potentially other noise is insufficient to help the system move out of an attractor basin. We relate this top-down proposal, related to the stochastic dynamics of neuronal networks, to new evidence that there may be overactivity in glutamatergic systems in obsessive-compulsive disorders, and consider the implications for treatment.

The background is that there has been interest for some time in the application of complex systems theory to understanding brain function and behaviour (Riley and Turvey 2001, Peled 2004, Heinrichs 2005, Lewis 2005). Dynamically realistic neuronal network models of working memory (Wang 1999, Compte, Brunel, Goldman-Rakic and Wang 2000, Durstewitz et al. 2000a, Durstewitz et al. 2000b, Brunel and Wang 2001, Wang 2001), decision-making (Wang 2002, Deco and Rolls 2006), and attention (Rolls and Deco 2002, Deco and Rolls 2005a, Deco and Rolls 2005b), and how they are influenced by neuromodulators such as dopamine (Durstewitz et al. 1999, Durstewitz et al. 2000a, Brunel and Wang 2001, Durstewitz and Seamans 2002, Deco and Rolls 2003) have been produced, and provide a foundation for the present approach (Rolls, Loh and Deco 2008a).

8.15.2 A hypothesis about the increased stability of attractor networks and the symptoms of obsessive-compulsive disorder

We hypothesize that cortical and related attractor networks become too stable in obsessive-compulsive disorder, so that once in an attractor state, the networks tend to remain there too long (Rolls, Loh and Deco 2008a). The hypothesis is that the depths of the basins of attraction become deeper, and that this is what makes the attractor networks more stable. We further hypothesize that part of the mechanism for the increased depth of the basins of attraction is increased glutamatergic transmission, which increases the depth of the basins of attraction by increasing the firing rates of the neurons, and by increasing the effective value of the synaptic weights between the associatively modified synapses that define the attractor, as is made evident in Equation 1.21. The synaptic strength is effectively increased if more glutamate is released per action potential at the synapse, or if in other ways the currents injected into the neurons through the NMDA (N-methyl-d-aspartate) and/or AMPA synapses are larger. In addition, if NMDA receptor function is increased, this could also increase the stability of the system because of the temporal smoothing effect of the long time constant of the NMDA receptors (Wang 1999). This increased stability of cortical and related attractor networks, and the associated higher neuronal firing rates, could occur in different brain regions, and thereby produce different symptoms, as follows. If these effects occurred in high order motor areas, the symptoms could include inability to move out of one motor pattern, resulting for example in repeated movements or actions. In parts of the cingulate cortex and dorsal medial prefrontal cortex, this could result in difficulty in switching between actions or strategies (Rushworth, Behrens, Rudebeck and Walton 2007a, Rushworth, Buckley, Behrens, Walton and Bannerman 2007b), as the system would be locked into one action or strategy. If an action was locked into a high order motor area due to increased stability of an attractor network, then lower order motor areas might thereby not be able to escape easily what they implement, such as a sequence of movements, so that the sequence would be repeated.

A similar account, of becoming locked in one action and having difficulty in switching to another action, can be provided for response inhibition deficits, which have been found in OCD. The response inhibition deficit has been found in tasks such as go/no-go and stop-signal reaction time (SSRT) which examine motor inhibitory processes, and also the Stroop task, a putative test of cognitive inhibition (Hartston and Swerdlow 1999, Bannon, Gonsalvez, Croft and Boyce 2002, Penades, Catalan, Andres, Salamero and Gasto 2005, Bannon, Gonsalvez, Croft and Boyce 2006, Chamberlain, Fineberg, Blackwell, Robbins and Sahakian 2006, Chamberlain, Fineberg, Menzies, Blackwell, Bullmore, Robbins and Sahakian 2007, Penades, Catalan, Rubia, Andres, Salamero and Gasto 2007). For example, response inhibition deficits have been reported in OCD patients when performing the SSRT, which measures the time taken to internally suppress pre-potent motor responses (Chamberlain et al. 2006). Unaffected first-degree relatives of OCD patients are also impaired on this task compared with unrelated healthy controls, suggesting that response inhibition may be an endophenotype (or intermediate phenotype) for OCD (Chamberlain et al. 2007, Menzies et al. 2008).

If occurring in the lateral prefrontal cortex (including the dorsolateral and ventrolateral parts), the increased stability of attractor networks could produce symptoms that include a difficulty in shifting attention and in cognitive set shifting. These are in fact important symptoms that can be found in obsessive-compulsive disorder (Menzies et al. 2008). Two different forms of shifting have been investigated: affective set shifting, where the affective or reward value of a stimulus changes over time (e.g. a rewarded stimulus is no longer rewarded) (intradimensional or ID set shifting); and attentional set shifting, where the stimulus dimension (e.g. shapes or colors) to which the subject must attend is changed (extradimensional or ED set shifting). Deficits of attentional set shifting in OCD have been found in several neurocognitive studies using the CANTAB ID/ED set shifting task (Veale, Sahakian, Owen and Marks 1996, Watkins, Sahakian, Robertson, Veale, Rogers, Pickard, Aitken and Robbins 2005, Chamberlain et al. 2006, Chamberlain et al. 2007). This deficit is most consistently reported at the ED stage (in which the stimulus dimension, e.g. shape, color or number, alters and subjects have to inhibit their attention to this dimension and attend to a new, previously irrelevant dimension). The ED stage is analogous to the stage in the Wisconsin Card Sorting Task where a previously correct rule for card sorting is changed and the subject has to respond to the new rule (Berg 1948). This ED shift impairment in OCD patients is considered to reflect a lack of cognitive or attentional flexibility and may be related to the repetitive nature of OCD symptoms and behaviours. Deficits in attentional set shifting are considered to be more dependent upon dorsolateral and ventrolateral prefrontal regions than the orbital prefrontal regions included in the orbitofronto-striatal model of OCD (Pantelis, Barber, Barnes, Nelson, Owen and Robbins 1999, Rogers, Andrews, Grasby, Brooks and Robbins 2000, Nagahama, Okada, Katsumi, Hayashi, Yamauchi, Oyanagi, Konishi, Fukuyama and Shibasaki 2001, Hampshire and Owen 2006), suggesting that cognitive deficits in OCD may not be underpinned exclusively by orbitofrontal cortex pathology. Indeed, intradimensional or affective set shifting may not be consistently impaired in OCD (Menzies et al. 2008).

Planning may also be impaired in patients with OCD (Menzies et al. 2008), and this could arise because there is too much stability of attractor networks in the dorsolateral prefrontal cortex concerned with holding in mind the different short-term memory representations that encode the different steps of a plan (Rolls 2008d). Indeed, there is evidence for dorsolateral prefrontal cortex (DLPFC) dysfunction in patients with OCD, in conjunction with impairment on a version of the Tower of London, a task often used to probe planning aspects of executive function (van den Heuvel, Veltman, Groenewegen, Cath, van Balkom, van Hartskamp, Barkhof and van Dyck 2005). Impairment on the Tower of London task has also been demonstrated in healthy first-degree relatives of OCD patients (Delorme, Goùsse, Roy,

Trandafir, Mathieu, Mouren-Simeoni, Betancur and Leboyer 2007).

An increased firing rate of neurons in the orbitofrontal cortex, and anterior cingulate cortex, produced by hyperactivity of glutamatergic transmitter systems, would increase emotionality, which is frequently found in obsessive-compulsive disorder. Part of the increased anxiety found in obsessive-compulsive disorder could be related to an inability to complete tasks or actions in which one is locked. But part of our unifying proposal is that part of the increased emotionality in OCD may be directly related to increased firing produced by the increased glutamatergic activity in brain areas such as the orbitofrontal and anterior cingulate cortex. The orbitofrontal cortex and anterior cingulate cortex are involved in emotion, in that they are activated by primary and secondary reinforcers that produce affective states (Rolls 2004a, Rolls 2005, Rolls 2008d, Rolls and Grabenhorst 2008, Rolls 2009a), and in that damage to these regions alters emotional behaviour and emotional experience (Rolls, Hornak, Wade and McGrath 1994a, Hornak, Rolls and Wade 1996, Hornak, Bramham, Rolls, Morris, O'Doherty, Bullock and Polkey 2003, Hornak, O'Doherty, Bramham, Rolls, Morris, Bullock and Polkey 2004, Berlin, Rolls and Kischka 2004, Berlin, Rolls and Iversen 2005). Indeed, negative emotions as well as positive emotions activate the orbitofrontal cortex, with the emotional states produced by negative events tending to be represented in the lateral orbitofrontal cortex and dorsal part of the anterior cingulate cortex (Kringelbach and Rolls 2004, Rolls 2005, Rolls 2008d, Rolls 2009a). We may note that stimulus-reinforcer reversal tasks (also known as intra-dimensional shifts or affective reversal) are not generally impaired in patients with OCD (Menzies et al. 2008), and this is as predicted, for the machinery for the reversal including the detection of non-reward (Rolls 2005) is present even if the system is hyperglutamatergic.

If the increased stability of attractor networks occurred in temporal lobe semantic memory networks, then this would result in a difficulty in moving from one thought to another, and possibly in stereotyped thoughts, which again may be a symptom of obsessive-compulsive disorder (Menzies et al. 2008). The obsessional states are thus proposed to arise because cortical areas concerned with cognitive functions have states that become too stable. The compulsive states are proposed to arise partly in response to the obsessional states, but also partly because cortical areas concerned with actions have states that become too stable. The theory provides a unifying computational account of both the obsessional and compulsive symptoms, in that both arise due to increased stability of cortical attractor networks, with the different symptoms related to overstability in different cortical areas. The theory is also unifying in that a similar increase in glutamatergic activity in the orbitofrontal and anterior cingulate cortex could increase emotionality, as described earlier.

8.15.3 Glutamate and increased depth of the basins of attraction of attractor networks

To demonstrate how alterations of glutamate as a transmitter for the connections between the neurons may influence the stability of attractor networks, Rolls, Loh and Deco (2008a) performed integrate-and-fire simulations of the type described in Chapters 3 and 4 and Section 8.14. A feature of these simulations is that we simulated the currents produced by activation of NMDA and AMPA receptors in the recurrent collateral synapses, and took into account the effects of the spiking-related noise, which is an important factor in determining whether the attractor stays in a basin of attraction, or jumps over an energy barrier into another basin (Loh, Rolls and Deco 2007a, Rolls, Loh, Deco and Winterer 2008b, Deco, Rolls and Romo 2009b). For our investigations, we selected $w_+ = 2.1$, which with the default values of the NMDA and GABA conductances yielded stable dynamics, that is, a stable spontaneous state if no retrieval cue was applied, and a stable state of persistent firing after a retrieval cue had been applied and removed.

To investigate the effects of changes (modulations) in the NMDA, AMPA and GABA conductances, we chose for demonstration purposes increases of 3% for the NMDA, and 10% for the AMPA and GABA synapses between the neurons in the network shown in Fig. 3.7, as these were found to be sufficient to alter the stability of the attractor network. Fig. 3.14 on page 112 shows the stability of the spontaneous and persistent attractor. We plot the percentage of the simulation runs on which the network during the last second of the simulation was in the high firing rate attractor state. Fig. 3.14 shows that for the persistent run simulations, in which the cue triggered the attractor into the high firing rate attractor state, the network was still in the high firing rate attractor state in the baseline condition on approximately 88% of the runs, and that this had increased to 98% when the NMDA conductances were increased by 3% (+NMDA). Thus increasing the NMDA receptor-activated synaptic currents increased the stability of the network. Fig. 3.14 also shows that increasing AMPA by 10% (+AMPA) could also increase the stability of the persistent high firing rate attractor state, as did the combination +NMDA +AMPA.

Fig. 3.14 on page 112 shows that in the baseline condition the spontaneous state was unstable on approximately 10% of the trials, that is, on 10% of the trials the spiking noise in the network caused the network run in the condition without any initial retrieval cue to end up in a high firing rate attractor state. This is of course an error that is related to the spiking noise in the network. In the +NMDA condition, the spontaneous state had jumped to the high firing rate attractor state on approximately 22% of the runs, that is the low firing rate spontaneous state was present at the end of a simulation on only approximately 78% of the runs. Thus increasing NMDA receptor activated currents can contribute to the network jumping from what should be a quiescent state of spontaneous activity into a high firing rate attractor state. We relate this to the symptoms of obsessive-compulsive disorders, in that the system can jump into a state with a dominant memory (which might be an idea or concern or action) even when there is no initiating input. Fig. 3.14 also shows that +AMPA can make the spontaneous state more likely to jump to a persistent high firing rate attractor state, as can the combination +NMDA +AMPA.

We next investigated to what extent alterations of the GABA-receptor mediated inhibition in the network could restore the system towards more normal activity even when NMDA conductances were high. Fig. 3.15 shows that increasing the GABA currents by 10% when the NMDA currents are increased by 3% (+NMDA +GABA) can move the persistent state away from overstability back to the normal baseline state. That is, instead of the system ending up in the high firing rate attractor state in the persistent state simulations on 98% of the runs, the system ended up in the high firing rate attractor state on approximately 88% of the runs, the baseline level. Increasing GABA has a large effect on the stability of the spontaneous state, making it less likely to jump to a high firing rate attractor state. The combination +NMDA +GABA produced a spontaneous state in which the +GABA overcorrected for the effect of +NMDA. That is, in the +NMDA +GABA condition the network was very likely to stay in the spontaneous firing rate condition in which it was started, or, equivalently, when tested in the spontaneous condition, the network was less likely than normally to jump to a high firing rate attractor. Increasing GABA thus corrected for the effect of increasing NMDA receptor activated synaptic currents on the persistent type of run when there was an initiating stimulus; and overcorrected for the effect of increasing NMDA on the spontaneous state simulations when there was no initiating retrieval stimulus, and the network should remain in the low firing rate state until the end of the simulation run.

The implications for symptoms are that agents that increase GABA conductances might reduce and normalize the tendency to remain locked into an idea or concern or action; and would make it much less likely that the quiescent resting state would be left by jumping because of the noisy spiking towards a state representing a dominant idea or concern or action. The effects of increasing GABA receptor activated currents alone was to make the persistent simulations less stable (less likely to end in a high firing rate state), and the spontaneous simulations to be more stable (more likely to end up in the spontaneous state).

We next investigated how an increase of NMDA currents might make the system less distractible, and might make it overstable in remaining in an attractor. This was investigated as shown in Fig. 3.8 by setting up a system with two high firing rate attractors, S1 and S2, then starting the network in an S1 attractor state with S1 applied at t=0–0.5 s, and then applying a distractor S2 at time t=1.0–1.5 s to investigate how strong S2 had to be to distract the network out of its S1 attractor. We assessed how often in 200 trials the average activity during the last second (2 s–3 s) stayed above 10 Hz in the S1 attractor (percentage sustained activity shown on the ordinate of Fig. 4.11). The strength of the distractor stimulus applied to S2 was an increase in firing rate above the 2.4 kHz background activity which is distributed among 800 synapses per neuron.

Fig. 4.11 on page 137 shows that increasing the NMDA receptor activated currents by 5% (+NMDA) means that larger distractor currents must be applied to S2 to move the system away from S1 to S2. That is, the +NMDA condition makes the system more stable in its high firing rate attractor, and less able to be moved to another state by another stimulus (in this case S2). We relate this to the symptoms of obsessive-compulsive disorder, in that once in an attractor state (which might reflect an idea or concern or action), it is very difficult to get the system to move to another state. Increasing AMPA receptor activated synaptic currents (by 10%, +AMPA) produces similar, but smaller, effects.

Fig. 4.12 on page 138 shows that increasing GABA activated synaptic conductances (by 10%, +GABA) can partly normalize the overstability and decrease of distractibility that is produced by elevating NMDA receptor activated synaptic conductances (by 3%, +NMDA). The investigation and conventions used the same protocol as for the simulations shown in Fig. 4.11. We assessed how often in 200 trials the average activity during the last second (2 s–3 s) stayed above 10 Hz in the S1 attractor, instead of being distracted by the S2 distractor. The strength of the distractor stimulus applied to S2 is an increase in firing rate above the 2.4 kHz background activity which is distributed among 800 synapses per neuron. The higher the sustained activity is in S1 the higher is the stability to S1 and the lower is the distractibility by S2. What was found (as shown in Fig. 4.12) is that the increase of GABA was able to move the curve in part back towards the normal state away from the overstable and less distractible state produced by increasing NMDA receptor activated synaptic currents. The effect was particularly evident at low distractor currents.

8.15.4 Synthesis on obsessive-compulsive disorder

This provides a new approach to the symptoms of obsessive-compulsive disorder, for it deals with the symptoms in terms of overstability of attractor networks in the cerebral cortex (Rolls, Loh and Deco 2008a). If the same generic change in stability were produced in different cortical areas, then we have indicated how different symptoms might arise. Of course, if these changes were more evident in some areas than in others in different patients, this would help to account for the different symptoms in different patients. Having proposed a generic hypothesis for the disorder, we recognize of course that the exact symptoms that arise if stability in some systems is increased will be subject to the exact effects that these will have in an individual patient, who may react to these effects, and produce explanatory accounts for the effects, and ways to deal with them, that may be quite different from individual to individual.

The integrate-and-fire simulations show that an increase of NMDA or AMPA synaptic currents can increase the stability of attractor networks. They can become so stable that the intrinsic stochastic noise caused by the spiking of the neurons is much less effective in moving the system from one state to another, whether spontaneously in the absence of a distracting, that is new, stimulus, or in the presence of a new (distracting) stimulus that would normally move the system from one attractor state to another.

The simulations also show that the stability of the spontaneous (quiescent) firing rate state is reduced by increasing NMDA or AMPA receptor activated synaptic currents. We relate this to the symptoms of obsessive-compulsive disorder in that the system is more likely than normally, under the influence of the spiking stochastic noise caused by the neuronal spiking in the system, to jump into one of the dominant attractor states, which might be a recurring idea or concern or action.

This simulation evidence, that an increase of glutamatergic synaptic efficacy can increase the stability of attractor networks and thus potentially provide an account for some of the symptoms of obsessive-compulsive disorder, is consistent with evidence that glutamatergic function may be increased in some brain systems in obsessive-compulsive disorder (Rosenberg, MacMaster, Keshavan, Fitzgerald, Stewart and Moore 2000, Rosenberg, MacMillan and Moore 2001, Rosenberg, Mirza, Russell, Tang, Smith, Banerjee, Bhandari, Rose, Ivey, Boyd and Moore 2004, Pittenger et al. 2006) and that cerebro-spinal-fluid glutamate levels are elevated (Chakrabarty, Bhattacharyya, Christopher and Khanna 2005). Consistent with this, agents with antiglutamatergic activity such as riluzole, which can decrease glutamate transmitter release, may be efficacious in obsessive-compulsive disorder (Pittenger et al. 2006, Bhattacharyya and Chakraborty 2007). Further evidence for a link between glutamate as a neurotransmitter and OCD comes from genetic studies. There is evidence for a significant association between the SLC1A1 glutamate transporter gene and OCD (Stewart, Fagerness, Platko, Smoller, Scharf, Illmann, Jenike, Chabane, Leboyer, Delorme, Jenike and Pauls 2007). This transporter is crucial in terminating the action of glutamate as an excitatory neurotransmitter and in maintaining extracellular glutamate concentrations within a normal range (Bhattacharyya and Chakraborty 2007). In addition, Arnold et al. postulated that N-methyl-d-aspartate (NMDA) receptors were involved in OCD, and specifically that polymorphisms in the $3'$ untranslated region of GRIN2B (glutamate receptor, ionotropic, N-methyl-d-aspartate 2B) were associated with OCD in affected families (Arnold, Rosenberg, Mundo, Tharmalingam, Kennedy and Richter 2004).

We have described results with a single attractor network. It is natural to think of such a network as being implemented by associatively modifiable recurrent collateral synaptic connections between the pyramidal cells within a few mm of each other in a cortical area. Such circuitry is characteristic of the cerebral neocortex, and endows it with many remarkable and fundamental properties, including the ability to implement short-term memories, which are fundamental to cognitive processes such as top-down attention, and planning (Rolls 2008d). However, when two cortical areas are connected by forward and backward connections, which are likely to be associatively modifiable, then this architecture, again characteristic of the cerebral neocortex (Rolls 2008d), also provides the basis for the implementation of an attractor network, but with now a different set of glutamatergic synapses in which overactivity could be relevant to the theory described here. Further there are likely to be interactions between connected attractor networks (Rolls and Deco 2002, Deco and Rolls 2003, Deco and Rolls 2006, Rolls 2008d), and these are also relevant to the way in which the theory of OCD described here would be implemented in the brain.

One of the interesting aspects of the present model of obsessive-compulsive disorder (Rolls, Loh and Deco 2008a) is that it is in a sense the opposite of a recent model of schizophrenia (Rolls 2005, Loh, Rolls and Deco 2007a, Rolls, Loh, Deco and Winterer 2008b). In that model, we suggest that decreased NMDA synaptic currents may, by decreasing the firing rates and effectively the synaptic strengths in cortical attractor networks, decrease the stability of cortical attractor networks. This, if effected in the dorsolateral prefrontal cortex, might produce some of the cognitive symptoms of schizophrenia, including distractibility, poor attention, and poor short-term memory. If the same generic effect were present in the temporal lobe semantic memory networks, this might make thoughts move too freely from one thought to another loosely associated thought, producing bizarre thoughts and some of the positive symptoms of schizophrenia (providing an alternative to the parasitic or spurious attractor hypothesis (Hoffman and McGlashan 2001)). If implemented in the orbitofrontal cortex, the decrease of NMDA synaptic currents would produce lower neuronal firing, and thus less affect, which is a negative symptom of schizophrenia. These two models, of obsessive-compulsive disorder and schizophrenia, in being formal opposites of each other, may underline the importance of attractor networks to cortical function (Rolls 2008d), and emphasize that any alteration in their stability, whether towards too little stability or too much, will have major consequences for the operation of the cerebral cortex.

The theory about how the symptoms of obsessive-compulsive disorder could arise in relation to the increased stability of cortical attractor networks has implications for possible pharmaceutical approaches to treatment. One is that treatments that reduce glutamatergic activity, for example by partially blocking NMDA receptors, might be useful. Another is that increasing the inhibition in the cortical system, for example by increasing GABA receptor activated synaptic currents, might be useful, both by bringing the system from a state where it was locked into an attractor back to the normal level, and by making the spontaneous state more stable, so that it would be less likely to jump to an attractor state (which might represent a dominant idea or concern or action) (see Fig. 3.15). However, we emphasize that the way in which the network effects we consider produce the symptoms in individual patients will be complex, and will depend on the way in which each person may deal cognitively with the effects. In this respect, we emphasize that what we describe is a theory of obsessive-compulsive disorder, that the theory must be considered in the light of empirical evidence yet to be obtained, and that cognitive behaviour therapy makes an important contribution to the treatment of such patients. Stochastic neurodynamics provides we believe an important new approach for stimulating further thinking and research in this area.

8.16   Predicting a decision before the evidence is applied

There is a literature on how early one can predict from neural activity what decision will be taken (Hampton and O'Doherty 2007, Haynes and Rees 2005a, Haynes and Rees 2005b, Haynes and Rees 2006, Haynes, Sakai, Rees, Gilbert, Frith and Passingham 2007, Lau, Rogers and Passingham 2006, Pessoa and Padmala 2005, Rolls, Grabenhorst and Franco 2009b). For example, when subjects held in mind in a delay period which of two tasks, addition or subtraction, they intended to perform, then it was possible to decode or predict with fMRI whether addition or subtraction would later be performed from medial prefrontal cortex activations with accuracies in the order of 70%, where chance was 50% (Haynes et al. 2007). A problem with such studies is that it is often not possible to know exactly when the decision was taken at the mental level, or when preparation for the decision actually started, so it is difficult to know whether neural activity that precedes the decision itself in any way predicts the decision that will be taken (Rolls 2010a). In these circumstances, is there anything rigorous that our understanding of the neural mechanisms involved in the decision-making can provide? It turns out that there is.

While investigating the speed of decision-making using a network of the type described in Chapter 6, A.Smerieri and E.T.Rolls studied the activity that preceded the onset of the

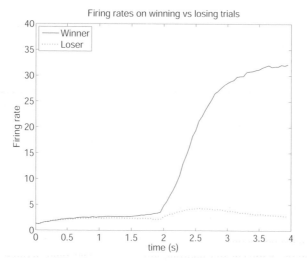

Fig. 8.7 Prediction of a decision before the evidence is applied. In this integrate-and-fire simulation of decision-making, the decision cues were turned on at t=2 s, with ΔI=0. The firing rate averaged over approximately 400 winning vs 400 losing trials for the attractor shows that the firing rate when the attractor will win is on average higher than that for when the attractor will lose at a time that starts in this case at approximately 1000 ms before the decision cues are applied. (At t=2 s with ΔI=0 the input firing rate on each of the 800 external input synapses onto every neuron of both of the selective attractor populations is increased from 3.00 to 3.04 Hz, as described in Chapter 6.)

decision cues. The results found in simulations in which the firing rate in the spontaneous firing period is measured before a particular attractor population won or lost the competition are illustrated in Fig. 8.7. The firing rate averaged over approximately 800 winning (correct) vs losing (error) trials for the attractor shows that the firing rate when the attractor will win is on average higher than that for when the attractor will lose at a time that starts in this case approximately 1000 ms before the decision cues are applied. Thus it is possible before the decision cues are applied to predict in this noisy decision-making attractor network something about what decision will be taken for periods in the order of 1 s before the decision cues are applied. (If longer time constants are used for some of the GABA inhibitory neurons in the network, the decision that will be taken can be predicted (probabilistically) for as much as 2 s before the decision cues are applied (Smerieri, Rolls and Feng 2010).)

What could be the mechanism? I suggest that the mechanism is as follows. There will be noise in the neuronal firing that will lead to low, but different, firing rates at different times in the period before the decision cues are applied of the two selective populations of neurons that represent the different decisions. If the firing rate of say population D1 (representing decision 1) is higher than that of the D2 population at a time just as the decision-cues are being applied, this firing will add to the effect of the decision cues, and make it more likely that the D1 population will win. These fluctuations in the spontaneous firing rate will have a characteristic time course that will be influenced by the time constants of the synapses etc in the system, so that if a population has somewhat higher firing at say 500 ms before the cues are applied, it will be a little more likely to also have higher firing some time later. By looking backwards in time one can see how long the effects of such statistical fluctuations can influence the decision that will be reached, and this is shown in Fig. 8.7 to be approximately 1000 ms in the network we have studied that is described in Chapter 6. I emphasize that this gradual increase of firing rate for the attractor that will win before the decision cues are applied is an effect found by averaging over very many trials, and that the fluctuations found on an individual trial (illustrated in Figs. 6.2 and 6.10) do not reveal obvious changes of the type illustrated in Fig. 8.7.

Thus we have a rigorous and definite answer and understanding of one way in which decisions that will be taken later are influenced by and can be probabilistically predicted from the prior state of the network. It is possible to make a probabilistic prediction of which decision will be taken from the prior activity of the system, before the decision cues are applied. This conclusion emerges from a fundamental understanding of how noise in the brain produces statistical fluctuations that can influence neural processes, a fundamental theme of this book.

8.17   Decision-making between interacting individuals

In this book, we have been considering the effects of noise in a system with interacting neurons. The interactions between the neurons are symmetrical (and assume there is no self-connectivity), and therefore the world of stable attractors, together with the effects of noise in such systems, provides a good model. (In the diluted connectivity case the system operates with similar properties (Rolls and Treves 1998).) What happens if there are interactions between individuals in a group, such as a flock of flying birds, or a swarm of social insects? Can a similar collective computation be invoked? Flocks of birds do seem to follow flight patterns that might be accounted for by interaction between neighbours. This apparent parallel has been followed up (Marshall, Bogacz, Planqué, Kovacs and Franks 2009, List, Elsholtz and Seeley 2009). However, although there are analogies, it is not clear that the interactions between the individuals are symmetric, or behave as if they were symmetric, so that the world of attractors and attractor dynamics may not apply in a formal sense. Nor is there a clear source of internal Poisson noise in the system that would make the systems analogous to those described here. However, there is clearly external noise in such systems, related for example to sensory capabilities, and there could be internal spiking-related noise in each individual, so the analogies may be developed further, especially with respect to noise.

8.18   Unifying principles of cortical design

In this section I draw together some unifying principles of cortical design.

The architecture of the cerebral cortex described in Section 1.8 shows that perhaps its most computationally prominent feature is short-range excitatory recurrent collateral connections between nearby pyramidal cells. This enables attractor states to be formed in what are autoassociation networks (Section 1.10). These stable attractor states provide the basis for short-term memory (Chapter 3). In the prefrontal cortex, this can be a short-term memory that is not dominated by sensory input, so can maintain items in short-term memory over intervening stimuli, and in this sense is an off-line store (Chapter 3).

Because the neocortical recurrent collateral connections are short-range, this allows several different items to be held in short-term memory simultaneously, as nearby attractor networks can operate independently of each other. This is important, for each attractor network can keep only one attractor state active easily (with graded firing rate representations that are found in the brain (Section 1.10)). Another important advantage of the short-range recurrent collateral connections is that the total memory capacity of the whole cerebral cortex if the system operated as one network with long-range connections would be that of any one attractor network, which is of the order of the number of recurrent connections per neuron, and might be 10,000 separate states in a typical neocortical network. This would be highly inefficient, and is too low a number (O'Kane and Treves 1992) (Section 1.8.4).

Part of the importance of a short-term memory that can hold several items in an off-line store is that this is a fundamental basis for what is required for planning, as it provides the ability to hold several steps of a plan in mind. Thus the ability to plan also depends on neocortical recurrent collateral connections.

The ability provided by short-term memory to hold even one item in mind provides the top-down source that maintains attention. The attentional short-term memory (largely in pre-frontal cortex) acts as a source of the top-down input for biased competition networks that help to implement consistent behaviour for a period. They do this not only by maintaining sensory attention (Chapter 4), but also by maintaining biases to select rules that can be rapidly switched for mapping sensory events to actions, as described in Section 3.7 and Fig. 3.6.

But the adaptive value of cortical recurrent networks goes beyond just holding several items in mind at the same time, for the implementation of short-term memory in the cortex can include holding the order in which items were presented or occurred as part of the short-term memory process, as described in Section 3.9. Part of the adaptive value of having a short-term memory in for example the prefrontal cortex that implements the order of the items is that short-term memory is an essential component of planning with multiple possible steps each corresponding to an item in a short-term memory, for plans require that the components be performed in the correct order. Similarly, if a series of events that has occurred results in an effect, the ability to remember the order in which those events occur can be very important in understanding the cause of the effect. This has great adaptive value in very diverse areas of human cognition, from interpreting social situations with several interacting individuals, to understanding mechanisms with a number of components. Part of the adaptive value of the operation of these cortical attractor networks in the noisy regime is that this regime allows slow transitions between attractors, and thus in several sequentially linked attractors or attractors with different time constants started at the same time to bridge significant periods of time, in the order of many seconds, as described in Section 3.9.

The ability to maintain several representations active at the same time may be a building block of language. Indeed, one might think of sentence comprehension as involving a trajectory through a state space of representations in which each representation provides a grammatical part of the sentence. For example, one cortical attractor network might store 10,000 nouns as potential subjects of a sentence, another cortical attractor network might store (up to) 10,000 verbs, and another cortical attractor network might store 10,000 nouns as potential objects in a sentence. These representations could be linked by stronger forward than backward synaptic connections. Each attractor would be activated with the set of neurons representing one of its items (e.g. one of its nouns) active at a high rate in the relevant part of the trajectory, but maintaining its firing (at a lower rate) because of the short-term memory properties of the attractor when the trajectory had moved to emphasise another, connected, attractor in the system.

The question of how syntax, the syntactic relations between the different representations, are managed in such a system is one of the great unresolved remaining mysteries in understanding brain function. The issue is that if several different populations of neurons are simultaneously active, how does one (or the brain) know how the firing of the different populations of neurons are related syntactically? With them all firing, which for example is the subject, and which the object? How does the next brain system that reads the information from the neuronal firing know which is the subject, and which the object? I suggest that part of the solution to this 'binding' problem (Rolls 2008d) may come from that fundamental aspect of cortical design used throughout the rest of the brain, spatial location. That is, the 'meaning' of a representation is encoded by which set of neurons, in a given cortical area, is firing. For example, the rest of the brain interprets firing in the visual cortical areas as representing a visual stimulus, in the auditory cortical areas an auditory stimulus, etc. This is place coding: which neurons fire encode which stimulus is present (Rolls 2008d). I suggest that a building block for syntax might be the same principle: place coding. The subject nouns are represented in one localised cortical attractor network in one small cortical region. The object nouns are represented in another localised cortical attractor network in another small cortical region. Et cetera. The syntactic role of the active representation would be place-encoded in the cortex according to this hypothesis. One could imagine that localised attractor networks closely coupled to a subject noun attractor network might include qualifying information about the subject, part of the underlying representation for a subject noun phrase, etc. Clearly there are major issues remaining to be understood here. There are further pointers, for example that in fact there are considerable constraints on language and syntax (Jackendoff 2002), and these may be understandable in the constraints that can occur in forward-connected sets of coupled but localised attractors in the neocortex.

A unifying principle is that these attractor dynamics implement what is a non-linear process, which involves because of the positive feedback implemented by the recurrent collateral connections a categorization of the inputs, with a clear winner present at the output, as described in Chapters 5 and 6. This categorization, a bifurcation in the dynamics, is the essence of memory recall. For memory recall, there may be a single dominant input, or a set of competing inputs. In both cases, one memory will be recalled. The same non-linear dynamics in cortical attractor networks is the basis for decision-making, as described in Chapters 5 and 6. The same non-linear dynamics is important in detecting a sensory input (Chapter 7), or for object recognition when a particular object must be recognised and categorised (Rolls 2008d). Thus the same fundamental architecture, of excitatory recurrent collateral connections to implement non-linear attractor dynamics, is used again and again in cortical computation, applied to what at first sight appear as very different processes: memory recall, the maintenance of a stable short-term memory, decision-making, and perception. The design of cortical architecture, once discovered by evolutionary processes, is then applied to many different functions, consistent with conservation of design, but not necessarily of function to which the design is applied, in evolution.

The relative strength of the recurrent collateral attractor effect to the effect of the incoming inputs to the network may be stronger the further one moves away from the primary sensory cortical area and up the hierarchy of processing towards memory, decision-making, attentional, action selection processes. This is consistent with the need to represent sensory stimuli faithfully and relatively linearly for early processing, and to later in processing start to make categorisations or decisions after all the relevant sensory evidence has been marshalled into the correct form (Rolls 2008d). Consistent with the greater dominance of recurrent collateral connections the further up the hierarchy one is from primary sensory areas, the dendritic spine system for the recurrent collateral synapses is especially well developed numerically on the dendrites of pyramidal neurons in the prefrontal cortex (Elston, Benavides-Piccione, Elston, Zietsch, Defelipe, Manger, Casagrande and Kaas 2006).

In these non-linear attractor processes, noise of course plays an important role, which, as described in this book, can be beneficial. The noise helps for example in decision-making, by making it probabilistic, which can help sample less favourable choices sometimes, which can be valuable in a changing environment (Section 8.7). The noise can also make behaviour unpredictable, which can be useful in for example escape from a predator, or in economic interactions (Section 8.7). The noise can also be useful in signal detection, in the process known as stochastic resonance (Section 7.4). The noise can also be useful in thought processes, by helping thoughts to be different on different occasions even from a similar starting points, leading to novel trajectories through the space of stored representations, leading often towards creativity (Section 8.9). There may be an optimal level of noise (relative to the depth of the attractor landscape) for this, and too much noise may be maladaptive and lead to thought processes that make associations too readily (in for example schizophrenia, as considered in Section 8.14), or which become locked into a state (in for example obsessive-compulsive disorder, as considered in Section 8.14). In this scenario, evolution may operate by producing differences in the signal-to-noise ratio between individuals, as this enables selection to operate to produce individuals who may, in a given environment, do particularly well by having a useful level of creativity or 'lateral thinking', but not too much (Section 8.9).

There is of course a major risk of having excitatory recurrent collateral connections in neocortical and hippocampal cortical design, with the cost for all the benefits described earlier being instability, which can be manifest in epilepsy. The cortex of course goes to great trouble to try to keep itself stable, but in its design it is inherently dynamically unstable. Some of the ways in which stability is promoted is by having long time constant excitatory receptors (NMDA, with a time constant of 100 ms) as well as AMPA receptors with a short time constant of 2–10 ms (Tegner, Compte and Wang 2002). If the relative efficacy of NMDA receptors is reduced, oscillations may break out, in for example the gamma range (40 Hz) (Buehlmann and Deco 2008). Having a diversity of inhibitory neurons also may help stability (Wang et al. 2004), and indeed in this case the balance of short time constant (5–10 ms) to long time constant (100 ms) GABA receptors can affect whether oscillations break out, in for example the theta (5–9 Hz) range (Smerieri, Rolls and Feng 2010).

In this scenario of the inherent risk of instability in the cerebral cortex due to positive feedback from the excitatory recurrent collateral connections, it is not surprising that oscillations can be found in the cerebral cortex, and that their frequency and amplitude can depend on the amount of incoming drive to a cortical area, as this will affect the dynamics of the circuitry, altering ion channel openings that will influence time constants. For example, if the eyes are closed and there is no input to the primary visual cortex, alpha waves (10 Hz) break out due to the lack of incoming drive, and these waves reflect abnormal functioning of the visual cortex, which is idling. Some have championed oscillations and related temporal phenomena as being functionally useful (Singer 1999, Buzsáki 2006, Lisman and Buzsaki 2008). One such temporal phenomenon is stimulus-dependent synchronization of neurons, which has been suggested to be involved in binding together neurons that fire in phase with each other (Malsburg 1999, Singer 1999). However, rigorous information theoretic analysis shows that temporal binding in the high order visual cortex where features are being bound to encode objects encode quantitatively very little information compared to that present in the firing rates (Aggelopoulos, Franco and Rolls 2005, Rolls 2008d). These results are found in macaques performing an attention-based visual search task in natural scenes, so do reflect the normal, natural, operation of the cerebral neocortex. Moreover, from a computational perspective, stimulus-dependent synchrony might in any case only be useful for grouping different neuronal populations, and would not easily provide a solution to the syntactic problem in vision in which the relations of features must be specified (Rolls 2008d).

p. 114 With respect to oscillations, it has been suggested that nested gamma and theta oscillations might provide a solution to temporal order effects in short-term memory (Lisman and Redish 2009), but as we have seen in Section 3.9, a much simpler solution consistent with recent experimental evidence is that temporal order memory can be implemented in the brain by neurons that encode time in their firing rates, by increasing their firing at particular parts of a delay period. This is a unifying and parsimonious approach in which firing rates are used to encode information, for this is the main code that has been established for many encoding systems in the brain (Rolls 2008d) (see Section 1.11).

However, given that oscillations can occur in cortical circuitry, we can ask whether there are beneficial effects that oscillations may have. In one modelling investigation, it was shown that reaction times can be faster if gamma oscillations are allowed to occur by decreasing NMDA efficacy (Buehlmann and Deco 2008). In another investigation, it was shown that adding some long time constant inhibitory GABA synapses to the decision-making network described by Deco and Rolls (2006) can introduce theta range (5–9 Hz) periodicity into the neuronal firing, and this can facilitate reaction times in the decision-making network (Smerieri, Rolls and Feng 2010). In this case, it was shown that the faster reaction times were being produced because sometimes, in association with the theta rhythm, the neurons were more depolarized, and this increase of excitability increased the speed of the decision (Smerieri, Rolls and Feng 2010), in the same way as altering synaptic currents can alter reaction times as described in Sections 8.14 and 8.15.

8.19 Apostasis

[2] Extraordinary experimental, conceptual, and theoretical progress has been made in the last 35 years in understanding how the brain might actually work. Thirty-five years ago, observations on how neurons respond in different brain areas during behaviour were only just starting, and there was very limited understanding of how any part of the brain might work or compute. Many of the observations appeared as interesting phenomena, but there was no conceptual framework in place for understanding how the neuronal responses might be generated, nor of how they were part of an overall computational framework. Through the developments that have taken place since then, some of them described in *Memory, Attention, and Decision-Making: A Unifying Computational Neuroscience Approach* (Rolls 2008d), we now have at least plausible and testable working models that are consistent with a great deal of experimental data about how many different types of memory, perception, attention, and decision-making may be implemented in the brain. We now have models, consistent with a great deal of neurophysiological data, of how perceptual systems build, by self-organizing learning and memory-related types of processing, representations of objects in the world in a form that is suitable as an input to memory systems, and how memory systems interact with perceptual systems (Rolls 2008d). We have models of how short-term memory is implemented in the brain (Rolls 2008d). We also, and quite closely related to these models, have detailed and testable models that make predictions about how attention works and is implemented in the brain (Rolls and Deco 2002). We also have testable models of long-term memory, including episodic memory and spatial memory, which again are based on, and are consistent with, what is being found to be implemented in the primate brain (Rolls 2008d). In addition, we have an understanding, at the evolutionary and adaptive level, of emotion, and at the computational level of how different brain areas implement the information processing underlying emotion (Rolls 2005, Rolls 2008d)).

Our understanding is developing to the stage where we can have not just a computational understanding of each of these systems separately, but also an understanding of how all these processes are linked and act together computationally. Even more than this, we have shown how many aspects of brain function, such as the operation of short-term memory systems, of attention, the effects of emotional states on memory and perception, and even decision-making and action selection, can now start to be understood in terms of the reciprocal *interactions* of connected neuronal systems (Rolls 2008d). Indeed, one of the implications of this research is that there are multiple memory systems in the brain (Rolls 2007c), each processing different types of information and using a range of computational implementations, and

---

[2] Apostasis — standing after.

that interactions between these memory systems are important in understanding the operation of perception, memory, attention, and decision-making in the brain.

This book adds to this approach by describing the important principle of how internal noise in the brain, produced for example by the random spiking times (for a given mean rate) of neurons, makes the brain into a stochastic dynamical system. We have shown how these stochastic dynamics can be analyzed and investigated, from the synaptic ion channel level, which generates some of the noise in the brain, to the level of the spiking neuron and its synaptic currents, through the effects found in finite-sized populations of such noisy neurons using approaches from theoretical physics, to functional neuroimaging signals and to behaviour. These stochastic dynamical principles have enabled us to approach the probabilistic operation of short-term memory and long-term memory systems in the brain, of attention, of decision-making, of signal detection, and even of creativity, and the fact that effectively we operate as non-deterministic systems. Indeed, we have seen how noise in the brain, far from being a problem, endows many advantages to brain information processing, including not only stochastic resonance, a useful property in sensory systems, but also probabilistic decision-making which enables us to keep exploring different possibilities and exploring the environment, and to be creative in science, art, and literature (Rolls 2010b). We have also seen how these new principles provide a new approach to understanding some disorders of the brain, which we argue are related to instability or overstability in the face of noise of the neuronal networks involved in many different functions in the brain. These disorders include some disorders of short-term memory, attention, and long-term memory, and are relevant to understanding inter alia aging, schizophrenia, and obsessive-compulsive disorder.

This leads us to emphasize that the understanding of how the brain actually operates is crucially dependent on a knowledge of the responses of single neurons, the computing elements of the brain, for it is through the connected operations in networks of single neurons, each with their own individual response properties, that the interesting collective computational properties of the brain arise in neuronal networks (Rolls 2008d), and that in finite-sized populations of such noisy neurons important stochastic computational processes emerge, as described in this book.

The approach taken in this book is towards how the dynamics of attractor networks used for decision-making and related functions is implemented neurobiologically, and in this respect is what we term a 'mechanistic' model. This may be contrasted with what we describe as 'phenomenological' models, which may model phenomena at some abstract level. Examples in the area of decision-making are the accumulator, counter, or race models of decision-making in which the noisy evidence for different choices accumulates by integration until some threshold is reached (Vickers 1979, Vickers and Packer 1982, Ratcliff and Rouder 1998, Ratcliff et al. 1999, Usher and McClelland 2001). In contrast to the 'mechanistic' approach taken in this book, such 'phenomenological' models do not describe a neurobiological mechanism, do not make specific predictions about how neuronal firing rates, synaptic currents, or BOLD signals will alter as decisions become easier, and do not make specific predictions about confidence in a decision, but do account for behavioural performance. In this context, it is of interest that the integrate-and-fire attractor approach as well as the other approaches can be analyzed as diffusion processes (Roxin and Ledberg 2008), and that our approach is not inconsistent with these accumulator or with Bayesian (Ma et al. 2006, Beck et al. 2008, Ma et al. 2008) approaches, but instead defines a mechanism, from which many important properties emerge, such as a representation of decision confidence on correct and error trials, as shown here, and specific predictions that can be tested with the experimental methods of neuroscience, again as described in this book. We emphasize that the empirical data alone are not sufficient to provide an understanding of how a complex process such as decision-making takes place in the brain, and that to understand the mechanism of decision-making (and related categorization processes), to make predictions, and to suggest treatments for disorders of the dynamical processes involved, the empirical data must be complemented by a mechanistic neurobiologically based theory and model.

In describing in this book and in *Memory, Attention, and Decision-Making* (Rolls 2008d) at least part of this great development in understanding in the last 35 years of how a significant number of parts of the brain actually operate, and how they operate together, we wish to make the point that we are at an exciting and conceptually fascinating time in the history of brain research. We are starting to see how a number of parts of the brain *could* work, and how stochastic dynamics is an important principle of brain function. There is much now that can be done and needs to be done to develop our understanding of how the brain actually works. But the work described in this book does give an indication of some of the types of information processing that take place in the brain, and an indication of the way in which we are now entering the age in which our conceptual understanding of brain function can be based on our developing understanding of how the brain computes, and does this noisily.

Understanding how the brain works normally is of course an important foundation for understanding its dysfunctions and its functioning when damaged. Some examples have been given in this book. We regard this as a very important long-term aim of the type of work described in this book.

Through noisy neural computation, decision-making and creativity

Appendix 1 Mean-field analyses, and stochastic dynamics

In this Appendix we describe a mean-field approach to understanding neuronal dynamics that is consistent with the integrate-and-fire simulation approach, in that the parameters that are used in the two approaches are similar. We thank Stefano Fusi for contributing to the derivation.

A.1 The Integrate-and-Fire model

First we describe the integrate-and-fire model for which a consistent mean-field approach will be derived.

The computational units of these networks are spiking neurons, which transform a large set of inputs, received from different neurons, into an output spike train, which constitutes the output signal of the neuron. This means that the spatiotemporal spike patterns produced by neural circuits convey information between neurons; this is the microscopic level at which the brain's representations and computations rest (Rolls and Deco 2002). We assume that the non-stationary temporal evolution of the spiking dynamics can be captured by one-compartment, point-like models of neurons, such as the leaky integrate-and-fire (LIF) model (Tuckwell 1988) used below. Other models relevant for systems neuroscience can be found in Dayan and Abbott (2001), Jirsa (2004) and Jirsa and McIntosh (2007).

In the LIF model, each neuron i can be fully described in terms of a single internal variable, namely the depolarization $V_i(t)$ of the neural membrane. The basic circuit of a leaky integrate-and-fire model consists of a capacitor C in parallel with a resistor R driven by a synaptic current (excitatory or inhibitory postsynaptic potential, EPSP or IPSP, respectively) (see Fig. 2.1). When the voltage across the capacitor reaches a threshold θ, the circuit is shunted (reset) and a δ-pulse (spike) is generated and transmitted to other neurons. The sub-threshold membrane potential of each neuron evolves according to a simple RC-circuit, with a time constant $\tau = RC$ given by the following equation:

$$\tau \frac{dV_i(t)}{dt} = -[V_i(t) - V_L] + RI_i(t), \tag{A.1}$$

where $I_i(t)$ is the total synaptic current flow into the cell i and V_L is the leak or resting potential of the cell in the absence of external afferent inputs. In order to simplify the analysis, we neglect the dynamics of the afferent neurons (see Brunel and Wang (2001) for extensions considering detailed synaptic dynamics including AMPA, NMDA, and GABA). The total synaptic current coming into the cell i is therefore given by the sum of the contributions of δ-spikes produced at presynaptic neurons. Let us assume that N neurons synapse onto cell i and that J_{ij} is the efficacy of synapse j, then the total synaptic afferent contribution (current driven through a resistance) is given by

$$RI_i(t) = \tau \sum_{j=1}^{N} J_{ij} \sum_k \delta(t - t_j^{(k)}), \tag{A.2}$$

where $t_j^{(k)}$ is the emission time of the kth spike from the jth presynaptic neuron. The sub-threshold dynamical Equation (A.1), given the input contribution (Equation A.2) can be integrated, and yields

$$V_i(t) = V_L + \sum_{j=1}^{N} J_{ij} \int_0^t e^{-s/\tau} \sum_k \delta \left(t - s - t_j^{(k)} \right) ds, \tag{A.3}$$

$$= V_L + \sum_{j=1}^{N} J_{ij} \, e^{-\left(t - t_j^{(k)} \right)/\tau} \sum_k H \left(t - t_j^{(k)} \right), \tag{A.4}$$

if the neuron i is initially ($t = 0$) at the resting potential ($V_i(0) = V_L$). In Equation A.4, $H(t)$ is the Heaviside function ($H(t) = 1$ if $t > 0$, and $H(t) = 0$ if $t < 0$). Thus, the incoming presynaptic δ-pulse from other neurons is basically low-pass filtered to produce an EPSP or IPSP in the postsynaptic cell. Nevertheless, the integrate-and-fire model is not only defined by the subthreshold dynamics but includes a reset after each spike generation, which makes the whole dynamics highly nonlinear. In what follows, we present a theoretical framework that is able to deal with this.

A.2   The population density approach

Realistic neuronal networks comprise a large number of neurons (e.g. a cortical column has $O(10^4)$-$O(10^8)$ neurons) which are massively interconnected (on average, a neuron makes contact with $O(10^4)$ other neurons). ($O(x)$ means in the order of x.) The underlying dynamics of such networks can be described explicitly by the set of coupled differential Equations A.1 above. Direct simulations of these equations yield a complex spatiotemporal pattern, covering the individual trajectory of the internal state of each neuron in the network. This type of direct simulation is computationally expensive, making it very difficult to analyze how the underlying connectivity relates to various dynamics. In fact, most key features of brain operation seem to emerge from the interplay of the components, rather than being generated by each component individually. One way to overcome these difficulties is by adopting the population density approach, using the Fokker–Planck formalism (e.g. De Groff, Neelakanta, Sudhakar and Aalo (1993)). The Fokker–Planck equation summarizes the flow and dispersion of states over phase-space in a way that is a natural summary of population dynamics in genetics (e.g. Feller (1951)) and neurobiology (e.g. Ricciardi and Sacerdote (1979) and Lansky, Sacerdote and Tomassetti (1995)).

In what follows, we derive the Fokker–Planck equation for neuronal dynamics that are specified in terms of spiking neurons. This derivation is a little dense but illustrates the approximating assumptions and level of detail that can be captured by density dynamics. The approach on which we focus was introduced by Knight, Manin and Sirovich (1996) (see also Omurtag, Knight and Sirovich (2000) and Knight (2000)). In this approach, individual integrate-and-fire neurons are grouped together into populations of statistically similar neurons. A statistical description of each population is given by a probability density function that expresses the distribution of neuronal states (i.e. membrane potentials) over the population. In general, neurons with the same state $V(t)$ at a given time t have a different history because of random fluctuations in the input current $I(t)$. The main source of randomness is from fluctuations in recurrent currents (resulting from '*quenched*' randomness in the connectivity and transmission delays) and fluctuations in the external currents. The key assumption in the population density approach is that the afferent input currents impinging on neurons in one population are uncorrelated. Thus, neurons sharing the same state $V(t)$ in a population are indistinguishable. Consequently, the dynamics of the membrane potential are described by the evolution of the probability density function:

$$p(v,t)dv = \text{Prob}\{V(t) \in [v, v + dv]\}, \tag{A.5}$$

which expresses the *population density*, which is the fraction of neurons at time t that have a membrane potential $V(t)$ in the interval $[v, v + dv]$.

The temporal evolution of the population density is given by the Chapman–Kolmogorov equation

$$p(v,t+dt) = \int_{-\infty}^{+\infty} p(v-\epsilon,t)\rho(\epsilon|v-\epsilon)d\epsilon, \tag{A.6}$$

where $\rho(\epsilon|v) = \text{Prob}\{V(t+dt) = v + \epsilon | V(t) = v\}$ is the conditional probability that generates an infinitesimal change $\epsilon = V(t+dt) - V(t)$ in the infinitesimal interval dt. The Chapman–Kolmogorov equation can be written in a differential form by performing a Taylor expansion in $p(v',t)\rho(\epsilon|v')$ around $v' = v$; i.e.

$$p(v',t)\rho(\epsilon|v') = \sum_{k=0}^{\infty} \frac{(-\epsilon)^k}{k!} \frac{\partial^k}{\partial v'^k} [p(v',t)\rho(\epsilon|v')]\Big|_{v'=v}. \tag{A.7}$$

In the derivation of the preceding equation we have assumed that $p(v',t)$ and $\rho(\epsilon|v')$ are infinitely many times differentiable in v. Inserting this expansion in Equation A.6, and replacing the time derivative in v' by the equivalent time derivative in v, we obtain

$$p(v,t+dt) = p(v,t)\int_{-\infty}^{+\infty} \rho(\epsilon|v)d\epsilon - \frac{\partial}{\partial v}\left[p(v,t)\left(\int_{-\infty}^{+\infty} \epsilon\rho(\epsilon|v)d\epsilon\right)\right]$$

$$+ \frac{1}{2}\frac{\partial^2}{\partial v^2}\left[p(v,t)\left(\int_{-\infty}^{+\infty} \epsilon^2\rho(\epsilon|v)d\epsilon\right)\right] + \ldots, \tag{A.8}$$

$$= \sum_{k=0}^{\infty} \frac{(-1)^k}{k!} \frac{\partial^k}{\partial v^k} [p(v,t)\langle\epsilon^k\rangle_v], \tag{A.9}$$

where $\langle\ldots\rangle_v$ denotes the average with respect to $\rho(\epsilon|v)$ at a given v. Finally, taking the limit for $dt \to 0$, we obtain:

$$\frac{\partial p(v,t)}{\partial t} = \sum_{k=1}^{\infty} \frac{(-1)^k}{k!} \frac{\partial^k}{\partial v^k}\left[p(v,t)\lim_{dt\to 0}\frac{1}{dt}\langle\epsilon^k\rangle_v\right]. \tag{A.10}$$

Equation A.10 is known as the Kramers–Moyal expansion of the original integral Chapman–Kolmogorov Equation A.6. It expresses the time evolution of the population density in differential form.

A.3 The diffusion approximation

The temporal evolution of the population density as given by Equation A.10 requires the moments $\langle\epsilon^k\rangle_v$ due to the afferent current during the interval dt. These moments can be calculated by the mean-field approximation. In this approximation, the currents impinging on each neuron in a population have the same statistics, because as we mentioned earlier, the history of these currents is uncorrelated. *The mean-field approximation entails replacing the time-averaged discharge rate of individual cells with a common time-dependent population activity (ensemble average).* This assumes ergodicity for all neurons in the population. The mean-field technique allows us to discard the index denoting the identity of any single neuron and express the infinitesimal change, $dV(t)$, in the membrane potential of all neurons as:

$$dV(t) = \langle J \rangle_J NQ(t)dt - \frac{V(t) - V_L}{\tau}dt, \qquad (A.11)$$

where N is the number of neurons, and $Q(t)$ is the mean population firing rate. This is determined by the proportion of active neurons by counting the number of spikes $n_{\text{spikes}}(t, t + dt)$ in a small time interval dt and dividing by N and by dt (Gerstner 2000); i.e.

$$Q(t) = \lim_{dt \to 0} \frac{n_{\text{spikes}}(t, t + dt)}{Ndt}. \qquad (A.12)$$

In Equation A.11 $\langle J \rangle_J$ denotes the average of the synaptic weights in the population. The moments of the infinitesimal depolarization $\epsilon = dV(t)$ can now be calculated easily from Equation A.11. The first two moments in the Kramers–Moyal expansion are called drift and diffusion coefficients, respectively, and they are given by:

$$M^{(1)} = \lim_{dt \to 0} \frac{1}{dt}\langle \epsilon \rangle_v = \langle J \rangle_J NQ(t) - \frac{v - V_L}{\tau} = \frac{\mu(t)}{\tau} - \frac{v - V_L}{\tau}, \qquad (A.13)$$

and

$$M^{(2)} = \lim_{dt \to 0} \frac{1}{dt}\langle \epsilon^2 \rangle_v = \langle J^2 \rangle_J NQ(t) = \frac{\sigma(t)^2}{\tau}. \qquad (A.14)$$

In general, keeping only the leading term linear in dt, it is easy to prove that for $k > 1$,

$$\langle \epsilon^k \rangle_v = \langle J^k \rangle_J NQ(t)dt + O(dt^2), \qquad (A.15)$$

and hence,

$$M^{(k)} = \lim_{dt \to 0} \frac{1}{dt}\langle \epsilon^k \rangle_v = \langle J^k \rangle_J NQ(t). \qquad (A.16)$$

The diffusion approximation arises when we neglect high-order ($k > 2$) terms. The diffusion approximation is exact in the limit of infinitely large networks, i.e. $N \to \infty$, if the synaptic efficacies scale appropriately with network size, such that $J \to 0$ but $NJ^2 \to$ const. In other words, the diffusion approximation is appropriate, if the minimal 'kick step' J is very small but the overall firing rate is very large. In this case, all moments higher than two become negligible, in relation to the drift (μ) and diffusion (σ^2) coefficients.

The diffusion approximation allows us to omit all higher orders $k > 2$ in the Kramers–Moyal expansion. The resulting differential equation describing the temporal evolution of the population density is called the **Fokker–Planck equation**, and reads

$$\frac{\partial p(v, t)}{\partial t} = \frac{1}{2\tau}\sigma^2(t)\frac{\partial^2 p(v, t)}{\partial v^2} + \frac{\partial}{\partial v}\left[\left(\frac{v - V_L - \mu(t)}{\tau}\right)p(v, t)\right]. \qquad (A.17)$$

In the particular case that the drift is linear and the diffusion coefficient $\sigma^2(t)$ is given by a constant, the Fokker–Planck equation describes a well known stochastic process called the Ornstein-Uhlenbeck process (Risken 1996). Thus, under the diffusion approximation, the Fokker–Planck Equation A.17 expresses an Ornstein-Uhlenbeck process. The Ornstein-Uhlenbeck process describes the temporal evolution of the membrane potential $V(t)$ when the input afferent currents are given by

$$RI(t) = \mu(t) + \sigma\sqrt{\tau}w(t), \qquad (A.18)$$

where $w(t)$ is a white noise process. Under the diffusion approximation, Equation A.18 can also be interpreted (by means of the Central Limit Theorem) as the case in which the sum of many Poisson processes (A.2) becomes a normal random variable with mean $\mu(t)$ and variance σ^2.

A.4 The mean-field model

The simulation of a network of integrate-and-fire neurons allows one to study the dynamical behaviour of the neuronal spiking rates. Alternatively, the integration of the non-stationary solutions of the Fokker–Planck Equation (A.17) also describes the dynamical behaviour of the network, and this would allow the explicit simulation of neuronal and cortical activity (single cells, EEG, fMRI) and behaviour (e.g. performance and reaction time). However, these simulations are computationally expensive and their results probabilistic, which makes them unsuitable for systematic explorations of parameter space.

However, the stationary solutions of the Fokker–Planck Equation (A.17) represent the stationary solutions of the original integrate-and-fire neuronal system. This allows one to construct bifurcation diagrams to understand the nonlinear mechanisms underlying equilibrium dynamics. This is an essential role of the mean-field approximation: to simplify analyses through the stationary solutions of the Fokker–Planck equation for a population density under the diffusion approximation (Ornstein-Uhlenbeck process) in a self-consistent form. In what follows, we consider stationary solutions for ensemble dynamics.

The Fokker–Planck Equation (A.17) describing the Ornstein-Uhlenbeck process, with $\mu = \langle J \rangle_J NQ(t)$ and $\sigma^2 = \langle J^2 \rangle_J NQ(t)$, can be rewritten as a continuity equation:

$$\frac{\partial p(v,t)}{\partial t} = -\frac{\partial F(v,t)}{\partial v}, \tag{A.19}$$

where F is the flux of probability defined as follows:

$$F(v,t) = -\frac{v - V_L - \mu}{\tau} p(v,t) - \frac{\sigma^2}{2\tau} \frac{\partial p(v,t)}{\partial v}. \tag{A.20}$$

The stationary solution should satisfy the following boundary conditions:

$$p(\theta, t) = 0, \tag{A.21}$$

and

$$\frac{\partial p(\theta, t)}{\partial v} = -\frac{2Q\tau}{\sigma^2}, \tag{A.22}$$

which expresses the fact that the probability current at threshold gives, by a self-consistent argument, the average firing rate Q of the population. Furthermore, at $v \to -\infty$ the probability density vanishes fast enough to be integrable; i.e.

$$\lim_{v \to -\infty} p(v,t) = 0, \tag{A.23}$$

and

$$\lim_{v \to -\infty} vp(v,t) = 0. \tag{A.24}$$

In addition, the probability of the membrane potential exceeding the threshold at time t has to be re-inserted at the reset potential at time $t + t_{ref}$ (where t_{ref} is the refractory period of the neurons), which can be accommodated by rewriting Equation A.19 as follows:

$$\frac{\partial p(v,t)}{\partial t} = -\frac{\partial}{\partial v} \left[F(v,t) + Q(t - t_{ref})H(v - V_{reset}) \right], \tag{A.25}$$

where $H(.)$ is the Heaviside function. The solution of Equation A.25 satisfying the boundary conditions (A.21) – (A.24) is:

$$p_s(v) = \frac{2Q\tau}{\sigma}\exp(-\frac{v - V_L - \mu}{\sigma^2})\int_{\frac{v-V_L-\mu}{\sigma}}^{\frac{\theta-V_L-\mu}{\sigma}} H\left(x - \frac{V_{reset} - V_L - \mu}{\sigma}\right)e^{x^2}dx. \quad (A.26)$$

Taking into account the fraction of neurons Qt_{ref} in the refractory period and the normalization of the mass probability,

$$\int_{-\infty}^{\theta} p_s(v)dv + Qt_{ref} = 1. \quad (A.27)$$

Finally, substituting (A.26) into (A.27), and solving for Q, we obtain the *population transfer function* ϕ of Ricciardi (Ricciardi and Sacerdote 1979):

$$Q = \left[t_{ref} + \tau\sqrt{\pi}\int_{\frac{V_{reset}-V_L-\mu}{\sigma}}^{\frac{\theta-V_L-\mu}{\sigma}} e^{x^2}\{1 + erf(x)\}dx\right]^{-1} = \phi(\mu, \sigma), \quad (A.28)$$

where $erf(x) = 2/\sqrt{\pi}\int_0^x e^{-y^2}dy$.

The stationary dynamics of each population can be described by the **population transfer function**, which provides the average population rate as a function of the average input current. This can be generalized easily for more than one population: the network is partitioned into populations of neurons whose input currents share the same statistical properties and fire spikes independently at the same rate. The set of stationary, self-reproducing rates Q_x for different populations x in the network can be found by solving a set of coupled self-consistency equations, given by:

$$Q_x = \phi(\mu_x, \sigma_x), \quad (A.29)$$

To solve the equations defined by A.29 for all x we integrate the differential equation below, describing the *approximate dynamics* of the system, which has fixed-point solutions corresponding to Equations A.29:

$$\tau_x \frac{dQ_x}{dt} = -Q_x + \phi(\mu_x, \sigma_x). \quad (A.30)$$

This enables *a posteriori* selection of parameters, which induce the emergent behaviour that we are looking for. One can then perform full non-stationary simulations using these parameters in the full integrate-and-fire scheme to generate *true dynamics*. The mean field approach ensures that these dynamics will converge to a stationary attractor that is consistent with the steady-state dynamics we require (Del Giudice, Fusi and Mattia 2003, Brunel and Wang 2001). In our case, the derived transfer function ϕ corresponds consistently to the assumptions of the simple leaky integrate-and-fire model described in Section A.1.

Further extensions for more complex and realistic models are possible. For example, an extended mean-field framework, which is consistent with the integrate-and-fire and realistic synaptic equations, that considers both the fast and slow glutamatergic excitatory synaptic dynamics (AMPA and NMDA) and the dynamics of GABA inhibitory synapses, can be found in Brunel and Wang (2001) and Section 2.7.2, and is further described and used in many investigations (Deco and Rolls 2003, Deco and Rolls 2005b, Deco and Rolls 2006, Loh, Rolls and Deco 2007a, Rolls, Loh and Deco 2008a, Rolls, Grabenhorst and Deco 2010b).

In addition, in Sections A.5 and A.6 we show how this mean-field approach can be extended to incorporate some of the effects of noise in the system.

A.5 Introducing noise into a mean-field theory

In this section we describe theoretical work that has been able to extend the mean-field approach (described earlier in Appendix A) in such a way that both non-stationary behaviour, and spiking noise fluctuations, are taken into account (Renart, Brunel and Wang 2003, La Camera, Rauch, Luescher, Senn and Fusi 2004, Mattia and Del Giudice 2004, Deco, Scarano and Soto-Faraco 2007b).

If there is one slow variable (such as the time course of the effects produced through NMDA receptors), the dynamics of the mean field variable can be derived by the use of an adiabatic argument (see La Camera et al. (2004) for more details). In fact, the dynamical evolution of the rate activity can be described via a system of first order differential equations of the Wilson–Cowan type plus a noise term that takes into account the spiking noise. In this framework, the stochastic neurodynamics of a network of spiking neurons grouped into M populations, can be expressed in a reduced form by the following system of coupled firing rate equations (Wilson and Cowan 1973, Dayan and Abbott 2001):

$$\frac{d\nu_i(t)}{dt} = -\nu_i(t) + \phi\Big(\lambda_i + \sum_{j=1}^{M} w_{ij}\nu_j(t)\Big) + \xi_i(t), \quad i = 1, ..., M, \tag{A.31}$$

where in the usual interpretation ν_i denotes the firing rate of population i (but also see Machens and Brody (2008)), and w_{ij} the synaptic strength between populations i and j. External inputs to the population i are denoted by λ_i, and a nonlinear transfer response function by $\phi(\cdot)$ (derived in Section A.4). (Note that we now denote the population or pool of neurons by i instead of by x as used previously, to avoid confusion with a new variable that we soon introduce.)

Fluctuations are modelled via an additive Gaussian noise term denoted by ξ_i. Here $\langle\xi_i(t)\rangle = 0$ and $\langle\xi_i(t)\xi_i(t')\rangle = \beta 2\delta_{ij}\delta(t-t')$, where the brackets $\langle...\rangle$ denote the average over stochastic random variables. This noise term represents finite-size effects that arise due to the finite number of N neurons in the populations. We note that there are two sources of noise in such spiking networks: the randomly arriving external Poissonian spike trains, and the statistical fluctuations due to the spiking of the neurons within the finite sized network. Here, we concentrate on finite-size effects due to the fact that the populations are described by a finite number N of neurons (see Mattia and Del Giudice (2004) and Mattia and Del Giudice (2002)). In the mean-field framework (see Mattia and Del Giudice (2004) and Mattia and Del Giudice (2002)), 'incoherent' fluctuations due to quenched randomness in the neurons' connectivity and/or to external input are already taken into account in the variance, and 'coherent' fluctuations give rise to new phenomena. In fact, the number of spikes emitted in a time interval dt by the network is a Poisson variable with mean and variance $N\nu(t)dt$. The estimate of $\nu(t)$, is then a stochastic process $\nu_N(t)$, well described in the limit of large $N\nu$ by $\nu_N(t) \simeq \nu(t) + \sqrt{\nu(t)/N}\gamma(t)$, where $\gamma(t)$ is Gaussian white noise with zero mean and unit variance, and $\nu(t)$ is the probability of emitting a spike per unit time in the infinite network. Such finite-N fluctuations, which affect the global activity ν_N, are coherently felt by all neurons in the network. This approach leads to the additive Gaussian noise corrections adopted in Equation A.31.

The classical mean-field approximation (Amit and Brunel 1997, Brunel and Wang 2001) corresponds to the case where the fluctuations can be neglected (i.e. where there is an infinite number of neurons in each population). In this case, a standard bifurcation analysis of the fixed points can be carried out. In the absence of noise the fixed points of Equation (A.31) can be determined by setting the time derivative equal to zero and solving for ν_i.

In summary, the system of coupled stochastic Equations A.31 captures the full dynamical evolution of the underlying stochastic neurodynamics. A more detailed approach to the analysis of stochastic dynamical systems with this framework is elaborated in Section A.6.

A.6 Effective reduced rate-models of spiking networks: a data-driven Fokker–Planck approach

In this section, we introduce a novel *top-down* approach that allows an effective reduction or simplification of a high-dimensional spiking neurodynamical system so that mean-field-like approaches can then be applied (Deco, Marti, Ledberg, Reig and Sanchez Vives 2009a).

Deciphering the fundamental mechanisms underlying brain functions demands explicit description of the computation performed by the neuronal and synaptic substrate. Indeed, the use of explicit mathematical neurodynamical models of brain function makes precise and describes the complex dynamics arising from the involved neuronal networks. Traditionally, as we have seen in the preceding sections, theoretical neuroscience follows a *bottom-up* approach consisting of two main steps: 1) the construction and simulation of models based on detailed specification of local microscopic neuronal and synaptic operations which with large numbers of neurons in a specified (hypothesized) network architecture leads to the desired global behaviour of the whole system, and 2) reduction of the hypothesized models such that an in-depth analytical study is feasible, and a systematic relation between structure (parameters), dynamics, and functional behaviour can be solidly established. A mere reproduction of phenomena of a complex system, like the brain, just by simulating the same kind of artificial complex system, is most of the time not useful, because there are usually no explicit underlying first principles.

Nevertheless, this *bottom-up* approach has several shortcomings. First, the standard mean field approach neglects the temporal properties of the synapses, i.e. considers only delta-like spiking input currents. There are extensions of the mean-field framework that include synaptic equations for fast and slow glutamatergic excitatory synaptic dynamics (AMPA and NMDA), and GABA-inhibitory synaptic dynamics (Brunel and Wang 2001), but they introduce extra oversimplifications and approximations that do not always hold. Second, to consider realistic non-stationary situations (e.g. transition phenomena in bistable perception, decision-making, or reaction times), one is obliged to estimate the underlying fluctuations which bias or even drive the dynamics. These fluctuations are intrinsically expressed in the detailed integrate-and-fire simulations (in the form of 'finite-size noise'), but are washed out in the mean-field reduction. Even more, the level of fluctuations even at the integrate-and-fire level given by the size of the network (Mattia and Del Giudice 2004) is another free parameter that is usually disregarded in such extended mean-field approaches. And finally, the whole bottom-up philosophy depends on initiating the process with a 'good' hypothesis about the underlying neuronal network architecture.

A.6.1 A reduced rate-model of spiking networks

We introduce here a novel *top-down* methodology that allows an effective reduction of a high-dimensional spiking neurodynamical system in a data-driven fashion (Deco, Marti, Ledberg, Reig and Sanchez Vives 2009a). The idea is to depart from the high-dimensional data first by reducing the dimensionality with techniques such as principal component analysis or principal curves. After this reduction, one fits the underlying dynamics by a system of mean-field-like reduced Langevin equations. In particular, we use a system of Langevin equations that describes the stochastic dynamics by assuming an underlying potential (or energy function).

The advantage of using this type of potential-based stochastic system is that for this system, the associated Fokker–Planck equation, which expresses the evolution of the probabilistic distribution of the variable, can be solved analytically at least for stationary conditions, so that we have a closed form for the asymptotic probabilistic distribution of the reduced data. In this schema, the underlying fluctuations can be explicitly extracted. The resulting system can be interpreted as a nonlinear diffusion system, which can be used for linking the underlying cortical dynamics with behaviour. In fact, it is usual in psychophysics to describe the behavioural data by (linear) diffusion models, as described in Section 5.7.1. The top-down approach described here offers therefore an extension of this type of model, in the sense that the new diffusion models described here used for fitting behaviour are nonlinear (due to the positive feedback in the attractor network), and are derived from the underlying neurophysiological experimental data. The model thus differs from traditional linear diffusion models of decision-making used to account for example for human reaction time data (Luce 1986, Ratcliff and Rouder 1998, Ratcliff et al. 1999, Usher and McClelland 2001).

For simplicity, we restrict ourselves to the reduction of stationary high-dimensional spiking data. In particular, we will show results on a simulated high-dimensional spiking network displaying bistability. The effective top-down reduction will allow explicit extraction of the underlying form of the energy function associated with the bistable behaviour, and the level of fluctuations, and consequently will allow us to calculate the characteristic escaping times in a much more efficient way, due to the fact that with the reduced system they can be calculated semi-analytically.

We describe now how to extract the *effective energy function* underlying the neurodynamical firing activity of a high-dimensional system of spiking neurons. Furthermore, we show how to estimate the concomitant noise. By doing this, we will able to write the stochastic neurodynamical equation describing the evolution of the neural system (Deco, Marti, Ledberg, Reig and Sanchez Vives 2009a).

We assume that the neural system that we are considering is stationary. Further, we assume that we can cluster into populations neurons that show a statistically similar type of firing activity. Let us denotes by $\nu_i(t)$ the population firing rate activity of population i (for $i = 1, \ldots, n$) at time t [3]. One can further reduce the dimensionality of the system by applying principal component analysis (PCA) (Johnson and Wichern 1998) to the multidimensional data $\nu_i(t)$. In order to keep the analysis as simple as possible, let us focus on cases where we can reduce the neural system to one dimension. In other words, keeping the first principal component, which we will denote $x(t)$, we obtain the one-dimensional projection capturing the most essential part of the neuronal variability underlying the dynamics of the neural system at the stationary state that we are studying.

A.6.1.1 Langevin 1D equation

To keep the analysis as simple as possible, we focus on cases where the description of the neural system can be captured by one single dimension. In such cases, it suffices to project the original dynamical variables on the first principal component, that is, on the direction where the variables have highest variability. We denote with $x(t)$ such a projection. To derive the effective energy function underlying the stochastic dynamics of $x(t)$, we assume that the temporal evolution of $x(t)$ can be described by a one-dimensional Langevin equation generating a Brownian motion of a highly damped particle in a one-dimensional potential $U(x)$ (Gardiner 1991)

---

[3]Formally, we define the population rate as $\nu = \frac{n_a(t, t+\Delta t)}{N \Delta t}$ where N is the number of neurons in the population, $n_a(t, t + \Delta t)$ is the number of spikes (summed over all neurons in the population) that occur between t and $t + \Delta t$, and Δt is a small time interval.

$$\dot{x} = -\frac{dU(x)}{dx} + \sqrt{2D}\eta(t), \tag{A.32}$$

where $\eta(t)$ is a Gaussian fluctuation term with

$$\langle \eta(t) \rangle = 0, \tag{A.33}$$
$$\langle \eta(t)\eta(t') \rangle = \delta(t - t'). \tag{A.34}$$

The parameter D is the noise intensity. The time coordinate we use is dimensionless and normalized to the time constant, so that time derivatives are of order unity and noise intensities have dimensions of a variance. The probability density of the stochastic process described by Equations A.32–A.34 satisfies the Fokker–Planck equation:

$$\frac{\partial P(x,t)}{\partial t} = \frac{\partial}{\partial x}\left(\frac{dU(x)}{dx}P(x,t)\right) + D\frac{\partial 2}{\partial x^2}P(x,t), \tag{A.35}$$

where $P(x,t)$ is the probability density of the random variable x. Equation A.35 admits a nontrivial stationary solution given by

$$\lim_{t\to\infty} P(x,t) \equiv P_{st}(x) = \frac{\exp[-\phi(x)]}{\int_{-\infty}^{+\infty}\exp[-\phi(x)]dx}. \tag{A.36}$$

where $\phi(x) \equiv U(x)/D$ is the normalized potential. This solution is well defined as long as the integral in the denominator converges, as is the case when the potential $U(x)$ grows to infinity as x goes to $\pm\infty$. This limit is satisfied if we impose the condition that $\lim_{x\to\pm\infty} f(x) = \mp\infty$.

A.6.1.2 Estimation of the 1D potential

The effective potential $\phi(x)$ is obtained by fitting the estimated probability density of the reduced empirical data with the stationary solution, Equation A.36. We perform this estimation assuming a continuous piecewise quadratic potential. First, we partition the range of x into M subintervals $[a_{i-1}, a_i)$, $i = 1, \ldots, M$. In each interval the potential is given by a quadratic polynomial

$$\phi(x) = \alpha_i(x - \gamma_i)2 + \beta_i, \quad \text{for } a_{i-1} \le x < a_i, \tag{A.37}$$

where the $3M$ parameters $\{\alpha_i, \beta_i, \gamma_i\}_{i=1,\ldots,M}$ are to be determined. To ensure continuity and differentiability of the potential at the boundaries the subintervals, the parameters α_i and β_i must obey the recurrent relation

$$\alpha_{i+1} = \alpha_i\frac{(a_i - \gamma_i)}{(a_i - \gamma_{i+1})}, \qquad\qquad \alpha_1 = c_0,$$
$$\beta_{i+1} = \beta_i + \alpha_i(a_i - \gamma_i)(\gamma_{i+1} - \gamma_i), \qquad \beta_1 = 0,$$

for $i = 1, \ldots, M-1$. We are then left with M free parameters, we denote by $\vec{\theta} \equiv (\gamma_1, \ldots, \gamma_M)$. Given the experimental data $\vec{x} = \{x_1, \ldots, x_N\}$, the values of the free parameters $\vec{\theta}$ are set maximizing the logarithm of the likelihood function

$$\log L(\vec{\theta}) = \log P_{st}(\vec{x}|\vec{\theta}) = \log \prod_{k=1}^{N} P_{st}(x_k|\vec{\theta})$$
$$= \sum_{k=1}^{N} \phi(x_k) + N\log\left(\int_{-\infty}^{+\infty} e^{-\phi(x)}dx\right), \tag{A.38}$$

where we assume in the second equality that the data is drawn independently and identically from the distribution $P_{\mathrm{st}}(x)$, given by Equation A.36. The maximization over $\vec{\theta}$ is computed using a downhill simplex method (Nelder and Mead 1965, Lagarias, Reeds, Wright and Wright 1998).

A.6.1.3 Estimation of the noise intensity

In addition to the potential $U(x)$ the effective stochastic dynamical Equation A.32 and the Fokker–Planck counterpart, Equation A.35, involve the noise intensity D. This has also to be specified in order to describe the full stochastic system. We estimate D through the mean escape time of the system from a metastable state. The analytical form of the mean escape time from an interval $[L, R]$ for the stochastic process described by Equation A.32 starting at $x_0 \in [L, R]$ is (Gardiner 1991) (an alternative procedure was derived in Malakhov and Pankratov (1996) for this case using piecewise parabolic potential profiles):

$$
\begin{aligned}
\tau(x_0, L, R) &= \frac{2}{D} \left\{ \int_{x_0}^{R} \left[\int_{L}^{v} \exp\{\phi(v) - \phi(u)\} \, du \right] dv \right. \\
&\quad \left. - d(x_0) \int_{L}^{R} \left[\int_{L}^{v} \exp\{\phi(v) - \phi(u)\} \, du \right] dv \right\} \quad \text{(A.39)} \\
&\equiv \frac{1}{D} I(x_0, L, R),
\end{aligned}
$$

where in the first equality we have defined

$$
d(x) \equiv \frac{\int_{x}^{R} \exp\{\phi(v)\} \, dv}{\int_{L}^{R} \exp\{\phi(v)\} \, dv}.
$$

The mean escape time $\tau(x_0, L, R)$ can be estimated from empirical data, while $I(x_0, L, R)$ can be evaluated numerically once the potential has been approximated maximizing Equation A.38. If we denote by $\hat{I}$ and $\hat{\tau}$ the estimated values of $I(x_0, L, R)$ and $\tau(x_0, L, R)$, the noise intensity can be inferred from the relation A.39,

$$
D \approx \hat{I}/\hat{\tau}. \quad \text{(A.40)}
$$

Once the energy function $\phi(x)$ and the noise intensity D are determined, the effective description given by Equations A.32–A.34, or, equivalently, by the associated Fokker–Planck Equation A.35, is complete. Let us note that D fixes also the time scale of the problem. The stationary distribution of the data determines unequivocally the normalized potential ϕ but it does not specify the time scale of dynamical evolution. Only after fixing D can we also fit the time scale of the data. In fact, we simulated the distribution of the transition and residence times of the data analyzed by explicit simulation of the Langevin Equation A.32 with the optimally fitted noise intensity D that adjusts the time scale by the mean escape time $\tau(x_0, L, R)$ for a particular x_0, L, R.

We demonstrate in Section A.6.2 how this top-down methodology works.

A.6.2 One-dimensional rate model

Here we show how the method works for a one-dimensional rate model described by a Langevin equation,

$$
\dot{x} = f(x) + \sigma\eta(t), \quad \text{(A.41)}
$$

where $f(x)$ is a nonlinear function and $\sigma\eta(t)$ is Gaussian white noise of standard deviation σ. The method (Deco, Marti, Ledberg, Reig and Sanchez Vives 2009a) provides, given a sample

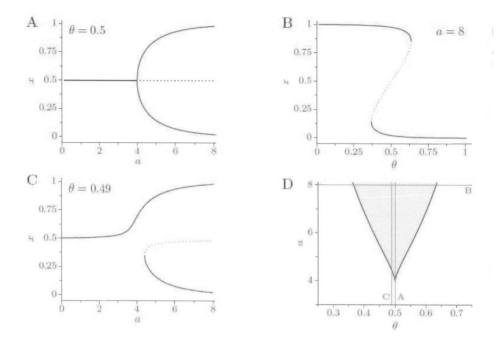

Fig. A.1 Bistability in a one-dimensional rate model. The rate model described by Equations A.41–A.43 has two stable fixed points for some values of the parameters a and θ. (A), (B), and (C) Location of the fixed points as a function of a, for θ=0.5 (A) and θ=0.49 (C); and as a function of θ, with a=8 (B). Solid curves correspond to stable fixed points, dotted curves to unstable fixed points. x is plotted on the ordinate. (D) Regions of monostability (white) and bistability (grey) in the parameter space (θ, a). The two vertical lines labelled A and C, and the horizontal line labelled A, show the sections in the parameter space represented in the bifurcation diagrams (A)–(C). (After Deco, Marti, Ledberg, Reig and Vives 2009a.)

of states $S = \{x^{(1)}, \ldots, x^{(N)}\}$, an effective description of the system of exactly the same type as Equation A.41. Our aim in this Section is to check that the piecewise approximation of the probability density described earlier recovers correctly the energy of the system, which is well-defined in this particular example. We also study the sensitivity of the estimation to the number of subintervals used in the piecewise quadratic approximation. We have chosen a simple system for illustrative purposes, and indeed, given that the system is one-dimensional, an energy function $U(x)$ satisfying $f(x) = -dU(x)/dx$ can be specified without the approximation method and can be used to check the accuracy of the method..

A.6.2.1 Fixed points and stability of the noiseless system

For the sake of concreteness, we choose $f(x)$ to be of the form:

$$f(x) = -x + \Phi(x), \tag{A.42}$$

where Φ is a sigmoidal activation function

$$\Phi(x) = 1/\left(1 + \exp[-a(x - \theta)]\right). \tag{A.43}$$

The parameter θ sets the location of the sigmoid's soft threshold, and a is a scale parameter. This specific choice of $f(x)$ is motivated by the form of the rate models used to describe the firing activity of neuronal assemblies (Wilson and Cowan 1972, Amit and Tsodyks 1991),

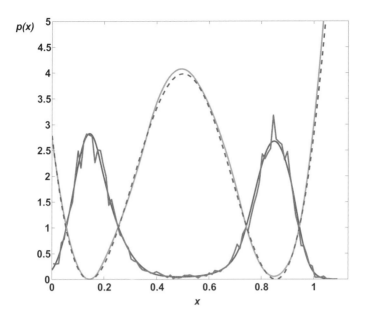

Fig. A.2 Simulation results for a one-dimensional bistable rate system. Stationary distribution of the rate variable $x(t)$. Black dashed: original energy function. Solid curve which follows it: estimated energy function. Irregular black line with two peaks: normalized histogram of simulated data, using the Langevin system in Equations A.41–A.43. Smooth black line with two peaks: maximum likelihood fit of the stationary distribution, using a piecewise quadratic approximation. (After Deco, Marti, Ledberg, Reig and Vives 2009a.)

although any other system satisfying the condition $\lim_{x \to \pm\infty} f(x) = \mp\infty$, which ensures the existence of at least one stable fixed point, would be equally valid (see earlier).

We first consider the deterministic system that results from switching noise off, i.e. setting $\sigma = 0$ in Equation A.41. The fixed points of the noiseless system satisfy the condition $f(x) = 0$, or equivalently $x = \Phi(x)$. Depending on the values of the parameters a and θ, there may be either one or three solutions to this equation. In the former case, the only intersection point will always correspond to a stable fixed point by construction, and the system will be monostable. When there exist instead three solutions, two of them correspond to stable fixed points, and the remaining one is an unstable fixed point; the system is thus bistable. We illustrate in Figs. A.1A–C the presence and number of fixed points as a function of the parameters. Note that bistability is only possible for high enough values of a, which dictates the degree of nonlinearity in the system. Also, in order for the system to be bistable, the value of θ should lie in some interval centered at $\theta = 0.5$. The endpoints of the interval can be derived from the fixed point condition, $f(x) = 0$, and the condition $f'(x) = 0$ that determines where pairs of fixed points appear or disappear. The interval turns out to be $\left[\ln\{(1/y_- - 1)e^{ay_-}\}/a, \ln\{(1/y_+ - 1)e^{ay_+}\}/a\right]$, where $y_\pm \equiv (1 \pm \sqrt{1 - 4/a})/2$, and is plotted as the region of bistability in Fig. A.1D.

It is straightforward to compute the energy function of the system:

$$U(x) = \int^x \{s - \Phi(s)\}\, ds$$
$$= \frac{x^2}{2} - x - \frac{1}{a}\ln\{1 + \exp[a(x - \theta)]\}. \tag{A.44}$$

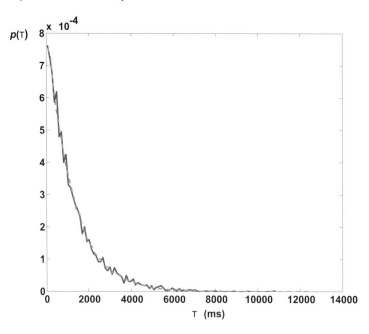

Fig. A.3 Simulation results for a one-dimensional bistable rate system. Distribution of the transition times (τ) between attractors. (Due to the symmetry, we show here only the transitions from the left to the right attractor of Fig. A.2). Dashed curve (red, smooth): original stochastic Equation A.41. Black solid curves: reconstructed (simulated) system. (After Deco, Marti, Ledberg, Reig and Vives 2009a.)

By construction, the local minima of the energy function $U(x)$ are at the stable fixed points of the system (A.41), while the local maxima are at the unstable fixed points.

A.6.2.2 Simulated data and extraction of the effective parameters

When there is bistability, the system will decay to either one of the two available stable states depending on the initial conditions and will remain there indefinitely. The picture changes dramatically when noise is added to the system: the variable x spends most of the time wandering around either one of the two fixed points until some rare and large fluctuation drives it to the neighbourhood of the other fixed point, where the process starts over. Therefore, rather than a pair of stable states, we have two *metastable* states that are long-lived in terms of the characteristic time scales of the system, but that are not truly stable at much longer time scales. Each of these two metastable *states* are better regarded as a unimodal *distribution* centered at one of the stable fixed points of the noiseless system. We loosely refer to each of these distributions as an *attractor*. Since the system spends most of the time in either one of the attractors, and there are quick, random, and rare switches between the two, the stationary distribution of x is bimodal.

As an example, let us choose the parameter values $a = 5$ and $\theta = 0.5$, which lie within the region of bistability depicted in Fig. A.1D. The analysis of stability of the deterministic limit of Equations A.41–A.43 shows that for $\theta = 0.5$ and $a = 5$ the stable fixed points are located at $x \approx 0.145$ and $x \approx 0.885$. By adding a moderate amount of noise to the system, stable fixed points turn into peaks of the stationary distribution, which is bimodal. Figure A.2 shows the stationary distribution of the state variable x when the stochastic system described by Equations A.41–A.43 is simulated long enough. Note that, in so doing, we are implicitly assuming that the sampling of the system over a long enough time can be identified

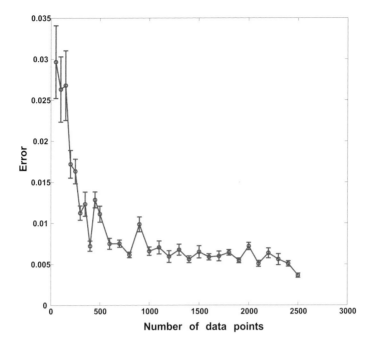

Fig. A.4 Simulation results for a one-dimensional bistable rate system. The error in estimating the true energy function as a function of the number of data points utilized. The error bars are the mean squared errors. Parameters: α=5, θ=0.5, σ=0.06. (After Deco, Marti, Ledberg, Reig and Vives 2009a.)

with the sampling of independent realizations of the same process (ergodicity). Figure A.2 also shows the maximum likelihood estimate of the stationary distribution $P_{st}(x)$ using a piecewise quadratic approximation, as well as the associated energy function. Note that the two peaks of the distribution are centered at the stable fixed points of the noiseless system. With the estimated probability density we can easily extract the underlying energy, $\phi(x)$, following the procedure described earlier. Figure A.2 also includes the true energy function of the original system, given by Equation A.44, for a comparison with the estimated energy function. There is good agreement between the two.

The noise intensity was estimated from the mean escape time from a metastable state, Equation A.39. In our example, we estimated the average time needed for the system initialized at the fixed point at $x_0 \approx 0.145$ (*left* attractor) to cross a boundary at some $x = \lambda$. The location λ of the boundary was chosen somewhere between the separatrix at $x = 0.5$ to the fixed point at $x \approx 0.885$ (*right* attractor), to make sure that the mean first-passage time corresponded to a real escape from one attractor to the other. The analytical form for the mean escape time from an interval $[L, R]$, when the system is initialized at $x_0 \in [L, R]$, is given by Equation A.39 above. The expression A.39 can be further simplified using the fact that $\phi(x)$ goes to infinity in the limit $L \to -\infty$, which ensures that the system can escape only through the right endpoint of the interval. Thus identifying L with $-\infty$ and R with λ, the mean escape time through λ reads

$$\tau(x_0, -\infty, \lambda) = \frac{2}{D} \int_{x_0}^{\lambda} \left[\int_{-\infty}^{v} \exp\{\phi(v) - \phi(u)\} \, du \right] dv$$

$$\equiv \frac{1}{D} I(x_0, -\infty, \lambda)$$

(A.45)

The noise intensity can then be estimated as $D = \hat{I}(x_0, -\infty, \lambda)/\hat{\tau}(x_0, -\infty, \lambda)$, where $\hat{I}$ is the numerical evaluation of $I(x_0, -\infty, \lambda)$ using the piecewise-quadratic approximation of the potential, and $\hat{\tau}$ is the sample mean of first-passage times. We obtained an estimated value for the noise intensity of $D \approx 3.8 \cdot 10^{-3}$, which is close to the value $D = \sigma2 = 3.6 \cdot 103$, used in the simulations.

We also checked the robustness of the method by checking the error in estimating the true energy function as a function of the number of data points utilized. Fig. A.4 shows the result plotting the error in estimating the true energy function obtained after optimization as a function of the number of data points used. Excellent results are already obtained for a relatively low number of data points (300).

In conclusion, we have demonstrated in this Section A.6.2 how the mean-field analysis with noise described in Section A.6.1 operates, and can make a useful contribution to understanding analytically the operation of an attractor system with noise, and can provide for example estimates of the escaping times for different amounts of noise. By analysing the escaping times found in a system such as a network in the brain, or behaviour, we can thus for example estimate the noise in the system. Analyses using this approach are now being applied to estimate the noise in neural data (Deco, Marti, Ledberg, Reig and Sanchez Vives 2009a).

References

Abbott, L. F. (1991). Realistic synaptic inputs for model neural networks, *Network* **2**: 245–258.

Abbott, L. F. and Blum, K. I. (1996). Functional significance of long-term potentiation for sequence learning and prediction, *Cerebral Cortex* **6**: 406–416.

Abbott, L. F. and Nelson, S. B. (2000). Synaptic plasticity: taming the beast, *Nature Neuroscience* **3**: 1178–1183.

Abbott, L. F. and Regehr, W. G. (2004). Synaptic computation, *Nature* **431**: 796–803.

Abbott, L. F., Rolls, E. T. and Tovee, M. J. (1996). Representational capacity of face coding in monkeys, *Cerebral Cortex* **6**: 498–505.

Abbott, L. F., Varela, J. A., Sen, K. and Nelson, S. B. (1997). Synaptic depression and cortical gain control, *Science* **275**: 220–224.

Abeles, M. (1991). *Corticonics: Neural Circuits of the Cerebral Cortex*, Cambridge University Press, Cambridge.

Abramson, N. (1963). *Information Theory and Coding*, McGraw-Hill, New York.

Aggelopoulos, N. C. and Rolls, E. T. (2005). Natural scene perception: inferior temporal cortex neurons encode the positions of different objects in the scene, *European Journal of Neuroscience* **22**: 2903–2916.

Aggelopoulos, N. C., Franco, L. and Rolls, E. T. (2005). Object perception in natural scenes: encoding by inferior temporal cortex simultaneously recorded neurons, *Journal of Neurophysiology* **93**: 1342–1357.

Aigner, T. G., Mitchell, S. J., Aggleton, J. P., DeLong, M. R., Struble, R. G., Price, D. L., Wenk, G. L., Pettigrew, K. D. and Mishkin, M. (1991). Transient impairment of recognition memory following ibotenic acid lesions of the basal forebrain in macaques, *Experimental Brain Research* **86**: 18–26.

Albantakis, L. and Deco, G. (2009). The encoding of alternatives in multiple-choice decision making, *Proceedings of the National Academy of Sciences USA* **106**: 10308–10313.

Aleman, A. and Kahn, R. S. (2005). Strange feelings: do amygdala abnormalities dysregulate the emotional brain in schizophrenia?, *Prog Neurobiol* **77**(5): 283–298.

Amaral, D. G. (1986). Amygdalohippocampal and amygdalocortical projections in the primate brain, *in* R. Schwarcz and Y. Ben-Ari (eds), *Excitatory Amino Acids and Epilepsy*, Plenum Press, New York, pp. 3–18.

Amaral, D. G. (1987). Memory: anatomical organization of candidate brain regions, *in* F. Plum and V. Mountcastle (eds), *Higher Functions of the Brain. Handbook of Physiology, Part I*, American Physiological Society, Washington, DC, pp. 211–294.

Amaral, D. G. and Price, J. L. (1984). Amygdalo-cortical projections in the monkey (Macaca fascicularis), *Journal of Comparative Neurology* **230**: 465–496.

Amaral, D. G., Price, J. L., Pitkanen, A. and Carmichael, S. T. (1992). Anatomical organization of the primate amygdaloid complex, *in* J. P. Aggleton (ed.), *The Amygdala*, Wiley-Liss, New York, chapter 1, pp. 1–66.

Amit, D. J. (1989). *Modelling Brain Function*, Cambridge University Press, New York.

Amit, D. J. (1995). The Hebbian paradigm reintegrated: local reverberations as internal representations, *Behavioral and Brain Sciences* **18**: 617–657.

Amit, D. J. and Brunel, N. (1997). Model of global spontaneous activity and local structured activity during delay periods in the cerebral cortex, *Cerebral Cortex* **7**: 237–252.

Amit, D. J. and Tsodyks, M. V. (1991). Quantitative study of attractor neural network retrieving at low spike rates. I. Substrate – spikes, rates and neuronal gain, *Network* **2**: 259–273.

Amit, D. J., Gutfreund, H. and Sompolinsky, H. (1987). Statistical mechanics of neural networks near saturation, *Annals of Physics (New York)* **173**: 30–67.

Andersen, P., Dingledine, R., Gjerstad, L., Langmoen, I. A. and Laursen, A. M. (1980). Two different responses of hippocampal pyramidal cells to application of gamma-aminobutyric acid, *Journal of Physiology* **307**: 279–296.

Andersen, R. A. (1995). Coordinate transformations and motor planning in the posterior parietal cortex, *in* M. S. Gazzaniga (ed.), *The Cognitive Neurosciences*, MIT Press, Cambridge, MA, chapter 33, pp. 519–532.

Arnold, P. D., Rosenberg, D. R., Mundo, E., Tharmalingam, S., Kennedy, J. L. and Richter, M. A. (2004). Association of a glutamate (NMDA) subunit receptor gene (GRIN2B) with obsessive-compulsive disorder: a preliminary study, *Psychopharmacology (Berl)* **174**: 530–538.

Arnsten, A. F. and Li, B. M. (2005). Neurobiology of executive functions: catecholamine influences on prefrontal cortical functions, *Biological Psychiatry* **57**: 1377–1384.

Artola, A. and Singer, W. (1993). Long term depression: related mechanisms in cerebellum, neocortex and hippocampus, *in* M. Baudry, R. F. Thompson and J. L. Davis (eds), *Synaptic Plasticity: Molecular, Cellular and Functional Aspects*, MIT Press, Cambridge, MA, chapter 7, pp. 129–146.

Asaad, W. F., Rainer, G. and Miller, E. K. (2000). Task-specific neural activity in the primate prefrontal cortex, *Journal of Neurophysiology* **84**: 451–459.

Baddeley, A. (1986). *Working Memory*, Oxford University Press, New York.

Baddeley, R. J., Abbott, L. F., Booth, M. J. A., Sengpiel, F., Freeman, T., Wakeman, E. A. and Rolls, E. T. (1997). Responses of neurons in primary and inferior temporal visual cortices to natural scenes, *Proceedings of the Royal Society B* **264**: 1775–1783.

Bagal, A. A., Kao, J. P. Y., Tang, C.-M. and Thompson, S. M. (2005). Long-term potentiation of exogenous glutamate responses at single dendritic spines., *Proc Natl Acad Sci U S A* **102**(40): 14434–9.

Bannon, S., Gonsalvez, C. J., Croft, R. J. and Boyce, P. M. (2002). Response inhibition deficits in obsessive-compulsive disorder, *Psychiatry Research* **110**: 165–174.

Bannon, S., Gonsalvez, C. J., Croft, R. J. and Boyce, P. M. (2006). Executive functions in obsessive-compulsive disorder: state or trait deficits?, *Aust N Z Journal of Psychiatry* **40**: 1031–1038.

Barlow, H. B. (1972). Single units and sensation: a neuron doctrine for perceptual psychology, *Perception* **1**: 371–394.

Barnes, C. A. (2003). Long-term potentiation and the ageing brain, *Philosophical Transactions of the Royal Society of London B* **358**: 765—772.

Barto, A. G. (1985). Learning by statistical cooperation of self-interested neuron-like computing elements, *Human Neurobiology* **4**: 229–256.

Bartus, R. T. (2000). On neurodegenerative diseases, models, and treatment strategies: lessons learned and lessons forgotten a generation following the cholinergic hypothesis, *Experimental Neurology* **163**: 495—529.

Battaglia, F. and Treves, A. (1998). Stable and rapid recurrent processing in realistic autoassociative memories, *Neural Computation* **10**: 431–450.

Baxter, R. D. and Liddle, P. F. (1998). Neuropsychological deficits associated with schizophrenic syndromes, *Schizophrenia Research* **30**: 239–249.

Baylis, G. C. and Rolls, E. T. (1987). Responses of neurons in the inferior temporal cortex in short term and serial recognition memory tasks, *Experimental Brain Research* **65**: 614–622.

Baylis, G. C., Rolls, E. T. and Leonard, C. M. (1985). Selectivity between faces in the responses of a population of neurons in the cortex in the superior temporal sulcus of the monkey, *Brain Research* **342**: 91–102.

Baylis, G. C., Rolls, E. T. and Leonard, C. M. (1987). Functional subdivisions of temporal lobe neocortex, *Journal of Neuroscience* **7**: 330–342.

Baynes, K., Holtzman, H. and Volpe, V. (1986). Components of visual attention: alterations in response pattern to visual stimuli following parietal lobe infarction, *Brain* **109**: 99–144.

Bear, M. F. and Singer, W. (1986). Modulation of visual cortical plasticity by acetylcholine and noradrenaline, *Nature* **320**: 172–176.

Beck, J. M., Ma, W. J., Kiani, R., Hanks, T., Churchland, A. K., Roitman, J., Shadlen, M. N., Latham, P. E. and Pouget, A. (2008). Probabilistic population codes for Bayesian decision making, *Neuron* **60**: 1142–1152.

Bender, W., Albus, M., Moller, H. J. and Tretter, F. (2006). Towards systemic theories in biological psychiatry, *Pharmacopsychiatry* **39 Suppl 1**: S4–S9.

Berg, E. (1948). A simple objective technique for measuring flexibility in thinking, *Journal of General Psychology* **39**: 15–22.

Berlin, H. and Rolls, E. T. (2004). Time perception, impulsivity, emotionality, and personality in self-harming borderline personality disorder patients, *Journal of Personality Disorders* **18**: 358–378.

Berlin, H., Rolls, E. T. and Kischka, U. (2004). Impulsivity, time perception, emotion, and reinforcement sensitivity in patients with orbitofrontal cortex lesions, *Brain* **127**: 1108–1126.

Berlin, H., Rolls, E. T. and Iversen, S. D. (2005). Borderline Personality Disorder, impulsivity, and the orbitofrontal cortex, *American Journal of Psychiatry* **58**: 234–245.

Bhattacharyya, S. and Chakraborty, K. (2007). Glutamatergic dysfunction–newer targets for anti-obsessional drugs, *Recent Patents CNS Drug Discovery* **2**: 47–55.

Bi, G.-Q. and Poo, M.-M. (1998). Activity-induced synaptic modifications in hippocampal culture, dependence on spike timing, synaptic strength and cell type, *Journal of Neuroscience* **18**: 10464–10472.

Bi, G.-Q. and Poo, M.-M. (2001). Synaptic modification by correlated activity: Hebb's postulate revisited, *Annual Review of Neuroscience* **24**: 139–166.

Blake, R. and Logothetis, N. K. (2002). Visual competition, *Nature Reviews Neuroscience* **3**: 13–21.

Bliss, T. V. P. and Collingridge, G. L. (1993). A synaptic model of memory: long-term potentiation in the hippocampus, *Nature* **361**: 31–39.

Block, N. (1995). On a confusion about a function of consciousness, *Behavioral and Brain Sciences* **18**: 22–47.

Bloomfield, S. (1974). Arithmetical operations performed by nerve cells, *Brain Research* **69**: 115–124.

Boorman, E. D., Behrens, T. E., Woolrich, M. W. and Rushworth, M. F. (2009). How green is the grass on the other side? Frontopolar cortex and the evidence in favor of alternative courses of action, *Neuron* **62**: 733–743.

Booth, M. C. A. and Rolls, E. T. (1998). View-invariant representations of familiar objects by neurons in the inferior temporal visual cortex, *Cerebral Cortex* **8**: 510–523.

Braitenberg, V. and Schutz, A. (1991). *Anatomy of the Cortex*, Springer-Verlag, Berlin.

Broadbent, D. E. (1958). *Perception and Communication*, Pergamon, Oxford.

Brody, C., Hernandez, A., Zainos, A. and Romo, R. (2003a). Timing and neural encoding of somatosensory parametric working memory in macaque prefrontal cortex, *Cerebral Cortex* **13**: 1196–1207.

Brody, C., Romo, R. and Kepecs, A. (2003b). Basic mechanisms for graded persistent activity: discrete attractors, continuous attractors, and dynamic representations, *Current Opinion in Neurobiology* **13**: 204–211.

Brown, D. A., Gahwiler, B. H., Griffith, W. H. and Halliwell, J. V. (1990). Membrane currents in hippocampal neurons, *Progress in Brain Research* **83**: 141–160.

Brunel, N. and Amit, D. (1997). Model of global spontaneous activity and local structured delay activity during delay periods in the cerebral cortex, *Cerebral Cortex* **7**: 237–252.

Brunel, N. and Hakim, V. (1999). Fast global oscillations in networks of integrate-and-fire neurons with low firing rates, *Neural Computation* **11**: 1621–1671.

Brunel, N. and Wang, X. J. (2001). Effects of neuromodulation in a cortical network model of object working memory dominated by recurrent inhibition, *Journal of Computational Neuroscience* **11**: 63–85.

Buehlmann, A. and Deco, G. (2008). The neuronal basis of attention: rate versus synchronization modulation, *Journal of Neuroscience* **28**: 7679–7686.

Burgess, P. W. (2000). Strategy application disorder: the role of the frontal lobes in human multitasking, *Psychological Research* **63**: 279–288.

Burke, S. N. and Barnes, C. A. (2006). Neural plasticity in the ageing brain, *Nature Reviews Neuroscience* **7**: 30–40.

Burton, M. J., Rolls, E. T. and Mora, F. (1976). Effects of hunger on the responses of neurones in the lateral hypothalamus to the sight and taste of food, *Experimental Neurology* **51**: 668–677.

Buxton, R. B. and Frank, L. R. (1997). A model for the coupling between cerebral blood flow and oxygen metabolism during neural stimulation, *Journal of Cerebral Blood Flow and Metabolism* **17**: 64–72.

Buxton, R. B., Wong, E. C. and Frank, L. R. (1998). Dynamics of blood flow and oxygenation changes during brain activation: the balloon model, *Magnetic Resonance in Medicine* **39**: 855–864.

Buzsáki, G. (2006). *Rhythms of the Brain*, Oxford University Press, Oxford.

Capuano, B., Crosby, I. T. and Lloyd, E. J. (2002). Schizophrenia: genesis, receptorology and current therapeutics, *Current Medicinal Chemistry* **9**: 521–548.

Carlson, N. R. (2006). *Physiology of Behavior*, 9th edn, Pearson, Boston.

Carlsson, A. (2006). The neurochemical circuitry of schizophrenia, *Pharmacopsychiatry* **39 Suppl 1**: S10–14.

Carpenter, R. H. S. and Williams, M. (1995). Neural computation of log likelihood in control of saccadic eye movements, *Nature* **377**: 59–62.

Carter, C. S., Perlstein, W., Ganguli, R., Brar, J., Mintun, M. and Cohen, J. D. (1998). Functional hypofrontality and working memory dysfunction in schizophrenia, *American Journal of Psychiatry* **155**: 1285–1287.

Castner, S. A., Williams, G. V. and Goldman-Rakic, P. S. (2000). Reversal of antipsychotic-induced working memory deficits by short-term dopamine D1 receptor stimulation, *Science* **287**: 2020–2022.

Chakrabarty, K., Bhattacharyya, S., Christopher, R. and Khanna, S. (2005). Glutamatergic dysfunction in OCD, *Neuropsychopharmacology* **30**: 1735–1740.

Chamberlain, S. R., Fineberg, N. A., Blackwell, A. D., Robbins, T. W. and Sahakian, B. J. (2006). Motor inhibition and cognitive flexibility in obsessive-compulsive disorder and trichotillomania, *American Journal of Psychiatry* **163**: 1282–1284.

Chamberlain, S. R., Fineberg, N. A., Menzies, L. A., Blackwell, A. D., Bullmore, E. T., Robbins, T. W. and Sahakian, B. J. (2007). Impaired cognitive flexibility and motor inhibition in unaffected first-degree relatives of patients with obsessive-compulsive disorder, *American Journal of Psychiatry* **164**: 335–338.

Chelazzi, L. (1998). Serial attention mechanisms in visual search: a critical look at the evidence, *Psychological Research* **62**: 195–219.

Chelazzi, L., Miller, E., Duncan, J. and Desimone, R. (1993). A neural basis for visual search in inferior temporal cortex, *Nature (London)* **363**: 345–347.

Christie, B. R. (1996). Long-term depression in the hippocampus, *Hippocampus* **6**: 1–2.

Churchland, A. K., Kiani, R. and Shadlen, M. N. (2008). Decision-making with multiple alternatives, *Nature Neuroscience* **11**: 693–702.

Clark, L., Cools, R. and Robbins, T. W. (2004). The neuropsychology of ventral prefrontal cortex: decision-making and reversal learning, *Brain and Cognition* **55**: 41–53.

Colby, C. L., Duhamel, J. R. and Goldberg, M. E. (1993). Ventral intraparietal area of the macaque - anatomic location and visual response properties, *Journal of Neurophysiology* **69**: 902–914.

Collingridge, G. L. and Bliss, T. V. P. (1987). NMDA receptors: their role in long-term potentiation, *Trends in Neurosciences* **10**: 288–293.

Compte, A., Brunel, N., Goldman-Rakic, P. and Wang, X. J. (2000). Synaptic mechanisms and network dynamics underlying spatial working memory in a cortical network model, *Cerebral Cortex* **10**: 910–923.

Compte, A., Constantinidis, C., Tegner, J., Raghavachari, S., Chafee, M. V., Goldman-Rakic, P. S. and Wang, X. J. (2003). Temporally irregular mnemonic persistent activity in prefrontal neurons of monkeys during a delayed response task, *Journal of Neurophysiology* **90**: 3441–3454.

Corchs, S. and Deco, G. (2002). Large-scale neural model for visual attention: integration of experimental single cell and fMRI data, *Cerebral Cortex* **12**: 339–348.

Corchs, S. and Deco, G. (2004). Feature-based attention in human visual cortex: simulation of fMRI data, *Neuroimage* **21**: 36–45.

Corrado, G. S., Sugrue, L. P., Seung, H. S. and Newsome, W. T. (2005). Linear-nonlinear-Poisson models of primate choice dynamics, *Journal of the Experimental Analysis of Behavior* **84**: 581–617.

Cover, T. M. and Thomas, J. A. (1991). *Elements of Information Theory*, Wiley, New York.

Coyle, J. T. (2006). Glutamate and schizophrenia: beyond the dopamine hypothesis, *Cellular and Molecular Neurobiology* **26**: 365–384.

Coyle, J. T., Tsai, G. and Goff, D. (2003). Converging evidence of NMDA receptor hypofunction in the pathophysiology of schizophrenia, *Annals of the New York Academy of Sciences* **1003**: 318–327.

Crick, F. H. C. (1984). Function of the thalamic reticular complex: the searchlight hypothesis, *Proceedings of the National Academy of Sciences USA* **81**: 4586–4590.

Critchley, H. D. and Rolls, E. T. (1996). Olfactory neuronal responses in the primate orbitofrontal cortex: analysis in an olfactory discrimination task, *Journal of Neurophysiology* **75**: 1659–1672.

Dan, Y. and Poo, M.-M. (2004). Spike-timing dependent plasticity of neural circuits, *Neuron* **44**: 23–30.

Dan, Y. and Poo, M.-M. (2006). Spike-timing dependent plasticity: from synapse to perception, *Physiological Reviews* **86**: 1033–1048.

Davis, M. (1992). The role of the amygdala in conditioned fear, *in* J. P. Aggleton (ed.), *The Amygdala*, Wiley-Liss, New York, chapter 9, pp. 255–306.

Davis, M. (2000). The role of the amygdala in conditioned and unconditioned fear and anxiety, *in* J. P. Aggleton (ed.), *The Amygdala: a Functional Analysis*, 2nd edn, Oxford University Press, Oxford, chapter 6, pp. 213–287.

Dawkins, M. S. (1995). *Unravelling Animal Behaviour*, 2nd edn, Longman, Harlow.

Dayan, P. and Abbott, L. F. (2001). *Theoretical Neuroscience*, MIT Press, Cambridge, MA.

De Groff, D., Neelakanta, P., Sudhakar, R. and Aalo, V. (1993). Stochastical aspects of neuronal dynamics: Fokker–Planck approach, *Biological Cybernetics* **69**: 155–164.

de Lafuente, V. and Romo, R. (2005). Neuronal correlates of subjective sensory experience, *Nature Neuroscience* **8**: 1698–1703.

de Lafuente, V. and Romo, R. (2006). Neural correlates of subjective sensory experience gradually builds up accross cortical areas, *Proceedings of the National Academy of Science USA* **103**: 14266–14271.

Deco, G. and Lee, T. S. (2002). A unified model of spatial and object attention based on inter-cortical biased competition, *Neurocomputing* **44–46**: 775–781.

Deco, G. and Lee, T. S. (2004). The role of early visual cortex in visual integration: a neural model of recurrent interaction, *European Journal of Neuroscience* **20**: 1089–1100.

Deco, G. and Marti, D. (2007a). Deterministic analysis of stochastic bifurcations in multi-stable neurodynamical systems, *Biological Cybernetics* **96**: 487–496.

Deco, G. and Marti, D. (2007b). Extended method of moments for deterministic analysis of stochastic multistable neurodynamical systems, *Physical Review E Statistical Nonlinear and Soft Matter Physics* **75**: 031913.

Deco, G. and Rolls, E. T. (2002). Object-based visual neglect: a computational hypothesis, *European Journal of Neuroscience* **16**: 1994–2000.

Deco, G. and Rolls, E. T. (2003). Attention and working memory: a dynamical model of neuronal activity in the prefrontal cortex, *European Journal of Neuroscience* **18**: 2374–2390.

Deco, G. and Rolls, E. T. (2004). A neurodynamical cortical model of visual attention and invariant object recognition, *Vision Research* **44**: 621–644.

Deco, G. and Rolls, E. T. (2005a). Attention, short term memory, and action selection: a unifying theory, *Progress in Neurobiology* **76**: 236–256.

Deco, G. and Rolls, E. T. (2005b). Neurodynamics of biased competition and cooperation for attention: a model with spiking neurons, *Journal of Neurophysiology* **94**: 295–313.

Deco, G. and Rolls, E. T. (2005c). Sequential memory: a putative neural and synaptic dynamical mechanism, *Journal of Cognitive Neuroscience* **17**: 294–307.

Deco, G. and Rolls, E. T. (2005d). Synaptic and spiking dynamics underlying reward reversal in the orbitofrontal cortex, *Cerebral Cortex* **15**: 15–30.

Deco, G. and Rolls, E. T. (2006). A neurophysiological model of decision-making and Weber's law, *European Journal of Neuroscience* **24**: 901–916.

Deco, G. and Zihl, J. (2004). A biased competition based neurodynamical model of visual neglect, *Medical Engineering and Physics* **26**: 733–743.

Deco, G., Pollatos, O. and Zihl, J. (2002). The time course of selective visual attention: theory and experiments, *Vision Research* **42**: 2925–2945.

Deco, G., Rolls, E. T. and Horwitz, B. (2004). 'What' and 'where' in visual working memory: a computational neurodynamical perspective for integrating fMRI and single-neuron data, *Journal of Cognitive Neuroscience* **16**: 683–701.

Deco, G., Ledberg, A., Almeida, R. and Fuster, J. (2005). Neural dynamics of cross-modal and cross-temporal associations, *Experimental Brain Research* **166**: 325–336.

Deco, G., Perez-Sanagustin, M., de Lafuente, V. and Romo, R. (2007a). Perceptual detection as a dynamical bistability phenomenon: A neurocomputational correlate of sensation, *Proceedings of the National Academy of Sciences (USA)* **104**: 20073–20077.

Deco, G., Scarano, L. and Soto-Faraco, S. (2007b). Weber's law in decision-making: integrating behavioral data in humans with a neurophysiological model, *Journal of Neuroscience* **27**: 11192–11200.

Deco, G., Marti, D., Ledberg, A., Reig, R. and Sanchez Vives, M. S. (2009a). Effective reduced diffusion-models: a data driven approach to the analysis of neuronal dynamics, *PLoS Computational Biology* **5**: e1000587.

Deco, G., Rolls, E. T. and Romo, R. (2009b). Stochastic dynamics as a principle of brain function, *Progress in Neurobiology* **88**: 1–16.

Deco, G., Rolls, E. T. and Romo, R. (2010). Synaptic dynamics and decision-making, *Proceedings of the National Academy of Sciences*.

Del Giudice, P., Fusi, S. and Mattia, M. (2003). Modeling the formation of working memory with networks of integrate-and-fire neurons connected by plastic synapses, *Journal of Physiology (Paris)* **97**: 659–681.

Delorme, R., Gousse, V., Roy, I., Trandafir, A., Mathieu, F., Mouren-Simeoni, M. C., Betancur, C. and Leboyer, M. (2007). Shared executive dysfunctions in unaffected relatives of patients with autism and obsessive-compulsive disorder, *European Psychiatry* **22**: 32–38.

Dere, E., Easton, A., Nadel, L. and Huston, J. P. (eds) (2008). *Handbook of Episodic Memory*, Elsevier, Amsterdam.

Desimone, R. (1996). Neural mechanisms for visual memory and their role in attention, *Proceedings of the National Academy of Sciences USA* **93**: 13494–13499.

Desimone, R. and Duncan, J. (1995). Neural mechanisms of selective visual attention, *Annual Review of Neuroscience* **18**: 193–222.

Diamond, M. E., Huang, W. and Ebner, F. F. (1994). Laminar comparison of somatosensory cortical plasticity, *Science* **265**: 1885–1888.

Divac, I. (1975). Magnocellular nuclei of the basal forebrain project to neocortex, brain stem, and olfactory bulb. Review of some functional correlates, *Brain Research* **93**: 385–398.

Douglas, R. J. and Martin, K. A. C. (1990). Neocortex, in G. M. Shepherd (ed.), *The Synaptic Organization of the Brain*, 3rd edn, Oxford University Press, Oxford, chapter 12, pp. 389–438.

Douglas, R. J., Mahowald, M. A. and Martin, K. A. C. (1996). Microarchitecture of cortical columns, in A. Aertsen and V. Braitenberg (eds), *Brain Theory: Biological Basis and Computational Theory of Vision*, Elsevier, Amsterdam.

Douglas, R. J., Markram, H. and Martin, K. A. C. (2004). Neocortex, in G. M. Shepherd (ed.), *The Synaptic Organization of the Brain*, 5th edn, Oxford University Press, Oxford, chapter 12, pp. 499–558.

Duhamel, J. R., Colby, C. L. and Goldberg, M. E. (1992). The updating of the representation of visual space in parietal cortex by intended eye movements, *Science* **255**: 90–92.

Duncan, J. (1996). Cooperating brain systems in selective perception and action, in T. Inui and J. L. McClelland (eds), *Attention and Performance XVI*, MIT Press, Cambridge, MA, pp. 549–578.

Duncan, J. and Humphreys, G. (1989). Visual search and stimulus similarity, *Psychological Review* **96**: 433–458.

Durstewitz, D. and Seamans, J. K. (2002). The computational role of dopamine D1 receptors in working memory, *Neural Networks* **15**: 561–572.

Durstewitz, D., Kelc, M. and Gunturkun, O. (1999). A neurocomputational theory of the dopaminergic modulation of working memory functions, *Journal of Neuroscience* **19**: 2807–2722.

Durstewitz, D., Seamans, J. K. and Sejnowski, T. J. (2000a). Dopamine-mediated stabilization of delay-period activity in a network model of prefrontal cortex, *Journal of Neurophysiology* **83**: 1733–1750.

Durstewitz, D., Seamans, J. K. and Sejnowski, T. J. (2000b). Neurocomputational models of working memory, *Nature Neuroscience* **3 Suppl**: 1184–1191.

Easton, A. and Gaffan, D. (2000). Amygdala and the memory of reward: the importance of fibres of passage from the basal forebrain, in J. P. Aggleton (ed.), *The Amygdala: a Functional Analysis*, 2nd edn, Oxford University Press, Oxford, chapter 17, pp. 569–586.

Easton, A., Ridley, R. M., Baker, H. F. and Gaffan, D. (2002). Unilateral lesions of the cholinergic basal forebrain and fornix in one hemisphere and inferior temporal cortex in the opposite hemisphere produce severe learning impairments in rhesus monkeys, *Cerebral Cortex* **12**: 729–736.

Eccles, J. C. (1984). The cerebral neocortex: a theory of its operation, in E. G. Jones and A. Peters (eds), *Cerebral Cortex: Functional Properties of Cortical Cells*, Vol. 2, Plenum, New York, chapter 1, pp. 1–36.

Elliffe, M. C. M., Rolls, E. T. and Stringer, S. M. (2002). Invariant recognition of feature combinations in the visual system, *Biological Cybernetics* **86**: 59–71.

Elston, G. N., Benavides-Piccione, R., Elston, A., Zietsch, B., Defelipe, J., Manger, P., Casagrande, V. and Kaas, J. H. (2006). Specializations of the granular prefrontal cortex of primates: implications for cognitive processing, *Anatomical Record A Discov Mol Cell Evol Biol* **288**: 26–35.

Epstein, J., Stern, E. and Silbersweig, D. (1999). Mesolimbic activity associated with psychosis in schizophrenia. Symptom-specific PET studies, *Annals of the New York Academy of Sciences* **877**: 562–574.

Fahy, F. L., Riches, I. P. and Brown, M. W. (1993). Neuronal activity related to visual recognition memory and the encoding of recency and familiarity information in the primate anterior and medial inferior temporal and rhinal cortex, *Experimental Brain Research* **96**: 457–492.

Faisal, A., Selen, L. and Wolpert, D. (2008). Noise in the nervous system, *Nature Reviews Neuroscience* **9**: 292–303.

Fatt, P. and Katz, B. (1950). Some observations on biological noise, *Nature* **166**: 597–598.

Fazeli, M. S. and Collingridge, G. L. (eds) (1996). *Cortical Plasticity: LTP and LTD*, Bios, Oxford.

Feller, W. (1951). Diffusion equations in genetics, *In Proceedings of the Second Berkeley Symposium on Mathematical Statistics and Probability*, University of California Press, Berkeley, CA.

Field, D. J. (1994). What is the goal of sensory coding?, *Neural Computation* **6**: 559–601.

Foldiak, P. (2003). Sparse coding in the primate cortex, *in* M. A. Arbib (ed.), *Handbook of Brain Theory and Neural Networks*, 2nd edn, MIT Press, Cambridge, MA, pp. 1064–1608.

Franco, L., Rolls, E. T., Aggelopoulos, N. C. and Treves, A. (2004). The use of decoding to analyze the contribution to the information of the correlations between the firing of simultaneously recorded neurons, *Experimental Brain Research* **155**: 370–384.

Franco, L., Rolls, E. T., Aggelopoulos, N. C. and Jerez, J. M. (2007). Neuronal selectivity, population sparseness, and ergodicity in the inferior temporal visual cortex, *Biological Cybernetics* **96**: 547–560.

Franks, K. M., Stevens, C. F. and Sejnowski, T. J. (2003). Independent sources of quantal variability at single glutamatergic synapses, *Journal of Neuroscience* **23**: 3186–3195.

Fregnac, Y. (1996). Dynamics of cortical connectivity in visual cortical networks: an overview, *Journal of Physiology, Paris* **90**: 113–139.

Freud, S. (1900). *The Interpretation of Dreams*.

Frolov, A. A. and Medvedev, A. V. (1986). Substantiation of the "point approximation" for describing the total electrical activity of the brain with use of a simulation model, *Biophysics* **31**: 332–337.

Fuhrmann, G., Markram, H. and Tsodyks, M. (2002a). Spike frequency adaptation and neocortical rhythms, *Journal of Neurophysiology* **88**: 761–770.

Fuhrmann, G., Segev, I., Markram, H. and Tsodyks, M. (2002b). Coding of temporal information by activity-dependent synapses, *Journal of Neurophysiology* **87**: 140–148.

Funahashi, S., Bruce, C. and Goldman-Rakic, P. (1989). Mnemonic coding of visual space in monkey dorsolateral prefrontal cortex, *Journal of Neurophysiology* **61**: 331–349.

Furman, M. and Wang, X.-J. (2008). Similarity effect and optimal control of multiple-choice decision making, *Neuron* **60**: 1153–1168.

Fuster, J. M. (1973). Unit activity in prefrontal cortex during delayed-response performance: neuronal correlates of transient memory, *Joural of Neurophysiology* **36**: 61–78.

Fuster, J. M. (1989). *The Prefrontal Cortex*, 2nd edn, Raven Press, New York.

Fuster, J. M. (1997). *The Prefrontal Cortex*, 3rd edn, Lippincott-Raven, Philadelphia.

Fuster, J. M. (2000). *Memory Systems in the Brain*, Raven Press, New York.

Fuster, J. M. (2008). *The Prefrontal Cortex*, 4th edn, Academic Press, London.

Fuster, J. M., Bauer, R. H. and Jervey, J. P. (1982). Cellular discharge in the dorsolateral prefrontal cortex of the monkey in cognitive tasks, *Experimental Neurology* **77**: 679–694.

Fuster, J. M., Bodner, M. and Kroger, J. K. (2000). Cross-modal and cross-temporal association in neurons of frontal cortex, *Nature* **405**: 347–351.

Gardiner, C. (1991). *Handbook of Stochastic Methods (for Physics, Chemistry and the Natural Sciences)*, Springer-Verlag, Berlin.

Gazzaniga, M. S. (1988). Brain modularity: towards a philosophy of conscious experience, *in* A. J. Marcel and E. Bisiach (eds), *Consciousness in Contemporary Science*, Oxford University Press, Oxford, chapter 10, pp. 218–238.

Gazzaniga, M. S. (1995). Consciousness and the cerebral hemispheres, *in* M. S. Gazzaniga (ed.), *The Cognitive Neurosciences*, MIT Press, Cambridge, MA, chapter 92, pp. 1392–1400.

Gazzaniga, M. S. and LeDoux, J. (1978). *The Integrated Mind*, Plenum, New York.

Georgopoulos, A. P. (1995). Motor cortex and cognitive processing, *in* M. S. Gazzaniga (ed.), *The Cognitive Neurosciences*, MIT Press, Cambridge, MA, chapter 32, pp. 507–517.

Gerstner, W. (1995). Time structure of the activity in neural network models, *Physical Review E* **51**: 738–758.

Gerstner, W. (2000). Population dynamics of spiking neurons: fast transients, asynchronous states, and locking, *Neural Computation* **12**: 43–89.

Gerstner, W. and Kistler, W. (2002). *Spiking Neuron Models: Single Neurons, Populations and Plasticity*, Cambridge University Press, Cambridge.

Gerstner, W., Kreiter, A. K., Markram, H. and Herz, A. V. (1997). Neural codes: firing rates and beyond, *Proceedings of the National Academy of Sciences USA* **94**: 12740–12741.

Giocomo, L. M. and Hasselmo, M. E. (2007). Neuromodulation by glutamate and acetylcholine can change circuit dynamics by regulating the relative influence of afferent input and excitatory feedback, *Molecular Neurobiology* **36**: 184–200.

Glimcher, P. (2003). The neurobiology of visual-saccadic decision making, *Annual Review of Neuroscience* **26**: 133–179.

Glimcher, P. (2004). *Decisions, Uncertainty, and the Brain*, MIT Press, Cambridge, MA.

Glimcher, P. (2005). Indeterminacy in brain and behavior, *Annual Review of Psychology* **56**: 25–56.

Glover, G. H. (1999). Deconvolution of impulse response in event-related BOLD fMRI, *Neuroimage* **9**: 416–429.

Gnadt, J. W. and Andersen, R. A. (1988). Memory related motor planning activity in posterior parietal cortex of macaque, *Experimental Brain Research* **70**: 216–220.

Goff, D. C. and Coyle, J. T. (2001). The emerging role of glutamate in the pathophysiology and treatment of schizophrenia, *American Journal of Psychiatry* **158**: 1367–1377.

Gold, J. and Shadlen, M. (2000). Representation of a perceptual decision in developing oculomotor commands, *Nature* **404**: 390–394.

Gold, J. and Shadlen, M. (2002). Banburismus and the brain: decoding the relationship between sensory stimuli, decisions, and reward, *Neuron* **36**: 299–308.

Gold, J. I. and Shadlen, M. N. (2007). The neural basis of decision making, *Annual Review of Neuroscience* **30**: 535–574.

Goldbach, M., Loh, M., Deco, G. and Garcia-Ojalvo, J. (2008). Neurodynamical amplification of perceptual signals via system-size resonance, *Physica D* **237**: 316–323.

Goldberg, M. E. (2000). The control of gaze, *in* E. R. Kandel, J. H. Schwartz and T. M. Jessell (eds), *Principles of Neural Science*, 4th edn, McGraw-Hill, New York, chapter 39, pp. 782–800.

Goldman-Rakic, P. (1994). Working memory dysfunction in schizophrenia, *Journal of Neuropsychology and Clinical Neuroscience* **6**: 348–357.

Goldman-Rakic, P. (1995). Cellular basis of working memory, *Neuron* **14**: 477–485.

Goldman-Rakic, P. S. (1987). Circuitry of primate prefrontal cortex and regulation of behavior by representational memory, *Handbook of Physiology, Section 1, The Nervous System*, Vol. V. Higher Functions of the Brain, Part 1, American Physiological Society, Bethesda, MD, pp. 373–417.

Goldman-Rakic, P. S. (1996). The prefrontal landscape: implications of functional architecture for understanding human mentation and the central executive, *Philosophical Transactions of the Royal Society B* **351**: 1445–1453.

Goldman-Rakic, P. S. (1999). The physiological approach: functional architecture of working memory and disordered cognition in schizophrenia, *Biological Psychiatry* **46**: 650–661.

Goldman-Rakic, P. S., Castner, S. A., Svensson, T. H., Siever, L. J. and Williams, G. V. (2004). Targeting the dopamine D1 receptor in schizophrenia: insights for cognitive dysfunction, *Psychopharmacology (Berlin)* **174**: 3–16.

Goodale, M. A. (2004). Perceiving the world and grasping it: dissociations between conscious and unconscious visual processing, *in* M. S. Gazzaniga (ed.), *The Cognitive Neurosciences III*, MIT Press, Cambridge, MA, pp. 1159–1172.

Grabenhorst, F., Rolls, E. T. and Parris, B. A. (2008). From affective value to decision-making in the prefrontal cortex, *European Journal of Neuroscience* **28**: 1930–1939.

Grady, C. L. (2008). Cognitive neuroscience of aging, *Annals of the New York Academy of Sciences* **1124**: 127–144.

Green, D. and Swets, J. (1966). *Signal Detection Theory and Psychophysics*, Wiley, New York.

Green, M. F. (1996). What are the functional consequences of neurocognitive deficits in schizophrenia?, *American Journal of Psychiatry* **153**: 321–330.

Gregory, R. L. (1970). *The Intelligent Eye*, McGraw-Hill, New York.

Gutig, R. and Sompolinsky, H. (2006). The tempotron: a neuron that learns spike timing-based decisions, *Nature Neuroscience* **9**: 420–428.

Hafner, H., Maurer, K., Loffler, W., an der Heiden, W., Hambrecht, M. and Schultze-Lutter, F. (2003). Modeling the early course of schizophrenia, *Schizophrenia Bulletin* **29**: 325–340.

Hafting, T., Fyhn, M., Molden, S., Moser, M. B. and Moser, E. I. (2005). Microstructure of a spatial map in the entorhinal cortex, *Nature* **436**: 801–806.

Hamming, R. W. (1990). *Coding and Information Theory*, 2nd edn, Prentice-Hall, Englewood Cliffs, New Jersey.

Hampshire, A. and Owen, A. M. (2006). Fractionating attentional control using event-related fMRI, *Cerebral Cortex* **16**: 1679–1689.

Hampton, A. N. and O'Doherty, J. P. (2007). Decoding the neural substrates of reward-related decision making with functional MRI, *Proceedings of the National Academy of Sciences USA* **104**: 1377–1382.

Hampton, R. R. (2001). Rhesus monkeys know when they can remember, *Proceedings of the National Academy of Sciences of the USA* **98**: 5539–5362.

Hartston, H. J. and Swerdlow, N. R. (1999). Visuospatial priming and Stroop performance in patients with obsessive compulsive disorder, *Neuropsychology* **13**: 447–457.

Hasselmo, M. E. and Bower, J. M. (1993). Acetylcholine and memory, *Trends in Neurosciences* **16**: 218–222.

Hasselmo, M. E., Schnell, E. and Barkai, E. (1995). Learning and recall at excitatory recurrent synapses and cholinergic modulation in hippocampal region CA3, *Journal of Neuroscience* **15**: 5249–5262.

Hayden, B. Y., Nair, A. C., McCoy, A. N. and Platt, M. L. (2008). Posterior cingulate cortex mediates outcome-contingent allocation of behavior, *Neuron* **60**: 19–25.

Haynes, J. D. and Rees, G. (2005a). Predicting the orientation of invisible stimuli from activity in human primary visual cortex, *Nature Neuroscience* **8**: 686–691.

Haynes, J. D. and Rees, G. (2005b). Predicting the stream of consciousness from activity in human visual cortex, *Current Biology* **15**: 1301–1307.

Haynes, J. D. and Rees, G. (2006). Decoding mental states from brain activity in humans, *Nature Reviews Neuroscience* **7**: 523–534.

Haynes, J. D., Sakai, K., Rees, G., Gilbert, S., Frith, C. and Passingham, R. E. (2007). Reading hidden intentions in the human brain, *Current Biology* **17**: 323–328.

Hebb, D. O. (1949). *The Organization of Behavior: a Neuropsychological Theory*, Wiley, New York.

Heekeren, H. R., Marrett, S., Bandettini, P. A. and Ungerleider, L. G. (2004). A general mechanism for perceptual decision-making in the human brain, *Nature* **431**: 859–862.

Heekeren, H. R., Marrett, S. and Ungerleider, L. G. (2008). The neural systems that mediate human perceptual decision making, *Nature Reviews Neuroscience* **9**: 467–479.

Heinrichs, D. W. (2005). Antidepressants and the chaotic brain: implications for the respectful treatment of selves, *Philosophy, Psychiatry, and Psychology* **12**: 215–227.

Helmholtz, H. v. (1867). *Handbuch der physiologischen Optik*, Voss, Leipzig.

Hernandez, A., Zainos, A. and Romo, R. (2002). Temporal evolution of a decision-making process in medial premotor cortex, *Neuron* **33**: 959–972.

Hertz, J. A., Krogh, A. and Palmer, R. G. (1991). *Introduction to the Theory of Neural Computation*, Addison-Wesley, Wokingham, UK.

Hestrin, S., Sah, P. and Nicoll, R. (1990). Mechanisms generating the time course of dual component excitatory synaptic currents recorded in hippocampal slices, *Neuron* **5**: 247–253.

Heyes, C. (2008). Beast machines? Questions of animal consciousness, *in* L. Weiskrantz and M. Davies (eds), *Frontiers of Consciousness*, Oxford University Press, Oxford, chapter 9, pp. 259–274.

Hodgkin, A. L. and Huxley, A. F. (1952). A quantitative description of membrane current and its application to conduction and excitation in nerve, *Journal of Physiology* **117**: 500–544.

Hoffman, R. E. and McGlashan, T. H. (2001). Neural network models of schizophrenia, *Neuroscientist* **7**: 441–454.

Hoge, J. and Kesner, R. P. (2007). Role of ca3 and ca1 subregions of the dorsal hippocampus on temporal processing of objects, *Neurobiology of Learning and Memory* **88**: 225–231.

Hopfield, J. J. (1982). Neural networks and physical systems with emergent collective computational abilities, *Proceedings of the National Academy of Sciences USA* **79**: 2554–2558.

Hopfield, J. J. (1984). Neurons with graded response have collective computational properties like those of two-state neurons, *Proceedings of the National Academy of Sciences USA* **81**: 3088–3092.

Hopfield, J. J. and Herz, A. V. (1995). Rapid local synchronization of action potentials: toward computation with coupled integrate-and-fire neurons, *Proceedings of the National Academy of Sciences USA* **92**: 6655–6662.

Hornak, J., Rolls, E. T. and Wade, D. (1996). Face and voice expression identification in patients with emotional and behavioural changes following ventral frontal lobe damage, *Neuropsychologia* **34**: 247–261.

Hornak, J., Bramham, J., Rolls, E. T., Morris, R. G., O'Doherty, J., Bullock, P. R. and Polkey, C. E. (2003). Changes in emotion after circumscribed surgical lesions of the orbitofrontal and cingulate cortices, *Brain* **126**: 1691–1712.

Hornak, J., O'Doherty, J., Bramham, J., Rolls, E. T., Morris, R. G., Bullock, P. R. and Polkey, C. E. (2004). Reward-related reversal learning after surgical excisions in orbitofrontal and dorsolateral prefrontal cortex in humans, *Journal of Cognitive Neuroscience* **16**: 463–478.

Horne, J. (2006). *Sleepfaring : a journey through the science of sleep*, Oxford University Press, Oxford.

Horwitz, B. and Tagamets, M.-A. (1999). Predicting human functional maps with neural net modeling, *Human Brain Mapping* **8**: 137–142.

Horwitz, B., Tagamets, M.-A. and McIntosh, A. R. (1999). Neural modeling, functional brain imaging, and cognition, *Trends in Cognitive Sciences* **3**: 85–122.

Hubel, D. H. and Wiesel, T. N. (1962). Receptive fields, binocular interaction, and functional architecture in the cat's visual cortex, *Journal of Physiology* **160**: 106–154.

Hubel, D. H. and Wiesel, T. N. (1968). Receptive fields and functional architecture of monkey striate cortex, *Journal of Physiology, London* **195**: 215–243.

Ingvar, D. H. and Franzen, G. (1974). Abnormalities of cerebral blood flow distribution in patients with chronic schizophrenia, *Acta Psychiatrica Scandinavica* **50**: 425–462.

Insabato, A., Pannunzi, M., Rolls, E. T. and Deco, G. (2010). Confidence-related decision-making.

Ito, M. (1984). *The Cerebellum and Neural Control*, Raven Press, New York.

Ito, M. (1989). Long-term depression, *Annual Review of Neuroscience* **12**: 85–102.

Ito, M. (1993a). Cerebellar mechanisms of long-term depression, *in* M. Baudry, R. F. Thompson and J. L. Davis (eds), *Synaptic Plasticity: Molecular, Cellular and Functional Aspects*, MIT Press, Cambridge, MA, chapter 6, pp. 117–128.

Ito, M. (1993b). Synaptic plasticity in the cerebellar cortex and its role in motor learning, *Canadian Journal of Neurological Science* **Suppl. 3**: S70–S74.

Itti, L. and Koch, C. (2001). Computational modelling of visual attention, *Nature Reviews Neuroscience* **2**: 194–203.

Jackendoff, R. (2002). *Foundations of Language*, Oxford University Press, Oxford.

Jackson, B. S. (2004). Including long-range dependence in integrate-and-fire models of the high interspike-interval variability of cortical neurons, *Neural Computation* **16**: 2125–2195.

Jahr, C. and Stevens, C. (1990). Voltage dependence of NMDA-activated macroscopic conductances predicted by single-channel kinetics, *Journal of Neuroscience* **10**: 3178–3182.

Jirsa, V. K. (2004). Connectivity and dynamics of neural information processing, *Neuroinformatics* **2**: 183–204.

Jirsa, V. K. and McIntosh, A. R. (eds) (2007). *Handbook of Brain Activity*, Springer, Berlin.

Johansen-Berg, H., Gutman, D. A., Behrens, T. E., Matthews, P. M., Rushworth, M. F., Katz, E., Lozano, A. M. and Mayberg, H. S. (2008). Anatomical connectivity of the subgenual cingulate region targeted with deep brain stimulation for treatment-resistant depression, *Cerebral Cortex* **18**: 1374–1383.

Johnson, R. and Wichern, D. (1998). *Principal Components. In: Applied Multivariate Statistical Analysis*, Prentice Hall, New Jersey.

Jones, E. G. and Peters, A. (eds) (1984). *Cerebral Cortex, Functional Properties of Cortical Cells*, Vol. 2, Plenum, New York.

Jonsson, F. U., Olsson, H. and Olsson, M. J. (2005). Odor emotionality affects the confidence in odor naming, *Chemical Senses* **30**: 29–35.

Kable, J. W. and Glimcher, P. W. (2007). The neural correlates of subjective value during intertemporal choice, *Nature Neuroscience* **10**: 1625–1633.

Kacelnik, A. and Brito e Abreu, F. (1998). Risky choice and Weber's Law, *Journal of Theoretical Biology* **194**: 289–298.

Kadohisa, M., Rolls, E. T. and Verhagen, J. V. (2004). Orbitofrontal cortex neuronal representation of temperature and capsaicin in the mouth, *Neuroscience* **127**: 207–221.

Kadohisa, M., Rolls, E. T. and Verhagen, J. V. (2005a). Neuronal representations of stimuli in the mouth: the primate insular taste cortex, orbitofrontal cortex, and amygdala, *Chemical Senses* **30**: 401–419.

Kadohisa, M., Rolls, E. T. and Verhagen, J. V. (2005b). The primate amygdala: neuronal representations of the viscosity, fat texture, grittiness and taste of foods, *Neuroscience* **132**: 33–48.

Kahneman, D. (1973). *Attention and Effort*, Prentice-Hall, Englewood Cliffs, NJ.

Kahneman, D., Wakker, P. P. and Sarin, R. (1997). Back to Bentham? - Explorations of experienced utility, *Quarterly Journal of Economics* **112**: 375–405.

Kandel, E. R., Schwartz, J. H. and Jessell, T. H. (eds) (2000). *Principles of Neural Science*, 4th (5th in 2010) edn, Elsevier, Amsterdam.

Kanter, I. and Sompolinsky, H. (1987). Associative recall of memories without errors, *Physical Review A* **35**: 380–392.

Karno, M., Golding, J. M., Sorenson, S. B. and Burnam, M. A. (1988). The epidemiology of obsessive-compulsive disorder in five US communities, *Archives of General Psychiatry* **45**: 1094–1099.

Kelly, K. M., Nadon, N. L., Morrison, J. H., Thibault, O., Barnes, C. A. and Blalock, E. M. (2006). The neurobiology of aging, *Epilepsy Research* **68, Supplement** 1: S5–S20.

Kepecs, A., Uchida, N., Zariwala, H. A. and Mainen, Z. F. (2008). Neural correlates, computation and behavioural impact of decision confidence, *Nature* **455**: 227–231.

Kesner, R. P. and Rolls, E. T. (2001). Role of long term synaptic modification in short term memory, *Hippocampus* **11**: 240–250.

Khalil, H. (1996). *Nonlinear Systems*, Prentice Hall, Upper Saddle River, NJ.

Kiani, R. and Shadlen, M. N. (2009). Representation of confidence associated with a decision by neurons in the parietal cortex, *Science* **324**: 759–764.

Kievit, J. and Kuypers, H. G. J. M. (1975). Subcortical afferents to the frontal lobe in the rhesus monkey studied by means of retrograde horseradish peroxidase transport, *Brain Research* **85**: 261–266.

Kim, J. and Shadlen, M. (1999). Neural correlates of a decision in the dorsolateral prefrontal cortex of the macaque, *Nature Neuroscience* **2**: 176–185.

Kircher, T. T. and Thienel, R. (2005). Functional brain imaging of symptoms and cognition in schizophrenia, *Progress in Brain Research* **150**: 299–308.

Knight, B. (2000). Dynamics of encoding in neuron populations: some general mathematical features, *Neural Computation* **12**(3): 473–518.

Knight, B. W., Manin, D. and Sirovich, L. (1996). Dynamical models of interacting neuron populations, *in* E. C. Gerf (ed.), *Symposium on Robotics and Cybernetics: Computational Engineering in Systems Applications*, Cite Scientifique, Lille, France.

Koch, C. (1999). *Biophysics of Computation*, Oxford University Press, Oxford.

Koch, K. W. and Fuster, J. M. (1989). Unit activity in monkey parietal cortex related to haptic perception and temporary memory, *Experimental Brain Research* **76**: 292–306.

Kohonen, T. (1977). *Associative Memory: A System Theoretical Approach*, Springer, New York.

Kohonen, T. (1989). *Self-Organization and Associative Memory*, 3rd (1984, 1st edn; 1988, 2nd edn) edn, Springer-Verlag, Berlin.

Kolb, B. and Whishaw, I. Q. (2008). *Fundamentals of Human Neuropsychology*, 6th edn, Worth, New York.

Komorowski, R. W., Manns, J. R. and Eichenbaum, H. (2009). Robust conjunctive item-place coding by hippocampal neurons parallels learning what happens where, *Journal of Neuroscience* **29**: 9918–9929.

Kosslyn, S. M. (1994). *Image and Brain: The Resolution of the Imagery Debate*, MIT Press, Cambridge, MA.

Kramers, H. (1940). Brownian motion in a field of force and the diffusion model of chemical reactions, *Physicas* **7**: 284–304.

Krebs, J. R. and Davies, N. B. (1991). *Behavioural Ecology*, 3rd edn, Blackwell, Oxford.

Kringelbach, M. L. and Rolls, E. T. (2004). The functional neuroanatomy of the human orbitofrontal cortex: evidence from neuroimaging and neuropsychology, *Progress in Neurobiology* **72**: 341–372.

Kropff, E. and Treves, A. (2008). The emergence of grid cells: Intelligent design or just adaptation?, *Hippocampus* **18**: 1256–1269.

Kuhn, R. (1990). Statistical mechanics of neural networks near saturation, *in* L. Garrido (ed.), *Statistical Mechanics of Neural Networks*, Springer-Verlag, Berlin.

Kuhn, R., Bos, S. and van Hemmen, J. L. (1991). Statistical mechanics for networks of graded response neurons, *Physical Review A* **243**: 2084–2087.

La Camera, G., Rauch, A., Luescher, H., Senn, W. and Fusi, S. (2004). Minimal models of adapted neuronal response to in vivo-like input currents, *Neural Computation* **16**: 2101–2124.

Lagarias, J., Reeds, J., Wright, M. and Wright, P. (1998). Convergence properties of the Nelder-Mead simplex algorithm in low dimensions, *SIAM Journal on Optimization* **9**: 112–147.

Lamme, V. A. F. (1995). The neurophysiology of figure-ground segregation in primary visual cortex, *Journal of Neuroscience* **15**: 1605–1615.

Lansky, P., Sacerdote, L. and Tomassetti, F. (1995). On the comparison of Feller and Ornstein-Uhlenbeck models for neural activity, *Biological Cybernetics* **73**: 457–465.

Lanthorn, T., Storn, J. and Andersen, P. (1984). Current-to-frequency transduction in CA1 hippocampal pyramidal cells: slow prepotentials dominate the primary range firing, *Experimental Brain Research* **53**: 431–443.

Lau, H. C., Rogers, R. D. and Passingham, R. E. (2006). On measuring the perceived onsets of spontaneous actions, *Journal of Neuroscience* **26**: 7265–7271.

Law-Tho, D., Hirsch, J. and Crepel, F. (1994). Dopamine modulation of synaptic transmission in rat prefrontal cortex: An in vitro electrophysiological study, *Neuroscience Research* **21**: 151–160.

LeDoux, J. (2008). Emotional coloration of consciousness: how feelings come about, *in* L. Weiskrantz and M. Davies (eds), *Frontiers of Consciousness*, Oxford University Press, Oxford, pp. 69–130.

LeDoux, J. E. (1995). Emotion: clues from the brain, *Annual Review of Psychology* **46**: 209–235.

Lee, T. S., Mumford, D., Romero, R. D. and Lamme, V. A. F. (1998). The role of primary visual cortex in higher level vision, *Vision Research* **38**: 2429–2454.

Lee, T. S., Yang, C. Y., Romero, R. D. and Mumford, D. (2002). Neural activity in early visual cortex reflects behavioral experience and higher order perceptual saliency, *Nature Neuroscience* **5**: 589–597.

Lehky, S. R., Sejnowski, T. J. and Desimone, R. (2005). Selectivity and sparseness in the responses of striate complex cells, *Vision Research* **45**: 57–73.

Leuner, K. and Muller, W. E. (2006). The complexity of the dopaminergic synapses and their modulation by antipsychotics, *Pharmacopsychiatry* **39 Suppl 1**: S15–20.

Leung, H., Gore, J. and Goldman-Rakic, P. (2002). Sustained mnemonic response in the human middle frontal gyrus during on-line storage of spatial memoranda, *Journal of Cognitive Neuroscience* **14**: 659–671.

Levitt, J. B., Lund, J. S. and Yoshioka, T. (1996). Anatomical substrates for early stages in cortical processing of visual information in the macaque monkey, *Behavioural Brain Research* **76**: 5–19.

Levy, W. B. and Baxter, R. A. (1996). Energy efficient neural codes, *Neural Computation* **8**: 531–543.

Lewis, D. A., Hashimoto, T. and Volk, D. W. (2005). Cortical inhibitory neurons and schizophrenia, *Nature Reviews Neuroscience* **6**: 312–324.

Lewis, M. D. (2005). Bridging emotion theory and neurobiology through dynamic systems modeling, *Behavioral and Brain Sciences* **28**: 169–245.

Liddle, P. F. (1987). The symptoms of chronic schizophrenia: a re-examination of the positive-negative dichotomy, *British Journal of Psychiatry* **151**: 145–151.

Lieberman, J. A., Perkins, D., Belger, A., Chakos, M., Jarskog, F., Boteva, K. and Gilmore, J. (2001). The early stages of schizophrenia: speculations on pathogenesis, pathophysiology, and therapeutic approaches, *Biological Psychiatry* **50**: 884–897.

Lisman, J. E. and Buzsaki, G. (2008). A neural coding scheme formed by the combined function of gamma and theta oscillations, *Schizophrenia Bulletin* **34**: 974–980.

Lisman, J. E. and Redish, A. D. (2009). Prediction, sequences and the hippocampus, *Philosophical Transactions of the Royal Society of London B Biological Sciences* **364**: 1193–1201.

List, C., Elsholtz, C. and Seeley, T. D. (2009). Independence and interdependence in collective decision making: an agent-based model of nest-site choice by honeybee swarms, *Philosophical Transactions of the Royal Society of London B Biological Sciences* **364**: 755–762.

Little, W. A. (1974). The existence of persistent states in the brain, *Mathematical Bioscience* **19**: 101–120.

Liu, Y. and Wang, X.-J. (2001). Spike-frequency adaptation of a generalized leaky integrate-and-fire model neuron, *Journal of Computational Neuroscience* **10**: 25–45.

Liu, Y. H. and Wang, X.-J. (2008). A common cortical circuit mechanism for perceptual categorical discrimination and veridical judgment, *PLoS Computational Biology* p. e1000253.

Logothetis, N. K. (2008). What we can do and what we cannot do with fMRI, *Nature* **453**: 869–878.

Logothetis, N. K., Pauls, J., Augath, M., Trinath, T. and Oeltermann, A. (2001). Neurophysiological investigation of the basis of the fMRI signal, *Nature* **412**: 150–157.

Loh, M., Rolls, E. T. and Deco, G. (2007a). A dynamical systems hypothesis of schizophrenia, *PLoS Computational Biology* **3**: e228. doi:10.1371/journal.pcbi.0030228.

Loh, M., Rolls, E. T. and Deco, G. (2007b). Statistical fluctuations in attractor networks related to schizophrenia, *Pharmacopsychiatry* **40**: S78–84.

Longtin, A. (1993). Stochastic resonance in neuron models, *Journal of Statistical Physics* **70**(1/2): 309–327.

Lorente de No, R. (1933). Vestibulo-ocular reflex arc., *Archives of Neurology Psychiatry* **30**: 245–291.

Luce, R. D. (1986). *Response Time: Their Role in Inferring Elementary Mental Organization*, Oxford University Press, New York.

Luck, S. J., Chelazzi, L., Hillyard, S. A. and Desimone, R. (1997). Neural mechanisms of spatial selective attention in areas V1, V2, and V4 of macaque visual cortex, *Journal of Neurophysiology* **77**: 24–42.

Lund, J. S. (1984). Spiny stellate neurons, *in* A. Peters and E. Jones (eds), *Cerebral Cortex, Vol. 1, Cellular Components of the Cerebral Cortex*, Plenum, New York, chapter 7, pp. 255–308.

Lycan, W. G. (1997). Consciousness as internal monitoring, *in* N. Block, O. Flanagan and G. Guzeldere (eds), *The Nature of Consciousness: Philosophical Debates*, MIT Press, Cambridge, MA, pp. 755–771.

Lynch, G. and Gall, C. M. (2006). AMPAkines and the threefold path to cognitive enhancement, *Trends in Neuroscience* **29**: 554–562.

Lynch, M. A. (2004). Long-term potentiation and memory, *Physiological Reviews* **84**: 87–136.

Ma, W. J., Beck, J. M., Latham, P. E. and Pouget, A. (2006). Bayesian inference with probabilistic population codes, *Nature Neuroscience* **9**: 1432–1438.

Ma, W. J., Beck, J. M. and Pouget, A. (2008). Spiking networks for Bayesian inference and choice, *Current Opinion in Neurobiology* **18**: 217–222.

MacDonald, C. J. and Eichenbaum, H. (2009). Hippocampal neurons disambiguate overlapping sequences of non-spatial events, *Society for Neuroscience Abstracts* p. 101.21.

MacGregor, R. J. (1987). *Neural and Brain Modelling*, Academic Press, San Diego, CA.

Machens, C., Romo, R. and Brody, C. (2005). Flexible control of mutual inhibition: a neural model of two-interval discrimination, *Science* **307**: 1121–1124.

Machens, C. K. and Brody, C. D. (2008). Design of continuous attractor networks with monotonic tuning using a symmetry principle, *Neural Computation* **20**: 452–485.

Maier, A., Logothetis, N. K. and Leopold, D. A. (2005). Global competition dictates local suppression in pattern rivalry, *Journal of Vision* **5**: 668–677.

Malakhov, A. and Pankratov, A. (1996). Exact solution of kramers' problem for piecewise parabolic potential profiles, *Physica A* **229**: 109–126.

Malhotra, A. K., Pinals, D. A., Weingartner, H., Sirocco, K., Missar, C. D., Pickar, D. and Breier, A. (1996). NMDA receptor function and human cognition: the effects of ketamine in healthy volunteers, *Neuropsychopharmacology* **14**: 301–307.

Malsburg, C. v. d. (1999). The what and why of binding: the modeler's perspective, *Neuron* **24**: 95–104.

Mangia, S., Giove, F., Tkac, I., Logothetis, N. K., Henry, P. G., Olman, C. A., Maraviglia, B., Di Salle, F. and Ugurbil, K. (2009). Metabolic and hemodynamic events after changes in neuronal activity: current hypotheses, theoretical predictions and in vivo NMR experimental findings, *Journal of Cerebral Blood Flow and Metabolism* **29**: 441–463.

Manwani, A. and Koch, C. (2001). Detecting and estimating signals over noisy and unreliable synapses: information-theoretic analysis, *Neural Computation* **13**: 1–33.

Markram, H. and Siegel, M. (1992). The inositol 1,4,5-triphosphate pathway mediates cholinergic potentiation of rat hippocampal neuronal responses to NMDA, *Journal of Physiology* **447**: 513–533.

Markram, H. and Tsodyks, M. (1996). Redistribution of synaptic efficacy between neocortical pyramidal neurons, *Nature* **382**: 807–810.

Markram, H., Lubke, J., Frotscher, M. and Sakmann, B. (1997). Regulation of synaptic efficacy by coincidence of postsynaptic APs and EPSPs, *Science* **275**: 213–215.

Markram, H., Pikus, D., Gupta, A. and Tsodyks, M. (1998). Information processing with frequency-dependent synaptic connections, *Neuropharmacology* **37**: 489–500.

Marshall, J. A. R., Bogacz, R., Planqué, R., Kovacs, T. and Franks, N. R. (2009). On optimal decision making in brains and social insect colonies, *Journal of the Royal Society: Interface* **6**: 1065–1074.

Marti, D., Deco, G., Del Giudice, P. and Mattia, M. (2006). Reward-biased probabilistic decision-making: mean-field predictions and spiking simulations, *Neurocomputing* **39**: 1175–1178.

Marti, D., Deco, G., Mattia, M., Gigante, G. and Del Giudice, P. (2008). A fluctuation-driven mechanism for slow decision processes in reverberant networks, *PLoS ONE* **3**: e2534. doi:10.1371/journal.pone.0002534.

Martin, K. A. C. (1984). Neuronal circuits in cat striate cortex, *in* E. Jones and A. Peters (eds), *Cerebral Cortex, Vol. 2, Functional Properties of Cortical Cells*, Plenum, New York, chapter 9, pp. 241–284.

Martin, S. J., Grimwood, P. D. and Morris, R. G. (2000). Synaptic plasticity and memory: an evaluation of the hypothesis, *Annual Review of Neuroscience* **23**: 649–711.

Martinez-Trujillo, J. and Treue, S. (2002). Attentional modulation strength in cortical area MT depends on stimulus contrast, *Neuron* **35**: 365–370.

Mascaro, M. and Amit, D. J. (1999). Effective neural response function for collective population states, *Network* **10**: 351–373.

Mason, A. and Larkman, A. (1990). Correlations between morphology and electrophysiology of pyramidal neurones in slices of rat visual cortex. I. Electrophysiology, *Journal of Neuroscience* **10**: 1415–1428.

Masuda, N. and Aihara, K. (2003). Ergodicity of spike trains: when does trial averaging make sense?, *Neural Computation* **15**: 1341–1372.

Mattia, M. and Del Giudice, P. (2002). Population dynamics of interacting spiking neurons, *Physical Review E* **66**: 051917.

Mattia, M. and Del Giudice, P. (2004). Finite-size dynamics of inhibitory and excitatory interacting spiking neurons, *Physical Review E* **70**: 052903.

Maunsell, J. H. R. (1995). The brain's visual world: representation of visual targets in cerebral cortex, *Science* **270**: 764–769.

Maurer, A. P. and McNaughton, B. L. (2007). Network and intrinsic cellular mechanisms underlying theta phase precession of hippocampal neurons, *Trends in Neuroscience* **30**: 325–333.

Maynard Smith, J. (1982). *Evolution and the Theory of Games*, Cambridge University Press, Cambridge.

Maynard Smith, J. (1984). Game theory and the evolution of behaviour, *Behavioral and Brain Sciences* **7**: 95–125.

McAdams, C. and Maunsell, J. H. R. (1999). Effects of attention on orientation-tuning functions of single neurons in macaque cortical area V4, *Journal of Neuroscience* **19**: 431–441.

McClelland, J. L. and Rumelhart, D. E. (1988). *Explorations in Parallel Distributed Processing*, MIT Press, Cambridge, MA.

McCormick, D., Connors, B., Lighthall, J. and Prince, D. (1985). Comparative electrophysiology of pyramidal and sparsely spiny stellate neurons in the neocortex, *Journal of Neurophysiology* **54**: 782–806.

McCoy, A. N. and Platt, M. L. (2005). Risk-sensitive neurons in macaque posterior cingulate cortex, *Nature Neuroscience* **8**: 1220–1227.

McNaughton, B. L. and Morris, R. G. M. (1987). Hippocampal synaptic enhancement and information storage within a distributed memory system, *Trends in Neuroscience* **10**: 408–415.

Mel, B. W. and Fiser, J. (2000). Minimizing binding errors using learned conjunctive features, *Neural Computation* **12**: 731–762.

Mel, B. W., Ruderman, D. L. and Archie, K. A. (1998). Translation-invariant orientation tuning in visual "complex" cells could derive from intradendritic computations, *Journal of Neuroscience* **18**(11): 4325–4334.

Menzies, L., Chamberlain, S. R., Laird, A. R., Thelen, S. M., Sahakian, B. J. and Bullmore, E. T. (2008). Integrating evidence from neuroimaging and neuropsychological studies of obsessive-compulsive disorder: The orbitofronto-striatal model revisited, *Neuroscience and Biobehavioral Reviews* **32**: 525–549.

Mesulam, M.-M. (1990). Human brain cholinergic pathways, *Progress in Brain Research* **84**: 231–241.

Miller, E. K. and Desimone, R. (1994). Parallel neuronal mechanisms for short-term memory, *Science* **263**: 520–522.

Miller, E. K., Gochin, P. M. and Gross, C. G. (1993a). Suppression of visual responses of neurons in inferior temporal cortex of the awake macaque by addition of a second stimulus, *Brain Research* **616**: 25–29.

Miller, E. K., Li, L. and Desimone, R. (1993b). Activity of neurons in anterior inferior temporal cortex during a short-term memory task, *Journal of Neuroscience* **13**: 1460–1478.

Miller, E. K., Erickson, C. and Desimone, R. (1996). Neural mechanisms of visual working memory in prefrontal cortex of the macaque, *Journal of Neuroscience* **16**: 5154–5167.

Miller, G. A. (1956). The magic number seven, plus or minus two: some limits on our capacity for the processing of information, *Psychological Review* **63**: 81–93.

Miller, G. F. (2000). *The Mating Mind*, Heinemann, London.

Minai, A. A. and Levy, W. B. (1993). Sequence learning in a single trial, *International Neural Network Society World Congress of Neural Networks* **2**: 505–508.

Miyamoto, S., Duncan, G. E., Marx, C. E. and Lieberman, J. A. (2005). Treatments for schizophrenia: a critical review of pharmacology and mechanisms of action of antipsychotic drugs, *Molecular Psychiatry* **10**: 79–104.

Miyashita, Y. (1993). Inferior temporal cortex: where visual perception meets memory, *Annual Review of Neuroscience* **16**: 245–263.

Miyashita, Y. and Chang, H. S. (1988). Neuronal correlate of pictorial short-term memory in the primate temporal cortex, *Nature* **331**: 68–70.

Moller, H. J. (2005). Antipsychotic and antidepressive effects of second generation antipsychotics: two different pharmacological mechanisms?, *European Archives of Psychiatry and Clinical Neuroscience* **255**: 190–201.

Mongillo, G., Barak, O. and Tsodyks, M. (2008). Synaptic theory of working memory, *Science* **319**: 1543–1546.

Mora, F., Rolls, E. T. and Burton, M. J. (1976). Modulation during learning of the responses of neurones in the lateral hypothalamus to the sight of food, *Experimental Neurology* **53**: 508–519.

Moran, J. and Desimone, R. (1985). Selective attention gates visual processing in the extrastriate cortex, *Science* **229**: 782–784.

Moreno-Botel, R., Rinzel, J. and Rubin, N. (2007). Noise-induced alternations in an attractor network model of perceptual bistability, *Journal of Neurophysiology* **98**: 1125–1139.

Morris, R. G. M. (1989). Does synaptic plasticity play a role in information storage in the vertebrate brain?, *in* R. G. M. Morris (ed.), *Parallel Distributed Processing: Implications for Psychology and Neurobiology*, Oxford University Press, Oxford, chapter 11, pp. 248–285.

Morris, R. G. M. (2003). Long-term potentiation and memory, *Philosophical Transactions of the Royal Society of London B* **358**: 643–647.

Morris, R. G. M. (2006). Elements of a neurobiological theory of hippocampal function: the role of synaptic plasticity, synaptic tagging and schemas, *European Journal of Neuroscience* **23**: 2829–2846.

Morrow, L. and Ratcliff, G. (1988). The disengagement of covert attention and the neglect syndrome, *Psychobiology* **16**: 261–269.

Motter, B. (1994). Neural correlates of attentive selection for colours or luminance in extrastriate area V4, *Journal of Neuroscience* **14**: 2178–2189.

Motter, B. C. (1993). Focal attention produces spatially selective processing in visual cortical areas V1, V2, and V4 in the presence of competing stimuli, *Journal of Neurophysiology* **70**: 909–919.

Mountcastle, V. B. (1957). Modality and topographic properties of single neurons of cat's somatosensory cortex, *Journal of Neurophysiology* **20**: 408–434.

Mountcastle, V. B. (1984). Central nervous mechanisms in mechanoreceptive sensibility, *in* I. Darian-Smith (ed.), *Handbook of Physiology, Section 1: The Nervous System, Vol III, Sensory Processes, Part 2*, American Physiological Society, Bethesda, MD, pp. 789–878.

Mountcastle, V. B., Talbot, W. H., Darian-Smith, I. and Kornhuber, H. H. (1967). Neural basis of the sense of fluttervibration, *Science* **155**: 597–600.

Mountcastle, V. B., Talbot, W., Sakata, H. and Hyvarinen, J. (1969). Cortical neuronal mechanisms in flutter-vibration studied in unanesthetized monkeys. Neuronal periodicity and frequency discrimination, *Journal of Neurophysiology* **32**: 452–484.

Mueser, K. T. and McGurk, S. R. (2004). Schizophrenia, *Lancet* **363**: 2063–2072.

Muir, J. L., Everitt, B. J. and Robbins, T. W. (1994). AMPA-induced excitotoxic lesions of the basal forebrain: a significant role for the cortical cholinergic system in attentional function, *Journal of Neuroscience* **14**: 2313–2326.

Nagahama, Y., Okada, T., Katsumi, Y., Hayashi, T., Yamauchi, H., Oyanagi, C., Konishi, J., Fukuyama, H. and Shibasaki, H. (2001). Dissociable mechanisms of attentional control within the human prefrontal cortex, *Cerebral Cortex* **11**: 85–92.

Neisser, U. (1967). *Cognitive Psychology*, Appleton-Century-Crofts, New York.

Nelder, J. and Mead, R. (1965). A simplex method for function minimization, *The Computer Journal* **7**: 308.

Newcomer, J. W., Farber, N. B., Jevtovic-Todorovic, V., Selke, G., Melson, A. K., Hershey, T., Craft, S. and Olney, J. W. (1999). Ketamine-induced NMDA receptor hypofunction as a model of memory impairment and psychosis, *Neuropsychopharmacology* **20**: 106–118.

Newsome, W. T., Britten, K. H. and Movshon, J. A. (1989). Neuronal correlates of a perceptual decision, *Nature* **341**: 52–54.

Nicoll, R. A. (1988). The coupling of neurotransmitter receptors to ion channels in the brain, *Science* **241**: 545–551.

Nicoll, R. A. and Malenka, R. C. (1995). Contrasting properties of two forms of long-term potentiation in the hippocampus, *Nature* **377**: 115–118.

O'Doherty, J., Kringelbach, M. L., Rolls, E. T., Hornak, J. and Andrews, C. (2001). Abstract reward and punishment representations in the human orbitofrontal cortex, *Nature Neuroscience* **4**: 95–102.

O'Kane, D. and Treves, A. (1992). Why the simplest notion of neocortex as an autoassociative memory would not work, *Network* **3**: 379–384.

O'Keefe, J. and Recce, M. L. (1993). Phase relationship between hippocampal place units and the EEG theta rhythm, *Hippocampus* **3**: 317–330.

Okubo, Y., Suhara, T., Sudo, Y. and Toru, M. (1997a). Possible role of dopamine D1 receptors in schizophrenia, *Molecular Psychiatry* **2**: 291–292.

Okubo, Y., Suhara, T., Suzuki, K., Kobayashi, K., Inoue, O., Terasaki, O., Someya, Y., Sassa, T., Sudo, Y., Matsushima, E., Iyo, M., Tateno, Y. and Toru, M. (1997b). Decreased prefrontal dopamine D1 receptors in schizophrenia revealed by PET, *Nature* **385**: 634–636.

Olshausen, B. A. and Field, D. J. (1997). Sparse coding with an incomplete basis set: a strategy employed by V1, *Vision Research* **37**: 3311–3325.

Olshausen, B. A. and Field, D. J. (2004). Sparse coding of sensory inputs, *Current Opinion in Neurobiology* **14**: 481–487.

Omurtag, A., Knight, B. and Sirovich, L. (2000). On the simulation of large populations of neurons, *Journal of Computational Neuroscience* **8**: 51–53.

O'Neill, M. J. and Dix, S. (2007). AMPA receptor potentiators as cognitive enhancers, *IDrugs* **10**: 185–192.

O'Reilly, R. C. (2006). Biologically based computational models of high-level cognition, *Science* **314**: 91–94.

Pantelis, C., Barber, F. Z., Barnes, T. R., Nelson, H. E., Owen, A. M. and Robbins, T. W. (1999). Comparison of set-shifting ability in patients with chronic schizophrenia and frontal lobe damage, *Schizophrenia Research* **37**: 251–270.

Panzeri, S. and Treves, A. (1996). Analytical estimates of limited sampling biases in different information measures, *Network* **7**: 87–107.

Panzeri, S., Rolls, E. T., Battaglia, F. and Lavis, R. (2001). Speed of feedforward and recurrent processing in multilayer networks of integrate-and-fire neurons, *Network: Computation in Neural Systems* **12**: 423–440.

Peled, A. (2004). From plasticity to complexity: a new diagnostic method for psychiatry, *Medical Hypotheses* **63**: 110–114.

Penades, R., Catalan, R., Andres, S., Salamero, M. and Gasto, C. (2005). Executive function and nonverbal memory in obsessive-compulsive disorder, *Psychiatry Research* **133**: 81–90.

Penades, R., Catalan, R., Rubia, K., Andres, S., Salamero, M. and Gasto, C. (2007). Impaired response inhibition in obsessive compulsive disorder, *European Psychiatry* **22**: 404–410.

Personnaz, L., Guyon, I. and Dreyfus, G. (1985). Information storage and retrieval in spin-glass-like neural networks, *Journal de Physique Lettres (Paris)* **46**: 359–365.

Pessoa, L. and Padmala, S. (2005). Quantitative prediction of perceptual decisions during near-threshold fear detection, *Proceedings of the National Academy of Sciences USA* **102**: 5612–5617.

Peters, A. (1984a). Bipolar cells, *in* A. Peters and E. G. Jones (eds), *Cerebral Cortex, Vol. 1, Cellular Components of the Cerebral Cortex*, Plenum, New York, chapter 11, pp. 381–407.

Peters, A. (1984b). Chandelier cells, *in* A. Peters and E. G. Jones (eds), *Cerebral Cortex, Vol. 1, Cellular Components of the Cerebral Cortex*, Plenum, New York, chapter 10, pp. 361–380.

Peters, A. and Jones, E. G. (eds) (1984). *Cerebral Cortex, Vol. 1, Cellular Components of the Cerebral Cortex*, Plenum, New York.

Peters, A. and Regidor, J. (1981). A reassessment of the forms of nonpyramidal neurons in area 17 of the cat visual cortex, *Journal of Comparative Neurology* **203**: 685–716.

Peters, A. and Saint Marie, R. L. (1984). Smooth and sparsely spinous nonpyramidal cells forming local axonal plexuses, *in* A. Peters and E. G. Jones (eds), *Cerebral Cortex, Vol. 1, Cellular Components of the Cerebral Cortex*, New York, Plenum, chapter 13, pp. 419–445.

Pittenger, C., Krystal, J. H. and Coric, V. (2006). Glutamate-modulating drugs as novel pharmacotherapeutic agents in the treatment of obsessive-compulsive disorder, *NeuroRx* **3**(1): 69–81.

Posner, M., Walker, J., Friedrich, F. and Rafal, B. (1984). Effects of parietal injury on covert orienting of attention, *Journal of Neuroscience* **4**: 1863–1874.

Posner, M., Walker, J., Friedrich, F. and Rafal, R. (1987). How do the parietal lobes direct covert attention?, *Neuropsychologia* **25**: 135–146.

Postle, B. R. and D'Esposito, M. (1999). "What" – then – "Where" in visual working memory: An event-related fMRI study, *Journal of Cognitive Neuroscience* **11**: 585–597.

Postle, B. R. and D'Esposito, M. (2000). Evaluating models of the topographical organization of working memory function in frontal cortex with event-related fMRI, *Psychobiology* **28**: 132–145.

Powell, T. P. S. (1981). Certain aspects of the intrinsic organisation of the cerebral cortex, *in* O. Pompeiano and C. Ajmone Marsan (eds), *Brain Mechanisms and Perceptual Awareness*, Raven Press, New York, pp. 1–19.

Power, J. M. and Sah, P. (2008). Competition between calcium-activated K+ channels determines cholinergic action on firing properties of basolateral amygdala projection neurons, *Journal of Neuroscience* **28**: 3209–3220.

Preuschhof, C., Heekeren, H. R., Taskin, B., Schubert, T. and Villringer, A. (2006). Neural correlates of vibrotactile working memory in the human brain, *Journal of Neuroscience* **26**: 13231–13239.

Rafal, R. and Robertson, L. (1995). The neurology of visual attention, *in* M. S. Gazzaniga (ed.), *The Cognitive Neurosciences*, MIT Press, Cambridge, MA, pp. 625–648.

Rao, S. C., Rainer, G. and Miller, E. K. (1997). Integration of what and where in the primate prefrontal cortex, *Science* **276**: 821–824.

Ratcliff, R. and Rouder, J. F. (1998). Modeling response times for two-choice decisions, *Psychological Science* **9**: 347–356.

Ratcliff, R., Zandt, T. V. and McKoon, G. (1999). Connectionist and diffusion models of reaction time, *Psychological Reviews* **106**: 261–300.

Renart, A., Parga, N. and Rolls, E. T. (1999a). Associative memory properties of multiple cortical modules, *Network* **10**: 237–255.

Renart, A., Parga, N. and Rolls, E. T. (1999b). Backprojections in the cerebral cortex: implications for memory storage, *Neural Computation* **11**: 1349–1388.

Renart, A., Parga, N. and Rolls, E. T. (2000). A recurrent model of the interaction between the prefrontal cortex and inferior temporal cortex in delay memory tasks, *in* S. Solla, T. Leen and K.-R. Mueller (eds), *Advances in Neural Information Processing Systems*, Vol. 12, MIT Press, Cambridge, MA, pp. 171–177.

Renart, A., Moreno, R., Rocha, J., Parga, N. and Rolls, E. T. (2001). A model of the IT–PF network in object working memory which includes balanced persistent activity and tuned inhibition, *Neurocomputing* **38–40**: 1525–1531.

Renart, A., Brunel, N. and Wang, X.-J. (2003). Mean field theory of irregularly spiking neuronal populations and working memory in recurrent cortical networks, *in* J. Feng (ed.), *Computational Neuroscience: A Comprehen-*

sive Approach, Chapman and Hall, Boca Raton, FL, pp. 431–490.

Reynolds, J. and Desimone, R. (1999). The role of neural mechanisms of attention in solving the binding problem, *Neuron* **24**: 19–29.

Reynolds, J. and Desimone, R. (2003). Interacting roles of attention and visual saliency in V4, *Neuron* **37**: 853–863.

Reynolds, J. H., Chelazzi, L. and Desimone, R. (1999). Competitive mechanisms subserve attention in macaque areas V2 and V4, *Journal of Neuroscience* **19**: 1736–1753.

Reynolds, J. H., Pastemak, T. and Desimone, R. (2000). Attention increases sensitivity of V4 neurons, *Neuron* **26**: 703–714.

Riani, M. and Simonotto, E. (1994). Stochastic resonance in the perceptual interpretation of ambiguous figures: A neural network model, *Physical Review Letters* **72**(19): 3120–3123.

Ricciardi, L. and Sacerdote, L. (1979). The Ornstein-Uhlenbeck process as a model for neuronal activity. I. Mean and variance of the firing time, *Biological Cybernetics* **35**: 1–9.

Rieke, F., Warland, D., de Ruyter van Steveninck, R. R. and Bialek, W. (1996). *Spikes: Exploring the Neural Code*, MIT Press, Cambridge, MA.

Riley, M. A. and Turvey, M. T. (2001). The self-organizing dynamics of intentions and actions, *American Journal of Psychology* **114**: 160–169.

Risken, H. (1996). *The Fokker–Planck Equation: Methods of Solution and Applications*, Springer Verlag, Berlin.

Robertson, R. G., Rolls, E. T., Georges-Francois, P. and Panzeri, S. (1999). Head direction cells in the primate pre-subiculum, *Hippocampus* **9**: 206–219.

Robins, L. N., Helzer, J. E., Weissman, M. M., Orvaschel, H., Gruenberg, E., Burke, J. D. J. and Regier, D. A. (1984). Lifetime prevalence of specific psychiatric disorders in three sites, *Archives of General Psychiatry* **41**: 949–958.

Roelfsema, P. R., Lamme, V. A. and Spekreijse, H. (1998). Object-based attention in the primary visual cortex of the macaque monkey, *Nature* **395**: 376–381.

Rogers, R. D., Andrews, T. C., Grasby, P. M., Brooks, D. J. and Robbins, T. W. (2000). Contrasting cortical and sub-cortical activations produced by attentional-set shifting and reversal learning in humans, *Journal of Cognitive Neuroscience* **12**: 142–162.

Roitman, J. D. and Shadlen, M. N. (2002). Response of neurons in the lateral intraparietal area during a combined visual discrimination reaction time task, *Journal of Neuroscience* **22**: 9475–9489.

Roland, P. E. and Friberg, L. (1985). Localization of cortical areas activated by thinking, *Journal of Neurophysiology* **53**: 1219–1243.

Rolls, E. T. (1975). *The Brain and Reward*, Pergamon Press, Oxford.

Rolls, E. T. (1981a). Central nervous mechanisms related to feeding and appetite, *British Medical Bulletin* **37**: 131–134.

Rolls, E. T. (1981b). Processing beyond the inferior temporal visual cortex related to feeding, learning, and striatal function, *in* Y. Katsuki, R. Norgren and M. Sato (eds), *Brain Mechanisms of Sensation*, Wiley, New York, chapter 16, pp. 241–269.

Rolls, E. T. (1982). Neuronal mechanisms underlying the formation and disconnection of associations between visual stimuli and reinforcement in primates, *in* C. D. Woody (ed.), *Conditioning: Representation of Involved Neural Functions*, Plenum, New York, pp. 363–373.

Rolls, E. T. (1986a). Neural systems involved in emotion in primates, *in* R. Plutchik and H. Kellerman (eds), *Emotion: Theory, Research, and Experience*, Vol. 3: Biological Foundations of Emotion, Academic Press, New York, chapter 5, pp. 125–143.

Rolls, E. T. (1986b). Neuronal activity related to the control of feeding, *in* R. Ritter, S. Ritter and C. Barnes (eds), *Feeding Behavior: Neural and Humoral Controls*, Academic Press, New York, chapter 6, pp. 163–190.

Rolls, E. T. (1986c). A theory of emotion, and its application to understanding the neural basis of emotion, *in* Y. Oomura (ed.), *Emotions. Neural and Chemical Control*, Japan Scientific Societies Press; and Karger, Tokyo; and Basel, pp. 325–344.

Rolls, E. T. (1987). Information representation, processing and storage in the brain: analysis at the single neuron level, *in* J.-P. Changeux and M. Konishi (eds), *The Neural and Molecular Bases of Learning*, Wiley, Chichester, pp. 503–540.

Rolls, E. T. (1989a). Functions of neuronal networks in the hippocampus and neocortex in memory, *in* J. H. Byrne and W. O. Berry (eds), *Neural Models of Plasticity: Experimental and Theoretical Approaches*, Academic Press, San Diego, CA, chapter 13, pp. 240–265.

Rolls, E. T. (1989b). Parallel distributed processing in the brain: implications of the functional architecture of neuronal networks in the hippocampus, *in* R. G. M. Morris (ed.), *Parallel Distributed Processing: Implications for Psychology and Neurobiology*, Oxford University Press, Oxford, chapter 12, pp. 286–308.

Rolls, E. T. (1989c). The representation and storage of information in neuronal networks in the primate cerebral cortex and hippocampus, *in* R. Durbin, C. Miall and G. Mitchison (eds), *The Computing Neuron*, Addison-Wesley, Wokingham, England, chapter 8, pp. 125–159.

Rolls, E. T. (1990a). Theoretical and neurophysiological analysis of the functions of the primate hippocampus in memory, *Cold Spring Harbor Symposia in Quantitative Biology* **55**: 995–1006.

Rolls, E. T. (1990b). A theory of emotion, and its application to understanding the neural basis of emotion, *Cognition and Emotion* **4**: 161–190.

Rolls, E. T. (1992). Neurophysiology and functions of the primate amygdala, *in* J. P. Aggleton (ed.), *The Amygdala*, Wiley-Liss, New York, chapter 5, pp. 143–165.

Rolls, E. T. (1993). The neural control of feeding in primates, *in* D. Booth (ed.), *Neurophysiology of Ingestion*, Pergamon, Oxford, chapter 9, pp. 137–169.

Rolls, E. T. (1996). A theory of hippocampal function in memory, *Hippocampus* **6**: 601–620.

Rolls, E. T. (1999). *The Brain and Emotion*, Oxford University Press, Oxford.

Rolls, E. T. (2000a). Functions of the primate temporal lobe cortical visual areas in invariant visual object and face recognition, *Neuron* **27**: 205–218.

Rolls, E. T. (2000b). Neurophysiology and functions of the primate amygdala, and the neural basis of emotion, *in* J. P. Aggleton (ed.), *The Amygdala: Second Edition. A Functional Analysis*, Oxford University Press, Oxford, chapter 13, pp. 447–478.

Rolls, E. T. (2000c). The orbitofrontal cortex and reward, *Cerebral Cortex* **10**: 284–294.

Rolls, E. T. (2003). Consciousness absent and present: a neurophysiological exploration, *Progress in Brain Research* **144**: 95–106.

Rolls, E. T. (2004a). The functions of the orbitofrontal cortex, *Brain and Cognition* **55**: 11–29.

Rolls, E. T. (2004b). A higher order syntactic thought (HOST) theory of consciousness, *in* R. J. Gennaro (ed.), *Higher Order Theories of Consciousness*, John Benjamins, Amsterdam, chapter 7, pp. 137–172.

Rolls, E. T. (2005). *Emotion Explained*, Oxford University Press, Oxford.

Rolls, E. T. (2006a). Consciousness absent and present: a neurophysiological exploration of masking, *in* H. Ogmen and B. G. Breitmeyer (eds), *The First Half Second*, MIT Press, Cambridge, MA, chapter 6, pp. 89–108.

Rolls, E. T. (2006b). The neurophysiology and functions of the orbitofrontal cortex, *in* D. H. Zald and S. L. Rauch (eds), *The Orbitofrontal Cortex*, Oxford University Press, Oxford, chapter 5, pp. 95–124.

Rolls, E. T. (2007a). The affective neuroscience of consciousness: higher order syntactic thoughts, dual routes to emotion and action, and consciousness, *in* P. D. Zelazo, M. Moscovitch and E. Thompson (eds), *Cambridge Handbook of Consciousness*, Cambridge University Press, New York, chapter 29, pp. 831–859.

Rolls, E. T. (2007b). A computational neuroscience approach to consciousness, *Neural Networks* **20**: 962–982.

Rolls, E. T. (2007c). Memory systems: multiple systems in the brain and their interactions, *in* H. L. Roediger, Y. Dudai and S. M. Fitzpatrick (eds), *Science of Memory: Concepts*, Oxford University Press, New York, chapter 59, pp. 345–351.

Rolls, E. T. (2007d). The representation of information about faces in the temporal and frontal lobes of primates including humans, *Neuropsychologia* **45**: 124–143.

Rolls, E. T. (2007e). Sensory processing in the brain related to the control of food intake, *Proceedings of the Nutrition Society* **66**: 96–112.

Rolls, E. T. (2008a). Emotion, higher order syntactic thoughts, and consciousness, *in* L. Weiskrantz and M. Davies (eds), *Frontiers of Consciousness*, Oxford University Press, Oxford, chapter 4, pp. 131–167.

Rolls, E. T. (2008b). Face representations in different brain areas, and critical band masking, *Journal of Neuropsychology* **2**: 325–360.

Rolls, E. T. (2008c). Functions of the orbitofrontal and pregenual cingulate cortex in taste, olfaction, appetite and emotion, *Acta Physiologica Hungarica* **95**: 131–164.

Rolls, E. T. (2008d). *Memory, Attention, and Decision-Making. A Unifying Computational Neuroscience Approach*, Oxford University Press, Oxford.

Rolls, E. T. (2008e). The primate hippocampus and episodic memory, *in* E. Dere, A. Easton, L. Nadel and J. P. Huston (eds), *Handbook of Episodic Memory*, Elsevier, Amsterdam, chapter 4.2, pp. 417–438.

Rolls, E. T. (2008f). Top-down control of visual perception: attention in natural vision, *Perception* **37**: 333–354.

Rolls, E. T. (2009a). The anterior and midcingulate cortices and reward, *in* B. Vogt (ed.), *Cingulate Neurobiology and Disease*, Oxford University Press, Oxford, chapter 8, pp. 191–206.

Rolls, E. T. (2009b). From reward value to decision-making: neuronal and computational principles, *in* J.-C. Dreher and L. Tremblay (eds), *Handbook of Reward and Decision-Making*, Academic Press, New York, chapter 5, pp. 95–130.

Rolls, E. T. (2009c). The neurophysiology and computational mechanisms of object representation, *in* S. Dickinson, M. Tarr, A. Leonardis and B. Schiele (eds), *Object Categorization: Computer and Human Vision Perspectives*, Cambridge University Press, Cambridge, chapter 14, pp. 257–287.

Rolls, E. T. (2010a). Consciousness, decision-making, and neural computation, *in* V. Cutsuridis, A. Hussain, J. G. Taylor, D. Polani and N. Tishby (eds), *Perception-Reason-Action Cycle: models, algorithms and systems*, Springer, Berlin.

Rolls, E. T. (2010b). A neurobiological basis for affective feelings and aesthetics, *in* E. Schellekens and P. Goldie (eds), *Philosophy and Aesthetic Psychology*, Oxford University Press, Oxford.

Rolls, E. T. (2010c). Noise in the brain, decision-making, determinism, free will, and consciousness, *in* D. Collerton, E. Perry, F. LeBeau and H. Ashton (eds), *Conscious Nonconscious Connections: Neuroscience in Mind*, John Benjamin, Amsterdam.

Rolls, E. T. and Baylis, L. L. (1994). Gustatory, olfactory and visual convergence within the primate orbitofrontal cortex, *Journal of Neuroscience* **14**: 5437–5452.

Rolls, E. T. and Deco, G. (2002). *Computational Neuroscience of Vision*, Oxford University Press, Oxford.

Rolls, E. T. and Deco, G. (2006). Attention in natural scenes: neurophysiological and computational bases, *Neural Networks* **19**: 1383–1394.

Rolls, E. T. and Grabenhorst, F. (2008). The orbitofrontal cortex and beyond: from affect to decision-making, *Progress in Neurobiology* **86**: 216–244.

Rolls, E. T. and Kesner, R. P. (2006). A theory of hippocampal function, and tests of the theory, *Progress in Neurobiology* **79**: 1–48.

Rolls, E. T. and Stringer, S. M. (2000). On the design of neural networks in the brain by genetic evolution, *Progress in Neurobiology* **61**: 557–579.

Rolls, E. T. and Stringer, S. M. (2001). A model of the interaction between mood and memory, *Network: Computation in Neural Systems* **12**: 89–109.

Rolls, E. T. and Stringer, S. M. (2006). Invariant visual object recognition: a model, with lighting invariance, *Journal of Physiology – Paris* **100**: 43–62.

Rolls, E. T. and Tovee, M. J. (1994). Processing speed in the cerebral cortex and the neurophysiology of visual masking, *Proceedings of the Royal Society, B* **257**: 9–15.

Rolls, E. T. and Tovee, M. J. (1995). Sparseness of the neuronal representation of stimuli in the primate temporal visual cortex, *Journal of Neurophysiology* **73**: 713–726.

Rolls, E. T. and Treves, A. (1990). The relative advantages of sparse versus distributed encoding for associative neuronal networks in the brain, *Network* **1**: 407–421.

Rolls, E. T. and Treves, A. (1998). *Neural Networks and Brain Function*, Oxford University Press, Oxford.

Rolls, E. T. and Xiang, J.-Z. (2005). Reward–spatial view representations and learning in the primate hippocampus, *Journal of Neuroscience* **25**: 6167–6174.

Rolls, E. T. and Xiang, J.-Z. (2006). Spatial view cells in the primate hippocampus, and memory recall, *Reviews in the Neurosciences* **17**: 175–200.

Rolls, E. T., Burton, M. J. and Mora, F. (1976). Hypothalamic neuronal responses associated with the sight of food, *Brain Research* **111**: 53–66.

Rolls, E. T., Judge, S. J. and Sanghera, M. (1977). Activity of neurones in the inferotemporal cortex of the alert monkey, *Brain Research* **130**: 229–238.

Rolls, E. T., Sanghera, M. K. and Roper-Hall, A. (1979). The latency of activation of neurons in the lateral hypothalamus and substantia innominata during feeding in the monkey, *Brain Research* **164**: 121–135.

Rolls, E. T., Perrett, D. I., Caan, A. W. and Wilson, F. A. W. (1982). Neuronal responses related to visual recognition, *Brain* **105**: 611–646.

Rolls, E. T., Yaxley, S. and Sienkiewicz, Z. J. (1990). Gustatory responses of single neurons in the orbitofrontal cortex of the macaque monkey, *Journal of Neurophysiology* **64**: 1055–1066.

Rolls, E. T., Hornak, J., Wade, D. and McGrath, J. (1994a). Emotion-related learning in patients with social and emotional changes associated with frontal lobe damage, *Journal of Neurology, Neurosurgery and Psychiatry* **57**: 1518–1524.

Rolls, E. T., Tovee, M. J., Purcell, D. G., Stewart, A. L. and Azzopardi, P. (1994b). The responses of neurons in the temporal cortex of primates, and face identification and detection, *Experimental Brain Research* **101**: 474–484.

Rolls, E. T., Critchley, H. D. and Treves, A. (1996a). The representation of olfactory information in the primate orbitofrontal cortex, *Journal of Neurophysiology* **75**: 1982–1996.

Rolls, E. T., Critchley, H. D., Mason, R. and Wakeman, E. A. (1996b). Orbitofrontal cortex neurons: role in olfactory and visual association learning, *Journal of Neurophysiology* **75**: 1970–1981.

Rolls, E. T., Treves, A. and Tovee, M. J. (1997a). The representational capacity of the distributed encoding of information provided by populations of neurons in the primate temporal visual cortex, *Experimental Brain Research* **114**: 149–162.

Rolls, E. T., Treves, A., Tovee, M. and Panzeri, S. (1997b). Information in the neuronal representation of individual stimuli in the primate temporal visual cortex, *Journal of Computational Neuroscience* **4**: 309–333.

Rolls, E. T., Treves, A., Robertson, R. G., Georges-Francois, P. and Panzeri, S. (1998). Information about spatial view in an ensemble of primate hippocampal cells, *Journal of Neurophysiology* **79**: 1797–1813.

Rolls, E. T., Tovee, M. J. and Panzeri, S. (1999). The neurophysiology of backward visual masking: information analysis, *Journal of Cognitive Neuroscience* **11**: 335–346.

Rolls, E. T., Stringer, S. M. and Trappenberg, T. P. (2002). A unified model of spatial and episodic memory, *Proceedings of The Royal Society B* **269**: 1087–1093.

Rolls, E. T., Aggelopoulos, N. C. and Zheng, F. (2003a). The receptive fields of inferior temporal cortex neurons in natural scenes, *Journal of Neuroscience* **23**: 339–348.

Rolls, E. T., Franco, L., Aggelopoulos, N. C. and Reece, S. (2003b). An information theoretic approach to the contributions of the firing rates and the correlations between the firing of neurons, *Journal of Neurophysiology* **89**: 2810–2822.

Rolls, E. T., Verhagen, J. V. and Kadohisa, M. (2003c). Representations of the texture of food in the primate orbitofrontal cortex: neurons responding to viscosity, grittiness, and capsaicin, *Journal of Neurophysiology* **90**: 3711–3724.

Rolls, E. T., Aggelopoulos, N. C., Franco, L. and Treves, A. (2004). Information encoding in the inferior temporal visual cortex: contributions of the firing rates and the correlations between the firing of neurons, *Biological Cybernetics* **90**: 19–32.

Rolls, E. T., Xiang, J.-Z. and Franco, L. (2005). Object, space and object-space representations in the primate hippocampus, *Journal of Neurophysiology* **94**: 833–844.

Rolls, E. T., Franco, L., Aggelopoulos, N. C. and Jerez, J. M. (2006a). Information in the first spike, the order of spikes, and the number of spikes provided by neurons in the inferior temporal visual cortex, *Vision Research* **46**: 4193–4205.

Rolls, E. T., Stringer, S. M. and Elliot, T. (2006b). Entorhinal cortex grid cells can map to hippocampal place cells by competitive learning, *Network: Computation in Neural Systems* **17**: 447–465.

Rolls, E. T., Loh, M. and Deco, G. (2008a). An attractor hypothesis of obsessive-compulsive disorder, *European Journal of Neuroscience* **28**: 782–793.

Rolls, E. T., Loh, M., Deco, G. and Winterer, G. (2008b). Computational models of schizophrenia and dopamine modulation in the prefrontal cortex, *Nature Reviews Neuroscience* **9**: 696–709.

Rolls, E. T., McCabe, C. and Redoute, J. (2008c). Expected value, reward outcome, and temporal difference error representations in a probabilistic decision task, *Cerebral Cortex* **18**: 652–663.

Rolls, E. T., Tromans, J. M. and Stringer, S. M. (2008d). Spatial scene representations formed by self-organizing learning in a hippocampal extension of the ventral visual system, *European Journal of Neuroscience* **28**: 2116–2127.

Rolls, E. T., Critchley, H., Verhagen, J. V. and Kadohisa, M. (2009a). The representation of information about taste and odor in the primate orbitofrontal cortex, *Chemosensory Perception* p. Epub 24 September 2009.

Rolls, E. T., Grabenhorst, F. and Franco, L. (2009b). Prediction of subjective affective state from brain activations, *Journal of Neurophysiology* **101**: 1294–1308.

Rolls, E. T., Grabenhorst, F. and Parris, B. A. (2009c). Neural systems underlying decisions about affective odors, *Journal of Cognitive Neuroscience* p. Epub 25 March.

Rolls, E. T., Deco, G. and Loh, M. (2010a). A stochastic neurodynamics approach to the changes in cognition and memory in aging.

Rolls, E. T., Grabenhorst, F. and Deco, G. (2010b). Choice, difficulty, and confidence in the brain.

Rolls, E. T., Grabenhorst, F. and Deco, G. (2010c). Decision-making, errors, and confidence in the brain.

Romo, R. and Salinas, E. (2001). Touch and go: Decision-making mechanisms in somatosensation, *Annual Review Neuroscience* **24**: 107–137.

Romo, R. and Salinas, E. (2003). Flutter discrimination: Neural codes, perception, memory and decision making, *Nature Reviews Neuroscience* **4**: 203–218.

Romo, R., Hernandez, A., Zainos, A. and Salinas, E. (1998). Somatosensory discrimination based on cortical microstimulation, *Nature* **392**: 387–390.

Romo, R., Brody, C., Hernandez, A. and Lemus, L. (1999). Neuronal correlates of parametric working memory in the prefrontal cortex, *Nature* **339**: 470–473.

Romo, R., Hernandez, A., Zainos, A., Lemus, L. and Brody, C. (2002). Neural correlates of decision-making in secondary somatosensory cortex, *Nature Neuroscience* **5**: 1217–1225.

Romo, R., Hernandez, A., Zainos, A. and Salinas, E. (2003). Correlated neuronal discharges that increase coding efficiency during perceptual discrimination, *Neuron* **38**: 649–657.

Romo, R., Hernandez, A. and Zainos, A. (2004). Neuronal correlates of a perceptual decision in ventral premotor cortex, *Neuron* **41**: 165–173.

Rosenberg, D. R., MacMaster, F. P., Keshavan, M. S., Fitzgerald, K. D., Stewart, C. M. and Moore, G. J. (2000). Decrease in caudate glutamatergic concentrations in pediatric obsessive-compulsive disorder patients taking paroxetine, *Journal of the American Academy of Child and Adolescent Psychiatry* **39**: 1096–1103.

Rosenberg, D. R., MacMillan, S. N. and Moore, G. J. (2001). Brain anatomy and chemistry may predict treatment response in paediatric obsessive–compulsive disorder, *International Journal of Neuropsychopharmacology* **4**: 179–190.

Rosenberg, D. R., Mirza, Y., Russell, A., Tang, J., Smith, J. M., Banerjee, S. P., Bhandari, R., Rose, M., Ivey, J., Boyd, C. and Moore, G. J. (2004). Reduced anterior cingulate glutamatergic concentrations in childhood ocd and major depression versus healthy controls, *J Am Acad Child Adolesc Psychiatry* **43**: 1146–1153.

Roxin, A. and Ledberg, A. (2008). Neurobiological models of two-choice decision making can be reduced to a one-dimensional nonlinear diffusion equation, *PLoS Computational Biology* **4(3)**: e1000046.

Rumelhart, D. E. and McClelland, J. L. (1986). *Parallel Distributed Processing*, Vol. 1: Foundations, MIT Press, Cambridge, MA.

Rushworth, M. F. S., Walton, M. E., Kennerley, S. W. and Bannerman, D. M. (2004). Action sets and decisions in the medial frontal cortex, *Trends in Cognitive Sciences* **8**: 410–417.

Rushworth, M. F. S., Behrens, T. E., Rudebeck, P. H. and Walton, M. E. (2007a). Contrasting roles for cingulate and orbitofrontal cortex in decisions and social behaviour, *Trends in Cognitive Sciences* **11**: 168–176.

Rushworth, M. F. S., Buckley, M. J., Behrens, T. E., Walton, M. E. and Bannerman, D. M. (2007b). Functional organization of the medial frontal cortex, *Current Opinion in Neurobiology* **17**: 220–227.

Russchen, F. T., Amaral, D. G. and Price, J. L. (1985). The afferent connections of the substantia innominata in the monkey, Macaca fascicularis, *Journal of Comparative Neurology* **242**: 1–27.

Sah, P. (1996). Ca^{2+}-activated K^+ currents in neurones: types, physiological roles and modulation, *Trends in Neuroscience* **19**: 150–154.

Sah, P. and Faber, E. S. (2002). Channels underlying neuronal calcium-activated potassium currents, *Progress in Neurobiology* **66**: 345–353.

Salin, P. and Prince, D. (1996). Spontaneous GABA-A receptor mediated inhibitory currents in adult rat somatosensory cortex, *Journal of Neurophysiology* **75**: 1573–1588.

Salzman, D., Britten, K. and Newsome, W. (1990). Cortical microstimulation influences perceptual judgements of motion direction, *Nature* **364**: 174–177.

Sato, T. (1989). Interactions of visual stimuli in the receptive fields of inferior temporal neurons in macaque, *Experimental Brain Research* **77**: 23–30.

Sawaguchi, T. and Goldman-Rakic, P. S. (1991). D1 dopamine receptors in prefrontal cortex: Involvement in working memory, *Science* **251**: 947–950.

Sawaguchi, T. and Goldman-Rakic, P. S. (1994). The role of D1-dopamine receptor in working memory: local injections of dopamine antagonists into the prefrontal cortex of rhesus monkeys performing an oculomotor delayed-response task, *Journal of Neurophysiology* **71**: 515–528.

Schall, J. (2001). Neural basis of deciding, choosing and acting, *Nature Review Neuroscience* **2**: 33–42.

Scheuerecker, J., Ufer, S., Zipse, M., Frodl, T., Koutsouleris, N., Zetzsche, T., Wiesmann, M., Albrecht, J., Bruckmann, H., Schmitt, G., Moller, H. J. and Meisenzahl, E. M. (2008). Cerebral changes and cognitive dysfunctions in medication-free schizophrenia - An fMRI study, *Journal of Psychiatric Research* **42**: 469–476.

Schliebs, R. and Arendt, T. (2006). The significance of the cholinergic system in the brain during aging and in Alzheimer's disease, *Journal of Neural Transmission* **113**: 1625–1644.

Seamans, J. K. and Yang, C. R. (2004). The principal features and mechanisms of dopamine modulation in the prefrontal cortex, *Progress in Neurobiology* **74**: 1–58.

Seamans, J. K., Gorelova, N., Durstewitz, D. and Yang, C. R. (2001). Bidirectional dopamine modulation of GABAergic inhibition in prefrontal cortical pyramidal neurons, *Journal of Neuroscience* **21**: 3628–3638.

Seeman, P. and Kapur, S. (2000). Schizophrenia: more dopamine, more D2 receptors, *Proceedings of the National Academy of Sciences USA* **97**: 7673–7675.

Seeman, P., Weinshenker, D., Quirion, R., Srivastava, L. K., Bhardwaj, S. K., Grandy, D. K., Premont, R. T., Sotnikova, T. D., Boksa, P., El-Ghundi, M., O'Dowd, B. F., George, S.-R., Perreault, M. L., Mannisto, P. T., Robinson, S., Palmiter, R. D. and Tallerico, T. (2005). Dopamine supersensitivity correlates with D2 High states, implying many paths to psychosis, *Proceedings of the National Academy of Sciences USA* **102**: 3513–3518.

Seeman, P., Schwarz, J., Chen, J. F., Szechtman, H., Perreault, M., McKnight, G. S., Roder, J. C., Quirion, R., Boksa, P., Srivastava, L. K., Yanai, K., Weinshenker, D. and Sumiyoshi, T. (2006). Psychosis pathways converge via D2 high dopamine receptors, *Synapse* **60**: 319–346.

Seigel, M. and Auerbach, J. M. (1996). Neuromodulators of synaptic strength, *in* M. S. Fazeli and G. L. Collingridge (eds), *Cortical Plasticity*, Bios, Oxford, chapter 7, pp. 137–148.

Senn, W., Markram, H. and Tsodyks, M. (2001). An algorithm for modifying neurotransmitter release probability based on pre- and postsynaptic spike timing, *Neural Computation* **13**: 35–67.

Shadlen, M. and Newsome, W. (1996). Motion perception: seeing and deciding, *Proceedings of the National Academy of Science USA* **93**: 628–633.

Shadlen, M. N. and Newsome, W. T. (2001). Neural basis of a perceptual decision in the parietal cortex (area LIP) of the rhesus monkey, *Journal of Neurophysiology* **86**: 1916–1936.

Shallice, T. and Burgess, P. (1996). The domain of supervisory processes and temporal organization of behaviour, *Philosophical Transactions of the Royal Society B,* **351**: 1405–1411.

Shallice, T. and Burgess, P. W. (1991). Deficits in strategy application following frontal lobe damage in man, *Brain* **114**: 727–741.

Shang, Y., Claridge-Chang, A., Sjulson, L., Pypaert, M. and Miesenböck, G. (2007). Excitatory local circuits and their implications for olfactory processing in the fly antennal lobe, *Cell* **128**: 601–612.

Shannon, C. E. (1948). A mathematical theory of communication, *AT&T Bell Laboratories Technical Journal* **27**: 379–423.

Shepherd, G. M. (2004). *The Synaptic Organisation of the Brain*, 5th edn, Oxford University Press, Oxford.

Shergill, S. S., Brammer, M. J., Williams, S. C., Murray, R. M. and McGuire, P. K. (2000). Mapping auditory hallucinations in schizophrenia using functional magnetic resonance imaging, *Archives of General Psychiatry* **57**: 1033–1038.

Shiino, M. and Fukai, T. (1990). Replica-symmetric theory of the nonlinear analogue neural networks, *Journal of Physics A: Mathematical and General* **23**: L1009–L1017.

Shulman, G. L., Olliger, J. M., Linenweber, M., Petersen, S. and Corbetta, M. (2001). Multiple neural correlates of detection in the human brain, *Proceedings of the National Academy of Sciences USA* **98**: 313–318.

Sigala, N. and Logothetis, N. K. (2002). Visual categorisation shapes feature selectivity in the primate temporal cortex, *Nature* **415**: 318–320.

Sikstrom, S. (2007). Computational perspectives on neuromodulation of aging, *Acta Neurochirurgica, Supplement* **97**: 513–518.

Sillito, A. M. (1984). Functional considerations of the operation of GABAergic inhibitory processes in the visual cortex, *in* E. G. Jones and A. Peters (eds), *Cerebral Cortex, Vol. 2, Functional Properties of Cortical Cells*, Plenum, New York, chapter 4, pp. 91–117.

Singer, W. (1995). Development and plasticity of cortical processing architectures, *Science* **270**: 758–764.

Singer, W. (1999). Neuronal synchrony: A versatile code for the definition of relations?, *Neuron* **24**: 49–65.

Sjostrom, P. J., Turrigiano, G. G. and Nelson, S. B. (2001). Rate, timing, and cooperativity jointly determine cortical synaptic plasticity, *Neuron* **32**: 1149–1164.

Sloper, J. J. and Powell, T. P. S. (1979a). An experimental electron microscopic study of afferent connections to the primate motor and somatic sensory cortices, *Philosophical Transactions of the Royal Society of London, Series B* **285**: 199–226.

Sloper, J. J. and Powell, T. P. S. (1979b). A study of the axon initial segment and proximal axon of neurons in the primate motor and somatic sensory cortices, *Philosophical Transactions of the Royal Society of London, Series B* **285**: 173–197.

Smerieri, A., Rolls, E. T. and Feng, J. (2010). Theta-rhythm facilitates reaction time.

Smith, P. and Ratcliff, R. (2004). Psychology and neurobiology of simple decisions, *Trends in Neurosciences* **23**: 161–168.

Soltani, A. and Wang, X.-J. (2006). A biophysically based neural model of matching law behavior: melioration by stochastic synapses, *Journal of Neuroscience* **26**: 3731–3744.

Somogyi, P. and Cowey, A. C. (1984). Double bouquet cells, *in* A. Peters and E. G. Jones (eds), *Cerebral Cortex, Vol. 1, Cellular Components of the Cerebral Cortex*, Plenum, New York, chapter 9, pp. 337–360.

Somogyi, P., Kisvarday, Z. F., Martin, K. A. C. and Whitteridge, D. (1983). Synaptic connections of morphologically identified and physiologically characterized large basket cells in the striate cortex of the cat, *Neuroscience* **10**: 261–294.

Spitzer, H., Desimone, R. and Moran, J. (1988). Increased attention enhances both behavioral and neuronal performance, *Science* **240**: 338–340.

Spruston, N., Jonas, P. and Sakmann, B. (1995). Dendritic glutamate receptor channel in rat hippocampal CA3 and CA1 pyramidal neurons, *Journal of Physiology* **482**: 325–352.

Stephan, K. E., Baldeweg, T. and Friston, K. J. (2006). Synaptic plasticity and dysconnection in schizophrenia, *Biological Psychiatry* **59**: 929–939.

Stewart, S. E., Fagerness, J. A., Platko, J., Smoller, J. W., Scharf, J. M., Illmann, C., Jenike, E., Chabane, N., Leboyer, M., Delorme, R., Jenike, M. A. and Pauls, D. L. (2007). Association of the SLC1A1 glutamate transporter gene and obsessive-compulsive disorder, *American Journal of Medical Genetics B: Neuropsychiatric Genetics* **144**: 1027–1033.

Stocks, N. G. (2000). Suprathreshold stochastic resonance in multilevel threshold systems, *Physical Review Letters* **84**: 2310–2313.

Sugrue, L. P., Corrado, G. S. and Newsome, W. T. (2005). Choosing the greater of two goods: neural currencies for valuation and decision making, *Nature Reviews Neuroscience* **6**: 363–375.

Sutton, R. S. and Barto, A. G. (1981). Towards a modern theory of adaptive networks: expectation and prediction, *Psychological Review* **88**: 135–170.

Suzuki, W. A., Miller, E. K. and Desimone, R. (1997). Object and place memory in the macaque entorhinal cortex, *Journal of Neurophysiology* **78**: 1062–1081.

Szabo, M., Almeida, R., Deco, G. and Stetter, M. (2004). Cooperation and biased competition model can explain attentional filtering in the prefrontal cortex, *European Journal of Neuroscience* **19**: 1969–1977.

Szabo, M., Deco, G., Fusi, S., Del Giudice, P., Mattia, M. and Stetter, M. (2006). Learning to attend: Modeling the shaping of selectivity in infero-temporal cortex in a categorization task, *Biological Cybernetics* **94**: 351–365.

Szentagothai, J. (1978). The neuron network model of the cerebral cortex: a functional interpretation, *Proceedings of the Royal Society of London, Series B* **201**: 219–248.

Tagamets, M. and Horwitz, B. (1998). Integrating electrophysical and anatomical experimental data to create a large-scale model that simulates a delayed match-to-sample human brain study, *Cerebral Cortex* **8**: 310–320.

Talbot, W., Darian-Smith, I., Kornhuber, H. and Mountcastle, V. B. (1968). The sense of fluttervibration: comparison of the human capacity response patterns of mechanoreceptive afferents from the monkey hand, *Journal of Neurophysiology* **31**: 301–334.

Tanji, J. (2001). Sequential organization of multiple movements: involvement of cortical motor areas, *Annual Review of Neuroscience* **24**: 631–651.

Tegner, J., Compte, A. and Wang, X.-J. (2002). The dynamical stability of reverberatory neural circuits, *Biological Cybernetics* **87**: 471–481.

Thibault, O., Porter, N. M., Chen, K. C., Blalock, E. M., Kaminker, P. G., Clodfelter, G. V., Brewer, L. D. and

Landfield, P. W. (1998). Calcium dysregulation in neuronal aging and Alzheimers disease: history and new directions, *Cell Calcium* **24**: 417–433.

Thomson, A. M. and Deuchars, J. (1994). Temporal and spatial properties of local circuits in neocortex, *Trends in Neurosciences* **17**: 119–126.

Thorpe, S. J., Rolls, E. T. and Maddison, S. (1983). Neuronal activity in the orbitofrontal cortex of the behaving monkey, *Experimental Brain Research* **49**: 93–115.

Tovee, M. J. and Rolls, E. T. (1995). Information encoding in short firing rate epochs by single neurons in the primate temporal visual cortex, *Visual Cognition* **2**: 35–58.

Tovee, M. J., Rolls, E. T., Treves, A. and Bellis, R. P. (1993). Information encoding and the responses of single neurons in the primate temporal visual cortex, *Journal of Neurophysiology* **70**: 640–654.

Trantham-Davidson, H., Neely, L. C., Lavin, A. and Seamans, J. K. (2004). Mechanisms underlying differential D1 versus D2 dopamine receptor regulation of inhibition in prefrontal cortex, *Journal of Neuroscience* **24**: 10652–10659.

Treisman, A. (1982). Perceptual grouping and attention in visual search for features and for objects, *Journal of Experimental Psychology: Human Perception and Performance* **8**: 194–214.

Treves, A. (1990). Graded-response neurons and information encodings in autoassociative memories, *Physical Review A* **42**: 2418–2430.

Treves, A. (1991a). Are spin-glass effects relevant to understanding realistic auto-associative networks?, *Journal of Physics A* **24**: 2645–2654.

Treves, A. (1991b). Dilution and sparse encoding in threshold-linear nets, *Journal of Physics A* **23**: 1–9.

Treves, A. (1993). Mean-field analysis of neuronal spike dynamics, *Network* **4**: 259–284.

Treves, A. and Rolls, E. T. (1991). What determines the capacity of autoassociative memories in the brain?, *Network* **2**: 371–397.

Treves, A. and Rolls, E. T. (1992). Computational constraints suggest the need for two distinct input systems to the hippocampal CA3 network, *Hippocampus* **2**: 189–199.

Treves, A. and Rolls, E. T. (1994). A computational analysis of the role of the hippocampus in memory, *Hippocampus* **4**: 374–391.

Treves, A., Rolls, E. T. and Simmen, M. (1997). Time for retrieval in recurrent associative memories, *Physica D* **107**: 392–400.

Treves, A., Panzeri, S., Rolls, E. T., Booth, M. and Wakeman, E. A. (1999). Firing rate distributions and efficiency of information transmission of inferior temporal cortex neurons to natural visual stimuli, *Neural Computation* **11**: 611–641.

Tuckwell, H. (1988). *Introduction to Theoretical Neurobiology*, Cambridge University Press, Cambridge.

Turner, B. H. (1981). The cortical sequence and terminal distribution of sensory related afferents to the amygdaloid complex of the rat and monkey, *in* Y. Ben-Ari (ed.), *The Amygdaloid Complex*, Elsevier, Amsterdam, pp. 51–62.

Ungerleider, L. G. (1995). Functional brain imaging studies of cortical mechanisms for memory, *Science* **270**: 769–775.

Ungerleider, L. G. and Haxby, J. V. (1994). 'What' and 'Where' in the human brain, *Current Opinion in Neurobiology* **4**: 157–165.

Usher, M. and McClelland, J. (2001). On the time course of perceptual choice: the leaky competing accumulator model, *Psychological Reviews* **108**: 550–592.

van den Heuvel, O. A., Veltman, D. J., Groenewegen, H. J., Cath, D. C., van Balkom, A. J., van Hartskamp, J., Barkhof, F. and van Dyck, R. (2005). Frontal-striatal dysfunction during planning in obsessive-compulsive disorder, *Archives of General Psychiatry* **62**: 301–309.

Van Hoesen, G. W. (1981). The differential distribution, diversity and sprouting of cortical projections to the amygdala in the rhesus monkey, *in* Y. Ben-Ari (ed.), *The Amygdaloid Complex*, Elsevier, Amsterdam, pp. 77–90.

Veale, D. M., Sahakian, B. J., Owen, A. M. and Marks, I. M. (1996). Specific cognitive deficits in tests sensitive to frontal lobe dysfunction in obsessive-compulsive disorder, *Psychological Medicine* **26**: 1261–1269.

Verhagen, J. V., Rolls, E. T. and Kadohisa, M. (2003). Neurons in the primate orbitofrontal cortex respond to fat texture independently of viscosity, *Journal of Neurophysiology* **90**: 1514–1525.

Verhagen, J. V., Kadohisa, M. and Rolls, E. T. (2004). The primate insular taste cortex: neuronal representations of the viscosity, fat texture, grittiness, and the taste of foods in the mouth, *Journal of Neurophysiology* **92**: 1685–1699.

Vickers, D. (1979). *Decision Processes in Visual Perception*, Academic Press, New York.

Vickers, D. and Packer, J. (1982). Effects of alternating set for speed or accuracy on response time, accuracy and confidence in a unidimensional discrimination task, *Acta Psychologica* **50**: 179–197.

Vinje, W. E. and Gallant, J. L. (2000). Sparse coding and decorrelation in primary visual cortex during natural vision, *Science* **287**: 1273–1276.

Wang, S.-H. and Morris, R. G. M. (2010). Hippocampal-neocortical interactions in memory formation, consolidation, and reconsolidation, *Annual Review of Psychology* **61**: in press.

Wang, X. J. (1999). Synaptic basis of cortical persistent activity: the importance of NMDA receptors to working memory, *Journal of Neuroscience* **19**: 9587–9603.

Wang, X. J. (2001). Synaptic reverberation underlying mnemonic persistent activity, *Trends in Neurosciences* **24**: 455–463.

Wang, X. J. (2002). Probabilistic decision making by slow reverberation in cortical circuits, *Neuron* **36**: 955–968.

Wang, X.-J. (2008). Decision making in recurrent neuronal circuits, *Neuron* **60**: 215–234.

Wang, X. J., Tegner, J., Constantinidis, C. and Goldman-Rakic, P. S. (2004). Division of labor among distinct subtypes of inhibitory neurons in a cortical microcircuit of working memory, *Proceedings of the National Academy of Sciences USA* **101**: 1368–1373.

Warrington, E. K. and Weiskrantz, L. (1973). An analysis of short-term and long-term memory defects in man, *in* J. A. Deutsch (ed.), *The Physiological Basis of Memory*, Academic Press, New York, chapter 10, pp. 365–395.

Watkins, L. H., Sahakian, B. J., Robertson, M. M., Veale, D. M., Rogers, R. D., Pickard, K. M., Aitken, M. R. and Robbins, T. W. (2005). Executive function in Tourette's syndrome and obsessive-compulsive disorder, *Psychological Medicine* **35**: 571–582.

Weisenfeld, K. (1993). An introduction to stochastic resonance, *Annals of the New York Academy of Sciences* **706**: 13–25.

Weiskrantz, L. (1997). *Consciousness Lost and Found*, Oxford University Press, Oxford.

Weiss, A. P. and Heckers, S. (1999). Neuroimaging of hallucinations: a review of the literature, *Psychiatry Research* **92**: 61–74.

Weissman, M. M., Bland, R. C., Canino, G. J., Greenwald, S., Hwu, H. G., Lee, C. K., Newman, S. C., Oakley-Browne, M. A., Rubio-Stipec, M., Wickramaratne, P. J. et al. (1994). The cross national epidemiology of obsessive compulsive disorder. The Cross National Collaborative Group, *Journal of Clinical Psychiatry* **55 Suppl**: 5–10.

Welford, A. T. (ed.) (1980). *Reaction Times*, Academic Press, London.

Werner, G. and Mountcastle, V. (1965). Neural activity in mechanoreceptive cutaneous afferents: stimulus-response relations, Weber functions, and information transmission, *Journal of Neurophysiology* **28**: 359–397.

Wilson, F. A., O'Scalaidhe, S. P. and Goldman-Rakic, P. S. (1994). Functional synergism between putative gamma-aminobutyrate-containing neurons and pyramidal neurons in prefrontal cortex, *Proceedings of the Natlional Academy of Sciences U S A* **91**: 4009–4013.

Wilson, F. A. W. and Rolls, E. T. (1990a). Learning and memory are reflected in the responses of reinforcement-related neurons in the primate basal forebrain, *Journal of Neuroscience* **10**: 1254–1267.

Wilson, F. A. W. and Rolls, E. T. (1990b). Neuronal responses related to reinforcement in the primate basal forebrain, *Brain Research* **509**: 213–231.

Wilson, F. A. W. and Rolls, E. T. (1990c). Neuronal responses related to the novelty and familiarity of visual stimuli in the substantia innominata, diagonal band of broca and periventricular region of the primate, *Experimental Brain Research* **80**: 104–120.

Wilson, F. A. W. and Rolls, E. T. (1993). The effects of stimulus novelty and familiarity on neuronal activity in the amygdala of monkeys performing recognition memory tasks, *Experimental Brain Research* **93**: 367–382.

Wilson, F. A. W. and Rolls, E. T. (2005). The primate amygdala and reinforcement: a dissociation between rule-based and associatively-mediated memory revealed in amygdala neuronal activity, *Neuroscience* **133**: 1061–1072.

Wilson, F. A. W., O'Sclaidhe, S. P. and Goldman-Rakic, P. S. (1993). Dissociation of object and spatial processing domains in primate prefrontal cortex, *Science* **260**: 1955–1958.

Wilson, H. R. (1999). *Spikes, Decisions and Actions: Dynamical Foundations of Neuroscience*, Oxford University Press, Oxford.

Wilson, H. R. and Cowan, J. D. (1972). Excitatory and inhibitory interactions in localised populations of model neurons, *Biophysics Journal* **12**: 1–24.

Wilson, H. R. and Cowan, J. D. (1973). A mathematical theory of the functional dynamics of cortical and thalamic nervous tissue, *Kybernetik* **13**: 55–80.

Winterer, G. and Weinberger, D. R. (2004). Genes, dopamine and cortical signal-to-noise ratio in schizophrenia, *Trends in Neurosciences* **27**: 683–690.

Winterer, G., Ziller, M., Dorn, H., Frick, K., Mulert, C., Wuebben, Y., Herrmann, W. M. and Coppola, R. (2000). Schizophrenia: reduced signal-to-noise ratio and impaired phase-locking during information processing, *Clinical Neurophysiology* **111**: 837–849.

Winterer, G., Coppola, R., Goldberg, T. E., Egan, M. F., Jones, D. W., Sanchez, C. E. and Weinberger, D. R. (2004). Prefrontal broadband noise, working memory, and genetic risk for schizophrenia, *American Journal of Psychiatry* **161**: 490–500.

Winterer, G., Musso, F., Beckmann, C., Mattay, V., Egan, M. F., Jones, D. W., Callicott, J. H., Coppola, R. and Weinberger, D. R. (2006). Instability of prefrontal signal processing in schizophrenia, *American Journal of Psychiatry* **163**: 1960–1968.

Wolkin, A., Sanfilipo, M., Wolf, A. P., Angrist, B., Brodie, J. D. and Rotrosen, J. (1992). Negative symptoms and hypofrontality in chronic schizophrenia, *Archives of General Psychiatry* **49**: 959–965.

Wong, K. and Wang, X.-J. (2006). A recurrent network mechanism of time integration in perceptual decisions, *Journal of Neuroscience* **26**: 1314–1328.

Wong, K. F., Huk, A. C., Shadlen, M. N. and Wang, X.-J. (2007). Neural circuit dynamics underlying accumulation of time-varying evidence during perceptual decision making, *Frontiers in Computational Neuroscience* **1**: 6.

Xiang, Z., Huguenard, J. and Prince, D. (1998). GABA-A receptor mediated currents in interneurons and pyramidal cells of rat visual cortex, *Journal of Physiology* **506**: 715–730.

Zheng, P., Zhang, X.-X., Bunney, B. S. and Shi, W.-X. (1999). Opposite modulation of cortical N-methyl-D-aspartate receptor-mediated responses by low and high concentrations of dopamine, *Neuroscience* **91**: 527–535.

Zipser, K., Lamme, V. and Schiller, P. (1996). Contextual modulation in primary visual cortex, *Journal of Neuroscience* **16**: 7376–7389.

Index

Appendix 2 Colour Plates

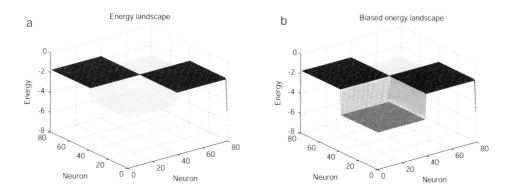

Fig. 2.6 (a) Energy landscape without any differential bias applied. The landscape is for a network with neurons 1–40 connected by strengthened synapses so that they form attractor 1, and neurons 41–80 connected by synapses strengthened (by the same amount) so that they form attractor 2. The energy basins in the two-dimensional landscape are calculated by Equation 2.17. In this two-dimensional landscape, there are two stable attractors, and each will be reached equally probably under the influence of noise. This scenario might correspond to decision-making where the input λ_1 to attractor 1 has the same value as the input λ_2 to attractor 2, and the network is equally likely under the influence of the noise to fall into attractor 1 representing decision 1 as into attractor 2 representing decision 2. (b) Energy landscape with bias applied to neurons 1–40. This make the basin of attraction deeper for attractor 1, as calculated with Equation 2.17. Thus, under the influence of noise caused by the randomness in the firing of the neurons, the network will reach attractor 1 more probably than it will reach attractor 2. This scenario might correspond to decision-making, where the evidence for decision 1 is stronger than for decision 2, so that a higher firing rate is applied as λ_1 to neurons 1–40. The scenario might also correspond to memory recall, in which memory 1 might be probabilistically more likely to be recalled than memory 2 if the evidence for memory 1 is stronger. Nevertheless, memory 2 will be recalled sometimes in what operates as a non-deterministic system.

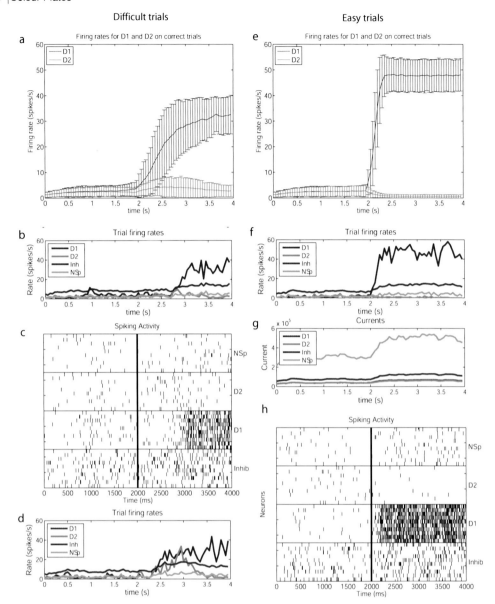

Fig. 6.2 (a) and (e) Firing rates (mean ± sd) for difficult (ΔI=0) and easy (ΔI=160) trials. The period 0–2 s is the spontaneous firing, and the decision cues were turned on at time = 2 s. The mean was calculated over 1000 trials. D1: firing rate of the D1 population of neurons on correct trials on which the D1 population won. D2: firing rate of the D2 population of neurons on the correct trials on which the D1 population won. A correct trial was one in which in which the mean rate of the D1 attractor averaged > 10 spikes/s for the last 1000 ms of the simulation runs. (b) The mean firing rates of the four populations of neurons on a difficult trial. Inh is the inhibitory population that uses GABA as a transmitter. NSp is the non-specific population of neurons (see Fig. 6.1). (c) Rastergrams for the trial shown in b. 10 neurons from each of the four pools of neurons are shown. (d) The firing rates on another difficult trial (ΔI=0) showing prolonged competition between the D1 and D2 attractors until the D1 attractor finally wins after approximately 1100 ms. (f) Firing rate plots for the 4 neuronal populations on a single easy trial (ΔI=160). (g) The synaptic currents in the four neuronal populations on the trial shown in f. (h) Rastergrams for the easy trial shown in f and g. 10 neurons from each of the four pools of neurons are shown. (After Rolls, Grabenhorst and Deco 2010b.)

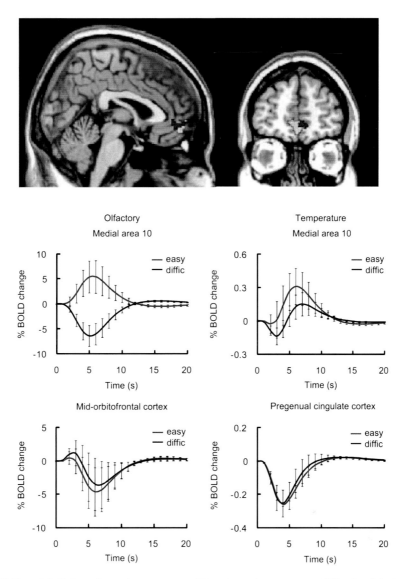

Fig. 6.8 Top: Medial prefrontal cortex area 10 activated on easy vs difficult trials in the olfactory pleasantness decision task (yellow) and the thermal pleasantness decision task (red). Middle: experimental data showing the BOLD signal in medial area 10 on easy and difficult trials of the olfactory affective decision task (left) and the thermal affective decision task (right). This medial area 10 was a region identified by other criteria (see text) as being involved in choice decision-making. Bottom: The BOLD signal for the same easy and difficult trials, but in parts of the pregenual cingulate and mid-orbitofrontal cortex implicated by other criteria (see text) in representing the subjective reward value of the stimuli on a continuous scale, but not in making choice decisions between the stimuli, or about whether to choose the stimulus in future. (After Rolls, Grabenhorst and Deco, 2010b.)

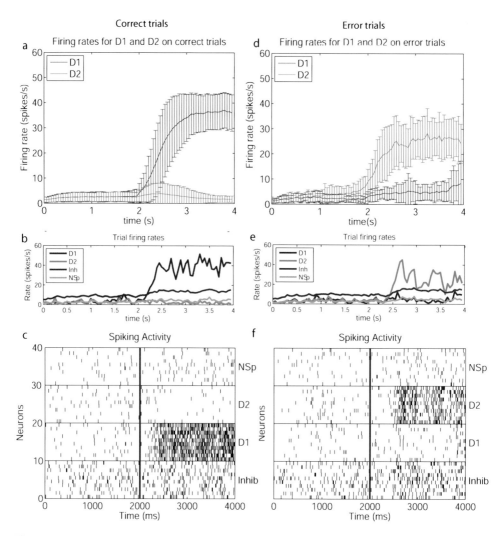

Fig. 6.10 (a and d) Firing rates (mean ± sd) for correct and error trials for an intermediate level of difficulty (ΔI=32). The period 0–2 s is the spontaneous firing, and the decision cues were turned on at time = 2 s. The means were calculated over 1000 trials. D1: firing rate of the D1 population of neurons, which is the correct population. D2: firing rate of the D2 population of neurons, which is the incorrect population. A correct trial was one in which in which the mean rate of the D1 attractor averaged > 10 spikes/s for the last 1000 ms of the simulation runs. (Given the attractor nature of the network and the parameters used, the network reached one of the attractors on >90% of the 1000 trials, and this criterion clearly separated these trials, as indicated by the mean rates and standard deviations for the last 1 s of the simulation as shown.) (b and e) The firing rates of the four populations of neurons on a single trial for a correct (b) and incorrect (e) decision. Inh is the inhibitory population that uses GABA as a transmitter. NSp is the non-specific population of neurons. (c and f) Rastergrams for the trials shown in (b) and (e). 10 neurons from each of the four pools of neurons are shown. (After Rolls, Grabenhorst and Deco 2010c.)

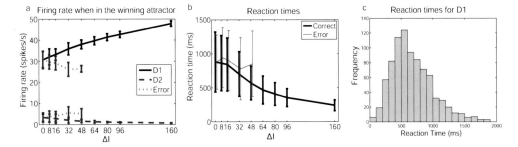

Fig. 6.11 (a) Firing rates (mean ± sd) on correct and error trials when in the winning attractor as a function of ΔI. $\Delta I=0$ corresponds to difficult, and $\Delta I=160$ corresponds to easy. The firing rates on correct trials for the winning population D1 are shown by solid lines, and for the losing population D2 by dashed lines. All the results are for 1000 simulation trials for each parameter value, and all the results shown are statistically highly significant. The results on error trials are shown by the dotted red lines, and in this case the D2 attractor wins, and the D1 attractor loses the competition. (There were no error trials for values of $\Delta I=64$ Hz or above.) (b) Reaction times (mean ± sd) for the D1 population to win on correct trials (thick line), and for D2 to win on error trials (thin line), as a function of the difference in inputs ΔI to D1 and D2. The plot for the error trials has been offset by a small amount so that its values can be seen clearly. $\Delta I=0$ corresponds to difficult, and $\Delta I=160$ corresponds to easy. (c) The distribution of reaction times for the model for $\Delta I=32$ illustrating the long tail of slow responses. Reaction times are shown for 837 correct trials, the level of performance was 95.7% correct, and the mean reaction time was 701 ms. (After Rolls, Grabenhorst and Deco, 2010c.)

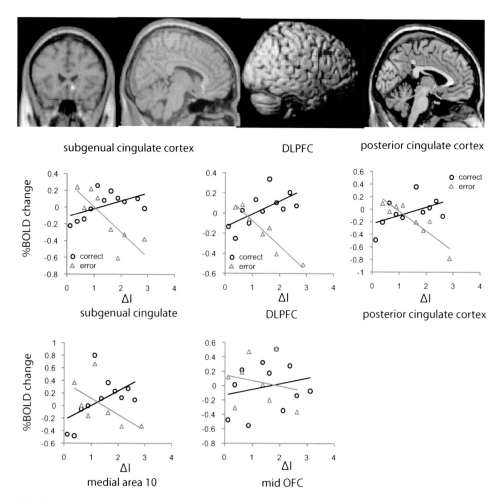

Fig. 6.14 Above: The subgenual cingulate cortex (with coronal and parasaggittal slices), dorsolateral prefrontal cortex (DLPFC) and posterior cingulate cortex regions with the percentage change in the BOLD signal positively correlated with ΔI on correct trials, and negatively correlated on error trials. Below: Separate regression plots for the relation between the BOLD signal and ΔI on correct and on error trials for these regions and for the medial prefrontal cortex area 10, and for a control region, the mid-orbitofrontal cortex (mid OFC) in which effects were not found. The regression plots show for discretized values of ΔI (averaged across subjects) the BOLD signal as a function of ΔI for correct and incorrect trials. For example, for the posterior cingulate cortex the signs of the correlations ($r = 0.51$ on correct trials, and $r = -0.91$ on error trials) are as predicted for a decision-making area of the brain, and are significantly different ($p<0.001$). The medial prefrontal cortex coordinates used for extracting the data were $[-4\ 54\ -6]$ with a 10 mm sphere around this location, the site in the same subjects for the peak of the contrast for easy – difficult trials (Rolls, Grabenhorst and Deco, 2010b). The same region $[2\ 50\ -12]$ showed more activity when binary decisions were made compared to when ratings were made in the same subjects (Rolls, Grabenhorst and Deco, 2010b). (After Rolls, Grabenhorst and Deco, 2010c.)

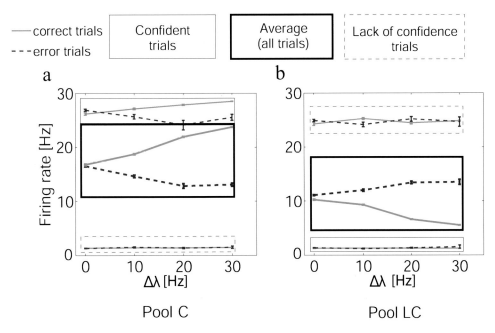

Fig. 6.20 Firing rates (mean ± s.e.m.) in the confidence decision-making network of the C (confident, 'positive outcome') (a), and LC (lack of confidence, 'negative outcome') (b) populations of neurons. In the solid thin boxes the firing rates of the C (left panel) and LC (right panel) pools are shown on trials on which the C (confidence) population won the competition. The lines in the solid thin boxes show for the C pool an average rate of close to 28 spikes/s which tends to increase with $\Delta\lambda$ when the first network is correct, and tends to decrease with $\Delta\lambda$ when the first network is in error. In the dashed thin boxes the firing rates of the C (left panel) and LC (right panel) pools are shown on trials on which the LC (lack of confidence) population won the competition. Again for the C population, if we take just the trials on which the C population lost the competition, the lines in the dashed thin boxes show for the C pool an average rate of close to 2 spikes/s. The activities of the C and LC populations of neurons, averaged over all trials, confident and lack of confidence, are shown in the thick boxes. As shown in Fig. 6.17, the confidence decision-making network was itself increasingly incorrect as $\Delta\lambda$ approached 0, and the firing rates in the thick lines reflect the fact that on some trials the C population won the competition, and on some trials it lost. (After Insabato, Pannunzi, Rolls and Deco, 2010.)

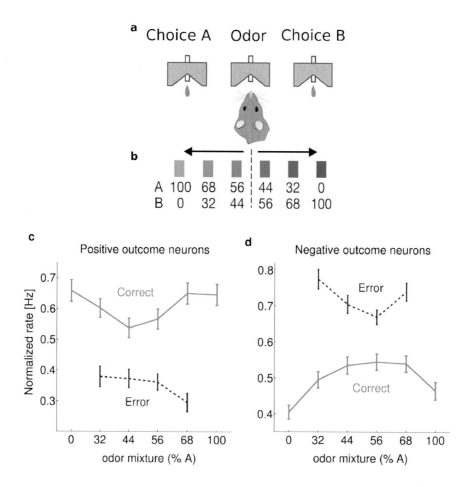

Fig. 6.21 (a) Odour mixture categorization task: when the rat enters the odour port, an odour mixture is delivered. Then the subject categorizes the mixture as A or B by moving left or right, according to the dominant odour. The stimulus is defined by the percentage of the two pure odourants A and B in the mixture as shown in (b). When odours are present in almost equal quantities, there is no bias to the decision process. Moving the stimulus towards pure odourants raises the bias towards one of the decisions and improves performance. (c) Mean normalized firing rate of the positive outcome selective neuronal population. These neurons fire faster on correct trials then on error trials, and increase their firing rate as the task becomes easier. On error trials, when the rat makes the wrong choice, these neurons fire more slowly, and decrease their firing rates more as the task becomes easier. If $\Delta\lambda$ is the discriminability of the stimuli, this is 0 for a 50% mixture of odours A and B, and $\Delta\lambda$ increases both to the left on the abscissa and to the right from 50% as the proportion of one of the odours in the mixture increases and the decision of which odour it is becomes easier. (d) Mean normalized firing rate of the negative outcome selective neuronal population in the orbitofrontal cortex as recorded by Kepecs et al (2008). These neurons have sustained activity on error trials, increasing their firing rate further when the decision is made easier. On correct trials they fire slower, and decrease their firing rates as the decision becomes easier. (Modified from Kepecs et al (2008)).